# Recent Advances in Hydrocarbon Production Processes from Geoenergy

# Recent Advances in Hydrocarbon Production Processes from Geoenergy

Guest Editors

**Lufeng Zhang**
**Fengrui Sun**
**Jie Wang**

MDPI

Basel • Beijing • Wuhan • Barcelona • Belgrade • Novi Sad • Cluj • Manchester

*Guest Editors*
Lufeng Zhang
Headquarters of China Petroleum & Chemical Corporation
Beijing
China

Fengrui Sun
State Key Laboratory of Petroleum Resources and Prospecting
China University of Geosciences (Beijing)
Beijing
China

Jie Wang
School of Petroleum Engineering
Yangtze University
Jingzhou
China

*Editorial Office*
MDPI AG
Grosspeteranlage 5
4052 Basel, Switzerland

This is a reprint of the Special Issue, published open access by the journal *Processes* (ISSN 2227-9717), freely accessible at: https://www.mdpi.com/journal/processes/special_issues/1I828M5AG9.

For citation purposes, cite each article independently as indicated on the article page online and as indicated below:

| Lastname, A.A.; Lastname, B.B. Article Title. *Journal Name* **Year**, *Volume Number*, Page Range. |
|---|

**ISBN 978-3-7258-5735-7 (Hbk)**
**ISBN 978-3-7258-5736-4 (PDF)**
**https://doi.org/10.3390/books978-3-7258-5736-4**

# Contents

# About the Editors

**Lufeng Zhang**

Lufeng Zhang, Senior Engineer, is currently employed at the headquarters of the China Petrochemical Corporation (Sinopec). He previously worked at the Sinopec Research Institute of Petroleum Exploration and Development. His research focuses primarily on acidization and acid fracturing technologies for unconventional reservoirs and their application in field operations. Zhang holds a PhD from the China University of Petroleum (Beijing) and is a recipient of the "Top Ten Outstanding University Students" award. He has authored multiple highly cited papers in prestigious petroleum journals including *SPE Journal*, *JPSE*, and *JNGSE*.

**Fengrui Sun**

Fengrui Sun is an associate professor and has studied, visited, and worked at the China University of Petroleum (East China), the University of Texas at Austin in the United States, the China University of Petroleum (Beijing), and the China University of Geosciences (Beijing). His research focuses on mineral prospecting and exploration, as well as petroleum engineering. For five consecutive years, from 2021 to 2025, he was honored as a "Highly Cited Researcher in China" (Elsevier). Fengrui Sun has had many research achievements in the fields of heavy oil thermal recovery and the numerical simulation of wellbore heat transfer.

**Jie Wang**

Jie Wang is an associate professor, currently working at the Yangtze University School of Petroleum Engineering. His main research focuses include enhanced oil recovery and reservoir stimulation technologies for unconventional reservoirs based on CCUS, formation, inhibition, and dissociation mechanisms of natural gas hydrates. He possesses extensive industry experience, having previously worked in Schlumberger China's Well Services department (the world's largest oilfield services company) and participating in over 100 well-site operations.

*processes*

MDPI

*Editorial*

# Special Issue on "Recent Advances in Hydrocarbon Production Processes from Geoenergy"

**Jie Wang [1,2,*], Lufeng Zhang [3] and Fengrui Sun [4]**

1 State Key Laboratory of Low Carbon Catalysis and Carbon Dioxide Utilization, Yangtze University, Wuhan 434023, China
2 School of Petroleum Engineering, Yangtze University, Wuhan 434023, China
3 Headquarters of China Petroleum & Chemical Corporation, Beijing 100728, China; zlfcupb@163.com
4 State Key Laboratory of Petroleum Resources and Prospecting, China University of Geosciences (Beijing), Beijing 100083, China; 13126682711@163.com
* Correspondence: wangjie@yangtzeu.edu.cn

## 1. Introduction

With the development of science and technology, the demand for oil and gas resources is increasing year by year. However, the recoverable reserves of conventional and readily accessible reservoirs are declining annually. It has become increasingly critical to enhance the development efficiency of unconventional oil and gas reservoirs in order to address growing energy demands and support sustainable development. Unconventional reservoirs typically refer to those with poor reservoir properties, such as low permeability and porosity, including shale oil and gas, tight oil and gas, coalbed methane, and deep reservoirs [1–3]. Commercial extraction from these reservoirs is generally attainable only through advanced technologies such as horizontal drilling, hydraulic fracturing, and acidization. Additionally, to maintain the productivity of unconventional wells during the mid-to-late development stages, timely water injection is necessary to replenish formation energy [4–6]. However, due to poor reservoir properties, the effect achieved by conventional water injection technology is often not satisfactory. This Special Issue of "Recent Advances in Hydrocarbon Production Processes from Geoenergy" in *Processes* has collected the latest research related to advanced stimulation and stabilization measures in the drilling and development of unconventional reservoirs, documenting data related to theoretical modeling, result analysis, and field applications of various technologies.

## 2. Advanced Analysis and Identification Technologies for "Sweet Spots" in Unconventional Hydrocarbon Reservoirs

The hydrocarbon characterization and monitoring analysis technology for unconventional reservoirs involves advancements in 3D seismic imaging, reservoir simulation, real-time monitoring systems, and artificial intelligence algorithms, etc., to enhance reservoir understanding and management, thereby achieving optimal production results [7,8]. Li et al. [9] studied the controlling factors of hydrocarbon enrichment in shale gas condensate reservoirs within rift lake basins. Through the application of geochemical pyrolysis logging, X-ray fluorescence (XRF) spectrum element logging, and SEM-based automated mineralogy, they found that thermal maturity thresholds, lithofacies assemblages, paleoenvironmental settings, and laminae and fracture systems are the main controlling factors for hydrocarbon enrichment in shale condensate gas reservoirs, and can be applied to the analysis and identification of "sweet spots" in unconventional reservoirs. Liu et al. [10]

 https://doi.org/10.3390/pr13113407

employed automatic mineral scanning equipment to obtain the microscopic physical property parameters of strata at continuous depths, and combined with the TOPSIS and AHP algorithms to optimize fracturing sweet spots in deep carbonate reservoirs. In addition, Li et al. [11] believe that the increase in oxygen content contributed to the biodiversity of the Fortunian stage in the Tarim Basin through the analysis of reservoir minerals and hydrocarbon compositions.

## 3. Advanced Hydraulic Fracturing Technologies for Unconventional Hydrocarbon Reservoirs

Hydraulic fracturing is one of the primary measures for production enhancement in unconventional reservoirs. With continuous research and development, current studies have gradually shifted from process optimization to mechanism research, focusing on related studies such as the optimization of hydraulic fracturing fracture parameters and the migration and placement laws of proppants in fractures, so as to improve the long-term conductivity of artificial fractures [12,13].

(1). Regarding the influencing factors of fracture propagation and hydraulic fracturing parameter optimization, Ma et al. [14] conducted simulation studies on the key factors affecting fracture propagation during hydraulic fracturing based on the displacement discontinuity method and a geological engineering integration method. The targeting reservoir types include shale, carbonate rock, volcanic rock, and deep coal seams.

(2). Regarding the influencing factors of fracture conductivity and optimization, Ci et al. [15] experimentally studied the impact of fracture angle, fracture number, proppant concentration, and closure pressure on fracture conductivity, providing theoretical support for constructing high-conductivity fracture networks in complex unconventional reservoirs.

(3). Regarding the research on proppant migration and placement behavior within fractures, Liang et al. [16] studied the migration and settlement laws of proppants in the sand-carrying fluid based on the power-law non-Newtonian fluid model, and the average relative error between the calculated settlement velocity and the actual settlement velocity was only 8.2%. Additionally, Zheng et al. [17] explored the application of $CO_2$ dry fracturing technology in shale reservoirs, contributing to the advancement of carbon neutrality and carbon storage initiatives within the petrochemical industry.

## 4. Stabilization Supporting Technologies for Unconventional Reservoir

To maintain the productivity of unconventional reservoirs during mid-to-late stages, emphasis has been placed on the flow laws of oil–water and gas–liquid two-phase flow, as well as the technology of large-volume water injection in low-permeability reservoirs to replenish formation energy [18,19]. Cheng et al. [20] employed both experimental and numerical simulation methods, integrated with a convolutional neural networks, to conduct comprehensive simulations on the flow behaviors of oil–water and gas–liquid two-phase flows in pipelines and annular geometries, thereby enhancing the efficiency of oil–water transportation and improving the accuracy of flow pattern recognition. To rapidly and efficiently inject fluids into the formation, replenish reservoir energy, and sustain oil and gas well production, Li et al. [21] investigated high-pressure water injection technology under near-fracturing pressure conditions in low-permeability reservoirs. This technology utilizes high pressure to alter reservoir physical properties and enhance the formation's water absorption capacity at the injection end, thereby significantly improving the waterflooding recovery factor in low-permeability reservoirs.

## 5. Conclusions

The papers in this Special Issue present the latest theoretical advancements on the stimulation and production stabilization technologies for unconventional reservoirs. Due to the complex geological structures and poor reservoir properties of unconventional reservoirs, a series of technological innovations in drilling and development are required to achieve commercial exploitation and long-term stable production, providing a sustained energy supply for high-quality development of humanity. We hope that this Special Issue can clarify current and emerging research activities related to several key technologies in unconventional reservoir development, promoting the efficient exploitation of such reservoirs.

**Author Contributions:** Investigation, L.Z.; writing—original draft preparation, J.W.; writing—review and editing, L.Z.; J.W., and F.S. All authors have read and agreed to the published version of the manuscript.

**Conflicts of Interest:** The authors declare no conflicts of interest.

**List of Contributions:**

1. Fakher, S.; Khlaifat, A.; Salib, A.M.; Elsayed, A. Development of Volumetric Adsorption Isotherms for Volcanic Fly Ash from Egypt for Carbon Dioxide Capture Under Elevated Pressure and Temperature. *Processes* **2025**, *13*, 1570.
2. Li, Y.; Bi, C.; Fu, C.; Xu, Y.; Yuan, Y.; Tong, L.; Tang, Y.; Wang, Q. Controls on the Hydrocarbon Production in Shale Gas Condensate Reservoirs of Rift Lake Basins. *Processes* **2025**, *13*, 1868.
3. Zhang, Y.; Kuang, J.; Zhang, H.; Zhong, Y.; Dong, S. Experimental Study on the Acid Fracturing Fracture Propagation Law of a Fractured Carbonate Reservoir in the Majiagou Formation. *Processes* **2025**, *13*, 695.
4. Zhao, K.; Xu, H.; Wang, J.; Jiang, H.; Zhang, L. Study on Optimization of Stimulation Technology of Heterogeneous Porous Carbonate Reservoir. *Processes* **2024**, *12*, 1191.
5. Liu, H.; Hong, J.; Shu, W.; Wang, X.; Ma, X.; Li, H.; Wang, Y. Study on Evolution of Stress Field and Fracture Propagation Laws for Re-Fracturing of Volcanic Rock. *Processes* **2025**, *13*, 2346.
6. Cheng, C.; Gan, Q.; Su, Y.; Cheng, Y.; Zhang, Y. Experimental and CFD-Integrated Investigation into the Intricate Oil–Water Two-Phase Flow Dynamics Within Blind Tees: Uncovering Flow Behaviors for Advanced Process Engineering Applications. *Processes* **2025**, *13*, 619.

## References

1. Lu, Y.; Zhao, G.; Ge, Z.; Jia, Y.; Tang, J.; Gong, T.; Huang, S.; Li, Z.; Fu, W.; Mi, J. Challenges and development direction of deep fragmented soft coalbed methane in China. *Earth Energy Sci.* **2025**, *1*, 38–64. [CrossRef]
2. Alagoz, E.; Tali, B. *Reserves and Resources Evaluation in Unconventional Resources*; Unconventional Resources; CRC Press: Boca Raton, FL, USA, 2025; pp. 421–500.
3. Abdulhadi, D.; Ali, J.A.; Hama, S.M. Advanced Techniques for Improving the Production of Natural Resources from Unconventional Reservoirs: A State-of-the-Art Review. *Energy Fuels* **2025**, *39*, 10853–10876. [CrossRef]
4. Temizel, C.; Canbaz, C.H.; Palabiyik, Y.; Hosgor, F.B.; Atayev, H.; Ozyurtkan, M.H.; Aydin, H.; Yurukcu, M.; Boppana, N. A review of hydraulic fracturing and latest developments in unconventional reservoirs. In Proceedings of the Offshore Technology Conference, OTC 2022: OTC-31942-MS, Houston, TX, USA, 2–5 May 2022.
5. Li, Q.; Xing, H.; Liu, J.; Liu, X. A review on hydraulic fracturing of unconventional reservoir. *Petroleum* **2015**, *1*, 8–15. [CrossRef]
6. Liao, Q.; Wang, B.; Chen, X.; Tan, P. Reservoir stimulation for unconventional oil and gas resources: Recent advances and future perspectives. *Adv. Geo-Energy Res.* **2024**, *13*, 7–9. [CrossRef]
7. Khanjar, H. Applications of Machine Learning in Sweet-Spots Identification: A Review. *SPE J.* **2024**, *29*, 6623–6639. [CrossRef]
8. Gong, D.; Zhu, H.; Zhang, Z.; Kong, F.; Liu, H.; Zhou, G.; Wang, J.; Zhang, L. Geology–Engineering Double Sweet Spot in Unconventional Reservoirs: A Comprehensive Review. *SPE J.* **2025**, 1–26. [CrossRef]
9. Liu, X.; Wang, X.; Chen, F.; Zhu, X.; Mao, Z.; Liu, X.; Ma, H. Three-Dimensional Geomechanical Modeling and Hydraulic Fracturing Parameter Optimization for Deep Coalbed Methane Reservoirs: A Case Study of the Daniudi Gas Field, Ordos Basin. *Processes* **2025**, *13*, 1699. [CrossRef]
10. Liu, Y.; Xie, G.; Zheng, H.; Ma, X.; Ren, G.; Feng, X.; Zhao, W.; Ma, H.; Lei, F. Optimization of Fracturing Sweet Spot in Deep Carbonate Reservoirs by Combining TOPSIS and AHP Algorithm. *Processes* **2025**, *13*, 2777. [CrossRef]

11. Li, W.; Chen, Y.; Wang, L. Carbon Isotope and Sterane Records of Biological Diversity in the Fortunian Stage of the Early Cambrian Tarim Basin, Northwest China. *Processes* **2025**, *13*, 1530. [CrossRef]
12. Li, G.; Jia, C.; Zhao, Q.; Zhou, T.; Gao, J. Coal-rock gas accumulation mechanism and the whole petroleum system of coal measures. *Pet. Explor. Dev.* **2025**, *52*, 33–49. [CrossRef]
13. Shao, C.; Chen, X. Experimental study on improving the recovery rate of low-pressure tight oil reservoirs using molecular deposition film technology. *Appl. Sci.* **2024**, *14*, 9197. [CrossRef]
14. Ma, P.; Tang, S. Simulation of key influencing factors of hydraulic fracturing fracture propagation in a shale reservoir based on the displacement discontinuity method (DDM). *Processes* **2024**, *12*, 1000. [CrossRef]
15. Ci, J.; Wang, L.; Xiong, F.; Zhao, Y. Investigation into the Variation Law of Network Fracture Conductivity in Unconventional Oil and Gas Reservoirs. *Processes* **2024**, *12*, 1803. [CrossRef]
16. Liang, T.; Deng, Z.; Wu, J.; Xu, F.; Zheng, L.; Yang, M.; Zhou, F. A Power-Law-Based Predictive Model for Proppant Settling Velocity in Non-Newtonian Fluid. *Processes* **2025**, *13*, 2631. [CrossRef]
17. Zheng, B.; Tang, W.; Wang, Y.; Li, Y.; Shen, B.; Wang, Y.; Hu, L.; Deng, Y.; Xi, S.; Liu, X. Further Investigation of $CO_2$ Quasi-Dry Fracturing in Shale Reservoirs—An Experimental Study. *Processes* **2024**, *12*, 912. [CrossRef]
18. Mohammadi, M.H.; Kulakhmetovna, Y.A.; Joia, R. An overview of oil recovery techniques: From primary to enhanced oil recovery methods. *J. Res. Appl. Sci. Biotechnol.* **2024**, *3*, 291–301. [CrossRef]
19. Ozowe, W.; Daramola, G.O.; Ekemezie, I.O. Innovative approaches in enhanced oil recovery: A focus on gas injection synergies with other EOR methods. *Magna Sci. Adv. Res. Rev.* **2024**, *11*, 311–324. [CrossRef]
20. Cheng, C.; Yang, W.; Feng, X.; Zhao, Y.; Su, Y. A Convolutional Neural Network-Based Method for Distinguishing the Flow Patterns of Gas-Liquid Two-Phase Flow in the Annulus. *Processes* **2024**, *12*, 2596. [CrossRef]
21. Li, Y.; Xu, H.; Fu, S.; Zhao, H.; Chen, Z.; Bai, X.; Li, J.; Xiu, C.; Zhang, L.; Wang, J. Analysis of the Effectiveness Mechanism and Research on Key Influencing Factors of High-Pressure Water Injection in Low-Permeability Reservoirs. *Processes* **2025**, *13*, 2664. [CrossRef]

*processes*

MDPI

*Article*

# Enhanced Oil Recovery Mechanism and Parameter Optimization of Huff-and-Puff Flooding with Oil Displacement Agents in the Baikouquan Oilfield

**Hui Tian [1], Jianye Mou [2,*], Kunlin Xue [1], Xingyu Yi [1], Hao Liu [2] and Budong Gao [2]**

1 Baikouquan Oil Plant, Xinjiang Oilfield Branch, PetroChina, Karamay 834000, China
2 College of Petroleum Engineering, China University of Petroleum (Beijing), Beijing 102249, China
* Correspondence: moujianye@cup.edu.cn

**Abstract:** The Baikouquan Oilfield edge expansion wells suffer from poor reservoir properties and limited connectivity, leading to low waterflooding sweep efficiency and insufficient reservoir energy. While oil displacement agents (ODAs) are currently employed in huff-and-puff flooding to enhance recovery, there is a lack of a solid basis for selecting these ODAs, and the dominant mechanisms of enhanced oil recovery (EOR) remain unclear. To address this issue, this study combines experimental work and reservoir numerical simulation to investigate the mechanisms of EOR by ODAs, optimize the selection of ODAs, and fine-tune the huff-and-puff flooding parameters. The results show that the selected nanoemulsion ODA (Nano ODA) significantly reduces the oil–water interfacial tension (IFT) by 97%, thereby increasing capillary number. Additionally, the ODA induces a shift from water–wet to neutral–wet conditions on rock surfaces, reducing capillary forces and weakening spontaneous imbibition. The Nano ODA demonstrates strong emulsification and oil-carrying ability, with an emulsification efficiency of 75%. Overall, the ODA increases the relative permeability of the oil phase, reduces residual oil saturation, and achieves a recovery improvement of more than 10% compared with conventional waterflooding. The injection volume and shut-in time were optimized for the target well, and the recovery enhancement from multiple cycles of huff-and-puff flooding was predicted. The research in this paper is expected to provide guidance for the design of huff-and-puff flooding schemes in low-permeability reservoirs.

**Keywords:** huff-and-puff flooding; nanoemulsion ODA; imbibition; numerical simulation; parameter optimization

## 1. Introduction

The Baikouquan Oilfield, particularly the Bai21 well area of the Kexia Formation, contains abundant unconventional oil and gas resources. The reservoir lithology is dominated by coarse-grained lithic sandstone with fine gravel, and vermicular kaolinite is the major clay mineral, accounting for 37.50% of the total clay content. Reservoir porosity is mainly distributed between 8% and 14%, with an average value of 11.40%, while permeability ranges from 1 mD to 30 mD, with an average of 23.71 mD. Due to the presence of a large number of nanoscale pores, conventional water flooding has limited sweep efficiency and cannot effectively mobilize residual oil [1–4]. To overcome these limitations, post-fracturing huff-and-puff flooding has been widely applied as an effective reservoir development strategy. This process involves injecting a large volume of water containing oil displacement agents (ODAs) into the reservoir after fracturing, shutting in the well to allow the injected

ODA to interact with the formation, and then resuming production to mobilize additional oil. Compared with conventional water flooding, huff-and-puff flooding can not only supplement formation energy and restore reservoir pressure but also improve crude oil mobility by altering interfacial properties [5].

During post-fracturing huff-and-puff flooding, ODAs play a crucial role in enhancing oil recovery. ODAs typically contain surfactants that can modify rock wettability and reduce oil–water interfacial tension (IFT). Numerous experimental and theoretical studies have confirmed the dual role of surfactants in wettability alteration and IFT reduction. Sheng et al. investigated the influence of surfactants on spontaneous imbibition through combined simulations, theoretical analysis, and experiments, and reported that surfactant molecules could change the initial wettability of the rock from oil-wet to water-wet, which significantly enhanced oil displacement [6,7]. Similarly, Standnes et al. demonstrated that cationic surfactants exhibit stronger wettability alteration ability than anionic surfactants in oil-wet limestone reservoirs [8], a conclusion that was further validated by the imbibition experiments conducted by Salehi et al. [9]. In parallel, surfactants also contribute to EOR by reducing oil–water IFT. Huang [10] showed that reservoirs with stronger hydrophilicity exhibit more pronounced flooding effects, while the reduction in IFT by surfactants further improved displacement efficiency. Hendraningrat et al. synthesized silica nanoparticles and confirmed that their addition to aqueous solutions reduced oil–water IFT with increasing nanofluid concentration, thereby enhancing oil recovery in low- and medium-permeability sandstone reservoirs [11]. Foster found that when the oil–water IFT is reduced to the level of $10^{-3}$ mN/m, the capillary number increases significantly, which in turn leads to a substantial improvement in recovery efficiency [12]. Youssif et al. reported that hydrophilic monodisperse silica nanoparticles with an average particle size of 22 nm enhanced oil recovery by 13.28% when used at 0.1% concentration in a three-stage recovery sequence following water flooding [13]. These results highlight that ODAs, through surfactant and nanoparticle action, can simultaneously alter wettability and reduce IFT, both of which are essential for improving recovery efficiency.

In addition to improving microscopic displacement efficiency, ODA flooding has also been demonstrated to replenish reservoir energy. Wellington and Richardson conducted displacement experiments using both cationic and anionic surfactants at 0.4% mass fraction and found that residual oil in the core was effectively displaced [14]. Field applications further confirmed that post-fracturing huff-and-puff flooding can significantly restore reservoir energy and production capacity, even enabling horizontal wells to recover to their initial production levels [15].

Beyond laboratory and field studies, numerical simulation has become an indispensable tool for understanding and optimizing huff-and-puff flooding processes. Liu developed a mathematical model for fractured reservoirs that considered spontaneous imbibition, and the results showed that oil recovery increased by 2.15% when imbibition was included in the simulation [16]. Similarly, Sandeep et al. and Viet Hoai Nguyen established mathematical models that also incorporated imbibition mechanisms, reinforcing the importance of accounting for capillary-driven flow in tight reservoirs [17,18]. Zhu applied the STARS simulator in CMG software to evaluate the effect of Nano ODA injection in tight reservoirs and concluded that oil production rates could be substantially enhanced with such treatments [19].

Although significant progress has been made in understanding oil recovery via post-fracturing huff-and-puff flooding, the selection of optimal ODAs for the Baikouquan Oilfield Kexia Formation remains a considerable challenge. This is due to its specific mineralogical composition, complex pore structure, and predominantly water-wet nature, which collectively complicate displacement processes. Furthermore, most existing studies have concentrated on general surfactant-based EOR mechanisms, while limited attention has been given to the comparative performance of different ODAs under the unique petrophysical conditions of the Kexia Formation. Against this background, the present study systematically investigates the physicochemical properties and displacement performance of two ODAs: GPNR-2 Nano ODA, a nanoemulsion developed by GePeto Ltd., and KPS ODA, a petroleum sulfonate-based surfactant. Spontaneous imbibition tests, relative permeability measurements, and core flooding experiments were carried out on core samples from the Kexia Formation. Based on these experimental insights, a numerical huff-and-puff flooding model was constructed to optimize injection parameters.

Importantly, while many previous studies have reported that nanoemulsions reduce IFT and alter wettability, they seldom address reservoir-specific conditions or clarify the dominant recovery mechanisms in water-wet states. This study demonstrates that under the water-wet conditions of the Kexia Formation, emulsification and IFT reduction dominate oil recovery, whereas spontaneous imbibition plays only a minor role. By combining laboratory experiments with numerical huff-and-puff simulations, this work provides both mechanistic insights and practical guidance for reservoir-specific optimization. The findings not only provide theoretical support and technical guidance for efficient development of tight oil reservoirs in the Baikouquan Oilfield but also contribute to the broader understanding of surfactant–nanoparticle-assisted EOR processes in unconventional reservoirs worldwide.

## 2. Experiments on ODA Performance

Five kinds of experiments were conducted to clarify the mechanism of EOR of ODAs. They are (1) oil–water interface property tests, including oil–water IFT tests and wettability tests; (2) emulsion tests of different ODAs with oil; (3) relative permeability tests; (4) spontaneous imbibition tests; and (5) oil displacement tests. The overall performance of the two ODAs is systematically evaluated, and the dominant enhanced oil recovery (EOR) mechanisms are identified based on experimental results. It should be emphasized that in practical applications, flooding agents are selected based on their performance under reservoir conditions—such as interfacial tension reduction, emulsification efficiency, wettability alteration, and displacement efficiency. In this study, effective concentration ranges were therefore evaluated through interfacial tension, emulsification, contact angle, and core flooding tests, rather than by measuring the CMC (critical micelle concentration). The lack of direct measurement of CMC may introduce uncertainty in quantitatively distinguishing micellar versus sub-micellar effects, particularly with respect to interfacial tension reduction and wettability alteration, but it does not compromise the validity of the present findings. Future work will incorporate CMC measurements under representative reservoir conditions to enhance the rigor and completeness of the results. A numerical simulation is conducted to optimize the operational parameters of huff-and-puff flooding for the Kexia Formation. The complete research workflow is illustrated in Figure 1.

### 2.1. Experimental Materials and Instruments

The main equipment used in the experiments includes the following: TX-700C rotating oil–water IFT meter (produced by Beijing Pinzhi Chuangsi, Beijing, China), contact angle measurement instrument (LAUDA Scientific, Lauda-Königshofen, Germany), high-temperature and high-pressure displacement testing platform (Figure 2), Amott imbibition bottle (Beijing Sanjiao Trading, Beijing, China), A30 high-speed emulsifier (Shanghai Ouhe Machinery Equipment Co., Ltd., Shanghai, China), and a constant-temperature oven.

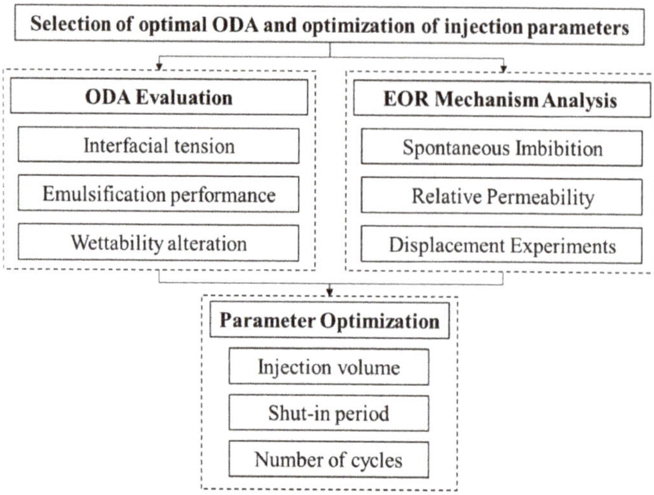


**Figure 1.** Workflow of ODA selection, EOR mechanism analysis, and injection parameter optimization.

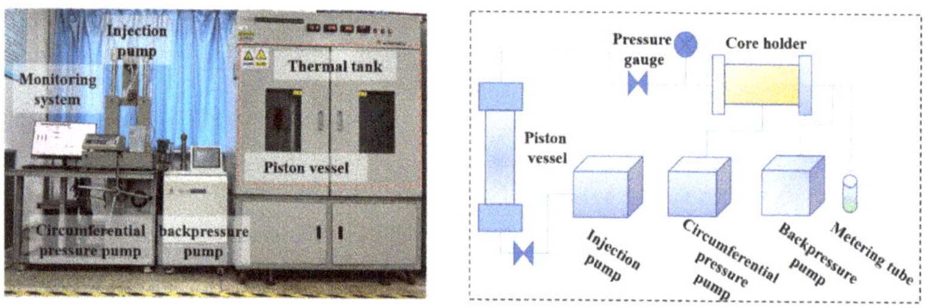


**Figure 2.** Oil displacement devices (**left**) and their schematic diagram (**right**).

The ODAs used in the experiments include a Nano ODA produced by Jiebeitong Petroleum Technology Co., Ltd., Chengdu, China, which primarily consists of organic solvents, diallylphenol ketone polyoxyethylene ether (surfactant), and water, with a surfactant content of 20–30%, and a KPS ODA provided by the Baikouquan Oil Plant, which mainly consists of petroleum sulfonate with an active content of 40%. The ODA concentration range of 0.1–0.3% was selected based on preliminary screening tests and literature evidence. Initial evaluations over a broader range (0.05–0.5 wt%) indicated that the most significant reduction in interfacial tension occurred up to 0.2%, beyond which further improvement was marginal, making this range both technically effective and economically practical. The water used in the experiment is the formation water from the Baikouquan Oil Plant, with a $NaHCO_3$ type composition and a salinity of 5276.9 mg/L. The crude oil used in the experiment is provided by the Baikouquan Oil Plant, Karamay, China, with a density of 0.833 g/cm$^3$ and a viscosity of 1.955 mPa·s. The core samples are also provided by the Baikouquan Oil Plant, with an average permeability of 1.18 mD and an average porosity of 14.8%. Table 1 shows the physical properties and purpose of each core sample.

**Table 1.** Core physical properties and its experimental purpose.

| Core Number | Length (cm) | Diameter (cm) | Permeability (mD) | Porosity (%) | Experimental Purpose | |
|---|---|---|---|---|---|---|
| 1-1 | 1.24 | 2.49 | 1.15 | 15.32 | Wettability test | Oil saturated |
| 1-2 | 1.25 | 2.50 | 1.15 | 15.32 | | Water flooded |
| 1-3 | 1.24 | 2.50 | 1.15 | 15.32 | | Nano ODA flooded |
| 1-4 | 1.23 | 2.50 | 1.15 | 15.32 | | KPS ODA flooded |
| 2 | 5.01 | 2.49 | 1.24 | 15.61 | Relative permeability test | Water |
| 3 | 5.00 | 2.50 | 0.98 | 13.12 | | Nano ODA |
| 4 | 4.99 | 2.50 | 1.42 | 15.28 | | KPS ODA |
| 5 | 5.00 | 2.49 | 0.98 | 14.62 | Imbibition test | Water+ Nano ODA |
| 6 | 5.01 | 2.49 | 1.15 | 16.15 | | Nano ODA |
| 7 | 5.00 | 2.50 | 1.02 | 14.24 | | Water + KPS ODA |
| 8 | 4.98 | 2.49 | 0.90 | 12.49 | | KPS ODA |
| 9 | 4.97 | 2.50 | 1.12 | 13.82 | Displacement test | Water |
| 10 | 4.99 | 2.49 | 1.25 | 14.53 | | 0.1% Nano ODA |
| 11 | 5.01 | 2.49 | 1.32 | 15.10 | | 0.2% Nano ODA |
| 12 | 5.00 | 2.49 | 1.28 | 13.94 | | 0.3% Nano ODA |
| 13 | 4.98 | 2.49 | 1.25 | 14.25 | | 0.1% KPS ODA |
| 14 | 4.99 | 2.50 | 1.38 | 16.18 | | 0.2% KPS ODA |
| 15 | 5.00 | 2.49 | 1.32 | 15.47 | | 0.3% KPS ODA |

Note: core disks 1-1 to 1-4 are cut from the same core plug.

## 2.2. Experimental Method

### 2.2.1. The IFT Test

The IFT test and wettability test are performed to measure the oil–water interface property. The IFT between oil and injected fluid acts as flow resistance when flooding the oil by injected fluid. The interfacial tensiometer was used to test the oil–water IFT between oil and ODA solution. Before the test, 2 μL of crude oil was injected into the quartz tube, which is filled with ODA solution. The quartz tube was carefully placed in the interfacial tensiometer. Then the interfacial tensiometer is started, and the IFT is measured according to the national standard of SY/T 5370—2018 [20].

### 2.2.2. Wettability Test

A 5 cm-long core sample (Core #1) saturated with crude oil was cut into four equal slices. Three slices were displaced with water, 0.2% Nano ODA, and 0.2% KPS ODA, respectively, until the water cut reached 98%. The slices were then polished, immersed in formation water, and crude oil droplets (1–2 mm in diameter) were applied to the mineral surface. Using an optical system, light was projected onto the droplets, magnified, and projected onto a screen to directly measure the contact angle. The experiment followed the "SY/T5153-2017 Method for Determining the Wettability of Reservoir Rocks" [21].

### 2.2.3. Emulsion Test

The emulsification performance of an ODA has a promotion effect on the oil displacement effect. The ODA to be tested was mixed with crude oil in an 8:2 volume ratio and added to a 10 mL stoppered graduated test tube. The mixture was then heated in a 40 °C constant-temperature oven for 20 min. After removal, the mixture was emulsified for 90 s at 20,000 r/min using a high-speed emulsifier to form a uniform system. The formation of water, oil, and emulsion layers was observed and quantified, and the emulsification rate was calculated as the emulsified oil volume divided by the total crude oil volume. The experiment followed the "Q/SY17583-2018 Technical Specification for Surfactants Used in Binary Composite Flooding" [22].

### 2.2.4. Relative Permeability Test

The distribution of oil–water saturation in porous media is a function of distance and time, and the relative permeability of oil and water saturation can be obtained by recording the production and pressure difference in each fluid according to the change in time. The relative permeability was measured through the constant velocity method. The testing procedures include (1) drying and vacuuming the core (2#, 3#, 4# core sample) for 8 h, (2) saturating the core with formation water and calculating the porosity, (3) displacing the core with oil until no water comes out from the outlet and calculating the relative permeability of the oil phase, and (4) displacing the oil-saturated core with formation water and calculating the relative permeability of the water phase. The relative permeability test follows the national standard of GB/T28912-2012 for the determination of the relative permeability of two-phase fluids in rocks [23].

#### 2.2.5. The Imbibition Experiment

The wetting phase can imbibe into the core under the capillary pressure to displace the non-wetting phase. Thus, to evaluate the imbibition ability of different ODA solutions, imbibition experiments have been performed. The saturated core samples (Core #5, #7) were placed in imbibition bottles and immersed in formation water at reservoir temperature, with oil production observed over time. The core mass was measured at intervals. After no oil was produced under formation water conditions for 4 days, the cores were weighed again and then replaced with Nano ODA solution (Core #5) and KPS ODA solution (Core #7), continuing the immersion until day 8. Oil displacement was observed, and mass measurements were taken at intervals. Core samples soaked for 8 days in Nano ODA (Core #6) and KPS ODA solution (Core #8) served as control groups, analyzing the effect of Nano ODA and KPS ODA solutions on spontaneous imbibition displacement. The oil displacement volume and oil recovery rate were calculated by measuring the mass difference in the cores under different solutions and soaking times.

#### 2.2.6. Oil Displacement Experiment

Oil displacement experiments were conducted to evaluate the performance of different oil displacement agents (ODAs) under controlled laboratory conditions, simulating the fluid flow behavior in the Kexia Formation. The detailed procedure is as follows:

(1) Core preparation: Cylindrical core plugs were dried in an oven at 80 °C until a constant weight was reached, then cooled to room temperature in a desiccator.
(2) Vacuum saturation: The dried cores were placed in a vacuum chamber and evacuated to a pressure below −0.08 MPa for at least 4 h to remove air from the pore spaces. Subsequently, formation brine was slowly introduced to saturate the cores under vacuum conditions, ensuring complete brine saturation.
(3) Establishing initial oil saturation: After brine saturation, the cores were mounted in a core holder. Crude oil was injected at a constant flow rate of 0.02 mL/min until no more water was produced from the outlet.
(4) Water/ODA flooding: The injection flow rate was maintained at 0.02 mL/min. Injection continued until the water cut reached 99% or 30 pore volumes (PV) were injected, whichever came first.
(5) Backpressure and confining pressure: A backpressure of 1 MPa was applied at the outlet to stabilize flow and maintain liquid-phase continuity. A confining pressure of 12 MPa was maintained to mimic in situ stress and avoid bypass.
(6) Pressure and production monitoring: Differential pressure across the core was continuously monitored using a precision pressure transducer. Oil and water production were recorded periodically throughout the flooding process.
(7) Post-flooding analysis: At the end of the displacement test, the final oil recovery factor and residual oil saturation were calculated based on cumulative oil production [24].

## 3. Experimental Results and Discussion

### 3.1. Effect of ODAs on Oil–Water IFT

The IFT test results are shown in Figure 3. The experiments demonstrate that both Nano ODA and KPS ODA reduce the oil–water IFT by over 97%. This is due to the presence of a large amount of surfactant in the displacement agent, which reduces direct contact between the two fluids and reduces the interfacial energy between oil and water, thereby lowering the IFT [25,26]. A comparison reveals that the Nano ODA is more effective in lowering the IFT than the KPS ODA. This could be attributed to nanoemulsion structure. The smaller particle size and larger specific surface area of the nanoemulsion allow it to disperse rapidly at the oil–water interface, forming multiple fine thin-layer structures that more significantly reduce the IFT [27]. Additionally, when the ODA concentration exceeds 0.2%, the rate of IFT reduction slows. The potential reason is that once the ODA concentration approaches a critical value, the interfacial sites become nearly saturated. Further increasing the surfactant concentration predominantly promotes micelle formation in the bulk solution rather than additional interfacial adsorption. Consequently, the rate at which the IFT decreases slows down beyond this critical concentration [28]. As the IFT decreases, the adhesion of crude oil to the rock surface also diminishes, converting more residual oil in the reservoir into movable oil, thereby enhancing recovery [29–31].

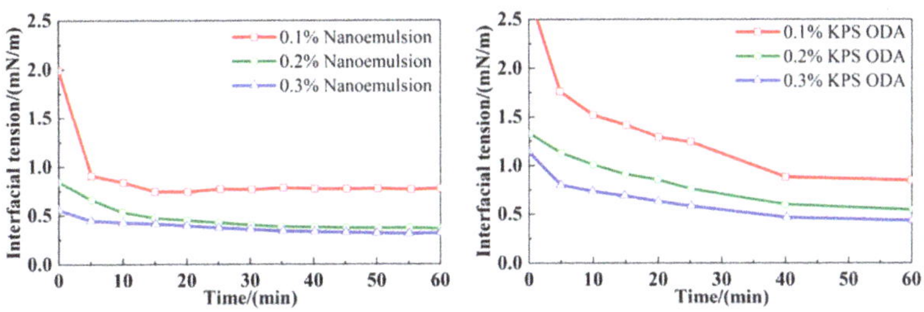


**Figure 3.** Nano ODA significantly reduces IFT, and when its concentration is beyond 0.2%, the IFT reduction rate decreases (**left**); KPS ODA reduces the IFT slowly, and compared with Nano ODA, the effect of KPS ODA in reducing IFT is slightly weaker (**right**).

### 3.2. Effect of ODAs on Wettability

During the development of low-permeability reservoirs, the wettability of the rock affects the imbibition process. When the reservoir is hydrophilic, capillary forces act as the driving force for imbibition. Displacement agents can alter the wettability of the rock, mainly through surfactant adsorption, which affects the surface affinity to water [32]. However, for strongly hydrophilic reservoirs, further increasing hydrophilicity is not always beneficial. Cui found that oil recovery is most effective when the rock exhibits weak hydrophilicity, close to neutral wettability [33]. To evaluate the effect of displacement agents on rock wettability in this study, core samples were prepared, including saturated oil core slices (Core 1-1), water-flooded core slices (Core 1-2), Nano ODA-flooded core slices (Core 1-3), and KPS ODA-flooded core slices (Core 1-4), for contact angle measurement. As shown in Figure 4, the contact angle for all four core samples was less than 90°, indicating that the cores were initially water-wet. After water flooding, the change in wettability was minimal. Following flooding with the ODAs, the measured contact angles increased slightly, demonstrating that the water-wetness weakened and shifted closer to a neutral condition. The Nano ODA exhibited a more pronounced effect in increasing the contact angle than the KPS ODA, indicating a greater tendency to weaken strong water-wetness. This change is directly evidenced by the measured contact angle data presented in Figure 4. The mention of surfactant adsorption explains the possible mechanism: surfactant molecules in the

ODAs can adsorb onto the rock surface and partially block hydrophilic mineral sites, thereby reducing the affinity between the rock and water. However, it should be noted that no direct adsorption measurements were conducted in this work; the inference is supported by the contact angle results and existing literature [34]. This moderate shift in wettability helps balance capillary and viscous forces, which is favorable for improving oil mobilization in low-permeability pore networks.

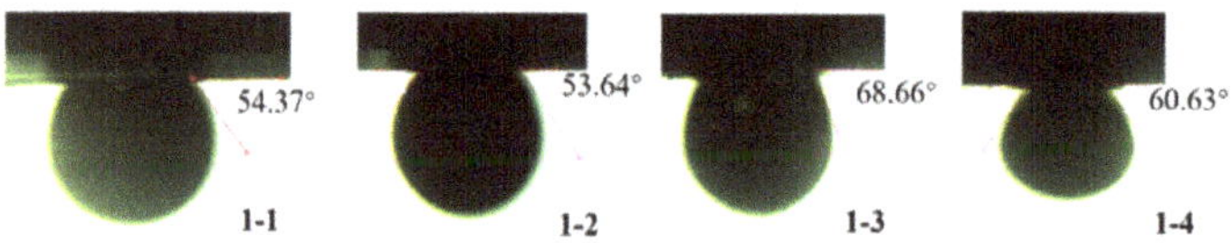


**Figure 4.** The contact angles after ODA flooding increased slightly, demonstrating that the water-wetness weakened and shifted closer to a neutral condition.

### 3.3. Emulsifying Efficiency of ODAs

By adding a displacement agent to the crude oil, a relatively stable water-in-oil (W/O) emulsion droplet is formed, which reduces the binding force of the reservoir on the released oil droplets. In addition, emulsification makes large oil clusters break into smaller droplets, reducing their adhesion to pore surfaces and lowering local capillary resistance. As a result, oil droplets become more mobile and can be more easily displaced by the injected fluid, thus improving oil recovery [35,36]. The changing process of the emulsified mixture is shown in Figure 5. The two mixed liquids, from top to bottom, include the pure oil phase (black), high oil-content phase (dark brown), and low oil-content phase (light brown), with the color gradually becoming lighter from top to bottom. The volume of emulsified crude oil can be determined by subtracting the amount of pure oil phase from the initial amount of crude oil added.

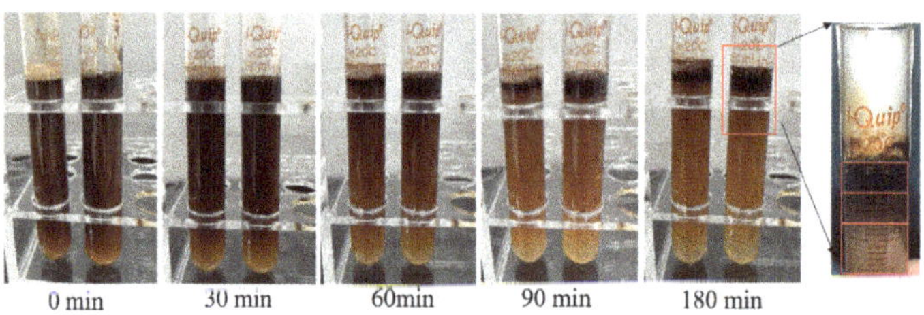


**Figure 5.** Emulsifying processes of Nano ODA (**left**) and KPS ODA (**right**).

The non-emulsified oil gradually rises to the top of the mixture due to buoyancy. After approximately 60 min, the separation process nearly ceases, and the volume of emulsified oil stabilizes. It is found that as the concentration of the displacement agent increases, the emulsification efficiency improves significantly (Figure 6). The emulsification efficiency for different concentrations and types of displacement agents is shown in Table 2. At the same concentration, the Nano ODA exhibits higher emulsification efficiency than the KPS ODA. The potential mechanism behind this phenomenon is that the Nano ODA produces finer droplets and denser surfactant coverage, improving emulsification performance beyond what petroleum sulfonate alone can achieve [37].

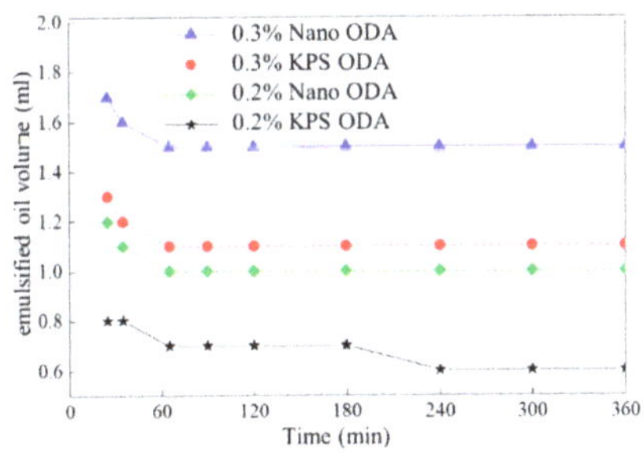


**Figure 6.** Variation in emulsified oil volume with time: as the concentration of the displacement agent increases, the emulsification efficiency improves significantly, and the Nano ODA exhibits higher emulsification efficiency than the KPS ODA.

**Table 2.** Emulsifying efficiency of ODAs of different concentrations.

| Number | 1 | 2 | 3 | 4 |
|---|---|---|---|---|
| ODA type | Nano ODA | Nano ODA | KPS ODA | KPS ODA |
| Concentration | 0.2% | 0.3% | 0.2% | 0.3% |
| Emulsifying efficiency | 53.0% | 68% | 32% | 57% |

### 3.4. Effect of ODAs on Oil–Water Relative Permeability

The relative permeability results are shown in Table 3 and Figure 7. A comparison reveals that the oil-phase relative permeability of the cores displaced by the ODAs increased, and the residual oil saturation was lower. Furthermore, the Nano ODA was more effective than the KPS ODA. This is because the smaller particle size of the nanoemulsion allows it to enter smaller pores, thereby displacing more crude oil from the narrow pore spaces. Meanwhile, the Nano ODA exhibits enhanced emulsification performance and generates stable, fine droplets, as confirmed by the emulsion tests. These droplets improve oil mobilization and contribute to increased oil-phase relative permeability. Additionally, the ODAs alter the rock surface wettability toward more neutral or oil-wet conditions, facilitating oil-phase flow. By reducing IFT and capillary forces, ODAs enable more efficient oil displacement and lower residual oil saturation. Therefore, the observed improvement in oil-phase relative permeability is attributed to the combined effects of emulsification, wettability alteration, and IFT reduction. This experimental result is also supported by other researchers' findings [38,39].

**Table 3.** Results of the relative permeability test.

| Injected Fluid (Core Number) | Oil Relative Permeability Under Water Saturation of 0.5 | Residual Oil Saturation (%) |
|---|---|---|
| Water (2#) | 0.222 | 0.378 |
| 0.2% Nano ODA (3#) | 0.369 | 0.325 |
| 0.2% KPS ODA (4#) | 0.285 | 0.346 |

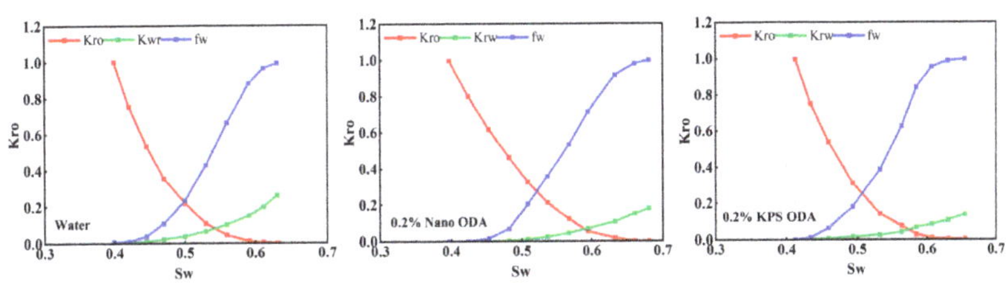


**Figure 7.** Relative permeability curve under different flooding fluids: the oil-phase permeability increased when ODA solution was used for displacement.

### 3.5. Effect of ODAs on Imbibition

In water-wet reservoirs, capillary forces are the main driving force for spontaneous imbibition. The surfactants in the ODA can influence both the oil–water IFT and the rock wettability, so it is essential to clarify the effect of the ODA on imbibition and displacement efficiency and subsequently determine its impact on the oil recovery mechanism of the target reservoir. Table 4 shows that the imbibition effect of formation water is better than that of the ODA, with a higher imbibition displacement efficiency of 27%, while the displacement efficiency of the ODA is less than 25%. Upon further comparison, the imbibition effect of the Nano ODA is weaker than that of the KPS ODA, with both Nano ODA and KPS ODA having a negative effect on imbibition displacement. A comprehensive analysis suggests that while the ODA can improve the mobility of crude oil in the pore spaces by emulsifying and lowering IFT, it also causes the originally water-wet rock to transition towards a neutral-wet state, and the reduction in IFT weakens the capillary forces driving imbibition. Overall, for water-wet reservoirs, the displacement agent has a certain negative effect on imbibition displacement, which contrasts with its positive effect in oil-wet reservoirs [40,41].

**Table 4.** Imbibition test results.

| Core Number | Fluid Type | Core Mass (g) | Saturated Oil Mass (g) | Core Mass After Imbibition (g) | Imbibition Efficiency (%) |
|---|---|---|---|---|---|
| 5# | Water + 0.2% Nano ODA | 49.61 | 2.37 | 51.34 (4 d)<br>51.31 (5 d–8 d) | 26.98<br>28.00 |
| 6# | 0.2% Nano ODA | 48.77 | 2.68 | 50.79 (8 d) | 24.55 |
| 7# | Water + 0.2% KPS ODA | 49.22 | 2.25 | 50.83 (4 d)<br>50.82 (5 d–8 d) | 28.01<br>28.91 |
| 8# | 0.2% KPS ODA | 48.15 | 2.07 | 49.68 (8 d) | 25.96 |

### 3.6. Effect of ODAs on Oil Recovery

To simulate the enhanced oil recovery effect of the ODA, core flooding experiments were conducted using formation water, Nano ODA, and KPS ODA. Various concentrations of the ODAs were tested to determine the optimal ODA concentration. Results (Figure 8) show that both Nano ODA and KPS ODA significantly improved the oil recovery, exceeding 60%, with lower residual oil saturation. The oil recovery increased with the ODA concentration, but the improvement slowed when the concentration exceeded 0.2%. Compared to KPS ODA, Nano ODA showed a stronger ability to enhance oil recovery. Based on the previous test results, the optimal ODA for the Baikouquan Oilfield Kexia Formation, is the Nano ODA with a concentration of 0.2%.

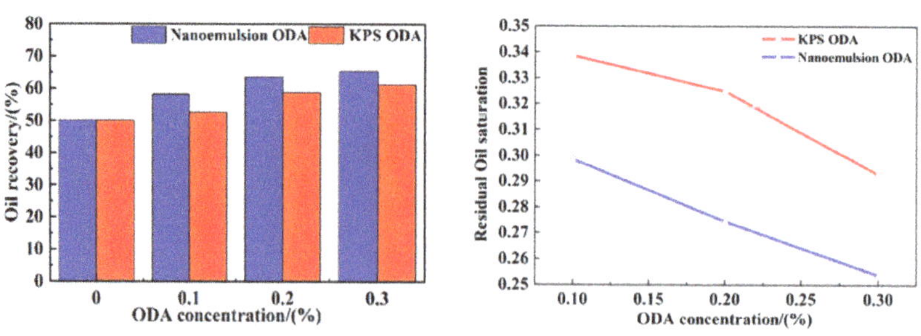


**Figure 8.** Oil recovery (**left**) and residual oil saturation (**right**) achieved by Nano ODA and KPS ODA flooding: both Nano ODA and KPS ODA can increase oil recovery and reduce residual oil saturation, but the effect of Nano ODA is superior.

## 4. Optimization of Huff-and-Puff Flooding Parameters

### 4.1. Model Setup

In the optimization of huff-and-puff flooding parameters, well 1090 in the Baikouquan Oilfield is selected as the target well. A geological model is established (1320 m × 750 m) based on the average reservoir physical properties of the Baikouquan Oilfield Kexia Formation, and numerical simulations are conducted using the CMG software (Version 2021). The reservoir pressure is set at 20.77 MPa, with a thickness of 19.7 m. The porosity ranges from 8% to 14%, with an average of 11.40%, and the permeability mainly ranges from 4 mD to 7 mD, with an average of 5.32 mD. The oil saturation ranges from 26.3% to 38.9%, with an average of 35.2%. The fracture length is 220 m, the fracture width is 0.01 m, and the fracture conductivity is 50.6 mD·m. In the simulation, relative permeability data for 0.2% Nano ODA is used as input, with an injection rate of 2 $m^3$/min.

### 4.2. Simulation Results

#### 4.2.1. Injection Volume

The huff-and-puff flooding process replenishes reservoir energy and restores formation pressure, with injection volume directly influencing both. A higher injection volume increases reservoir pressure but also extends injection time under a constant injection rate. As injection volume rises, more water is retained in the reservoir, leading to higher water cut in the produced fluids and thus reducing oil production efficiency. To determine the optimal injection volume, simulations were conducted for 2000 $m^3$, 3000 $m^3$, 4000 $m^3$, 5000 $m^3$, and 6000 $m^3$ with a shut-in time of 20 days. The results (Figure 9) show that while both cumulative and daily oil production initially increase with injection volume, the incremental gains diminish significantly at higher volumes. A distinct plateau is observed around 4000 $m^3$, indicating this as the optimal injection volume.

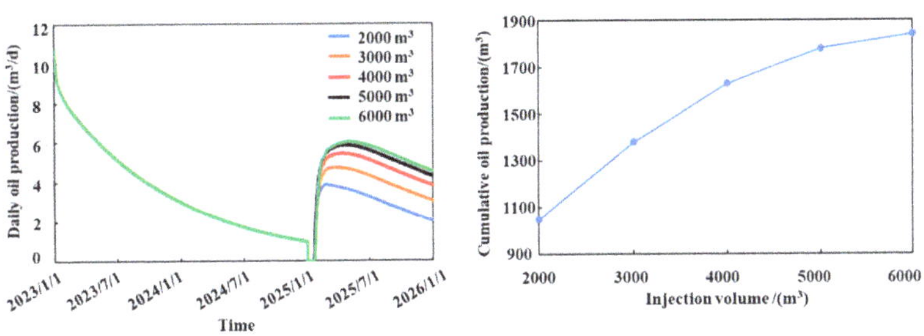


**Figure 9.** Daily production rate curves (**left**) and the cumulative production curve (**right**) under different injection volumes: both cumulative and daily oil production initially increase with injection volume; the incremental gains diminish significantly at higher volumes.

This trend is consistent with the concept of diminishing marginal returns, where each additional unit of injected fluid yields less additional oil due to reduced displacement efficiency and increasing water production. As Zhang et al. point out, beyond a certain injection threshold, further volume fails to proportionally increase the sweep area or oil mobility and instead leads to premature water breakthrough and inefficient use of injection energy [42,43]. Therefore, excessive injection not only offers limited production benefits but may also adversely affect economic and operational efficiency. Optimizing injection volume is thus essential to maximize recovery while minimizing unnecessary water handling and operational costs.

4.2.2. Shut-In Period

Well shut-in leverages the hydrophilicity of the reservoir, enhancing imbibition and oil displacement, redistributing oil–water saturation, and restoring pressure to facilitate oil recovery. The shut-in period significantly impacts the huff-and-puff flooding efficiency and ultimate oil recovery. To determine the optimal shut-in period, simulations were conducted with shut-in periods of 5, 10, 15, 20, 25, and 30 days under an injection volume of 4000 $m^3$. The results (Figure 10) show that both cumulative and daily oil production increase with shut-in time, exhibiting a roughly positive correlation. However, the growth rate slows after 15 days, with a clear plateau, indicating that the optimal shut-in time is around 15–20 days.

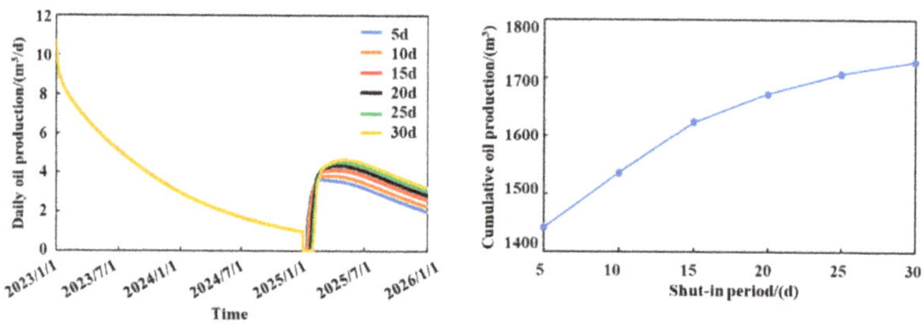


**Figure 10.** Daily production rate curves (**left**) and the cumulative production curve (**right**) under different shut-in periods: both cumulative and daily oil production increase with shut-in time, exhibiting a roughly positive correlation.

4.2.3. Huff-and-Puff Flooding Cycle

In the huff-and-puff flooding oil recovery process, the oil production increment from a single huff-and-puff flooding cycle is limited, necessitating multiple huff-and-puff flooding rounds. To evaluate the impact of huff-and-puff flooding cycles on oil recovery, simulations were conducted with an injection volume of 4000 $m^3$, shut-in period of 15 days, and a single cycle duration of 20 months. The oil production per cycle and cumulative oil production for three huff-and-puff flooding cycles were calculated. As shown in Figure 11, multiple huff-and-puff flooding cycles increase cumulative oil production. However, with each additional cycle, the oil increment per cycle decreases, primarily due to the reduction in oil saturation and diminishing effectiveness of each successive cycle.

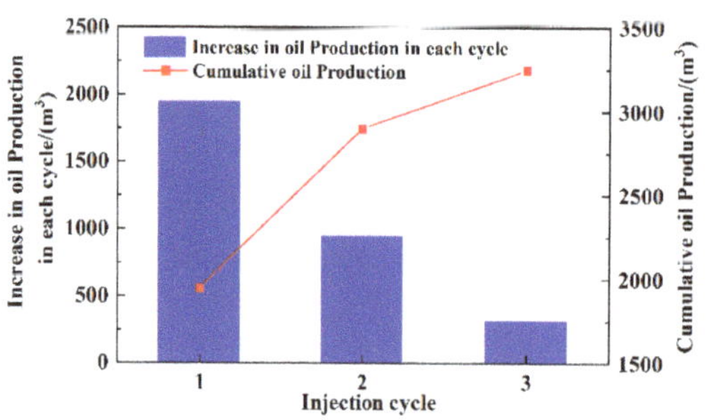


**Figure 11.** Increase in oil production in each huff-and-puff flooding cycle and cumulative oil production: the cumulative oil production increases with the number of huff-and-puff cycles, but the oil production per individual cycle shows a decreasing trend.

## 5. Discussions

While our study provides optimized injection volumes and shut-in periods for Nano ODA in the Kexia Formation, it is important to acknowledge several practical limitations for field implementation. Long-term injectivity may be affected by formation heterogeneity, pore-throat clogging, or potential adsorption of the nanoemulsion on rock surfaces. Chemical loss through adsorption or emulsification breakdown could reduce effective concentration in the reservoir. Moreover, operational challenges such as maintaining stable injection rates and handling large volumes of chemical agents must be considered. Environmental aspects, including potential toxicity, disposal of produced fluids, and regulatory compliance, also need to be addressed in large-scale applications. Future work should incorporate pilot-scale field tests and environmental risk assessments to ensure safe and effective deployment of Nano ODA.

## 6. Conclusions

This study investigated the huff-and-puff flooding mechanism in the Baikouquan Oilfield Kexia Formation through laboratory experiments and numerical simulations. Results show that in this water-wet, low-porosity lithic sandstone, ODAs enhance recovery via a synergistic mechanism of reservoir energy replenishment, significant IFT reduction (up to 97% for the nanoemulsion-based ODA), and strong emulsification (up to 75% efficiency) that mobilizes residual oil clusters and shifts relative permeability. Unlike most previous studies that attribute recovery mainly to wettability alteration and spontaneous imbibition, our findings reveal that in strongly water-wet formations, emulsification and IFT reduction dominate, with the nanoemulsion providing a particularly effective oil mobilization mechanism in nanoscale pores. Numerical simulations further indicate that 4000 $m^3$ injection with a 15–20 day shut-in is the optimal design for well 1090. Overall, this work not only provides a reservoir-specific flooding strategy for the Kexia Formation but also highlights a distinct recovery mechanism where nanoemulsion-driven emulsification plays the dominant role, advancing the understanding of surfactant–nanoparticle-assisted EOR in unconventional reservoirs.

**Author Contributions:** Conceptualization, J.M. and H.T.; methodology, J.M. and K.X.; software, H.L.; validation, B.G.; formal analysis, X.Y. and H.T.; investigation, H.L. and K.X.; writing—original draft preparation, H.L. and B.G.; writing—review and editing, J.M. and H.T. All authors have read and agreed to the published version of the manuscript.

**Funding:** This research received no external funding.

**Data Availability Statement:** The original contributions presented in this study are included in the article. Further inquiries can be directed to the corresponding author.

**Conflicts of Interest:** Hui Tian, Kunlin Xue and Xingyu Yi were employed by the PetroChina. The remaining authors declare that the research was conducted in the absence of any commercial or financial relationships that could be construed as a potential conflict of interest.

## References

1. Sun, W.; Zhi, H.; Li, J. Analysis of water flooding mechanism and development effect after water breakthrough in Ansete low permeability oilfield. *Exp. Petr. Geol.* **1999**, *3*, 256–260.
2. Thomas, A.; Kumar, A.; Rodrigues, K.; Sinclair, R.; Lackie, C.; Galipeault, A.; Blair, M. Understanding water flood response in tight oil formations: A case study of the lower Shaunavon. In Proceedings of the SPE/CSUR Unconventional Resources Conference, Calgary, AB, Canada, 30 September–2 October 2014; SPE-171671-MS. [CrossRef]
3. Wang, T. Influencing factors and analysis of carbon dioxide flooding effect. *China Petrol. Chem. Ind.* **2008**, *24*, 30–33.
4. Shang, Q. Sensitivity analysis of $CO_2$ flooding oil well productivity and influencing factors. *Petr. Dril. Tech.* **2011**, *39*, 83–88.
5. Wang, H.; Chen, Z.; Zhang, L. Discussion on injection huff and puff development model of water-wet sandstone reservoir. *Petrol. Explor. Dev.* **2004**, *5*, 86–88.
6. Li, G.; Zhang, S. Study on capillary imbibition effect of low permeability sandstone simulated by quartz glass. *J. Chongqing Univ. Sci. Technol. (Nat. Sci.)* **2010**, *12*, 1–3+7.

7. Sheng, J. What type of surfactants should be used to enhance spontaneous imbibition in shale and tight reservoirs? *J. Petrol. Sci. Eng.* **2017**, *159*, 635–643. [CrossRef]
8. Standnes, D.; Austad, T. Wettability alteration in chalk: Mechanism for wettability alteration from oil-wet to water-wet using surfactants. *J. Petrol. Sci. Eng.* **2000**, *28*, 123–143. [CrossRef]
9. Salehi, M.; Johnson, S.; Liang, J. Mechanistic study of wettability alteration using surfactants with applications in naturally fractured reservoirs. *Langmuir* **2008**, *24*, 14099–14107. [CrossRef] [PubMed]
10. Huang, D.; Xiang, D. Study on the mechanism of injection huff and puff oil recovery. *Petrol. Geol. Rec. Effi.* **2004**, *5*, 39–40+43.
11. Hendraningrat, L.; Torsæter, O. A study of water chemistry extends the benefits of using silica-based nanoparticles on enhanced oil recovery. *Appl. Nanosci.* **2016**, *6*, 83–95. [CrossRef]
12. Foster, W. A low-tension waterflooding process. *J. Petrol. Technol.* **1973**, *25*, 205–210. [CrossRef]
13. Youssif, M.; El-Maghraby, R.; Saleh, S.; Elgibaly, A. Silica nanofluid flooding for enhanced oil recovery in sandstone rocks. *Egypt J. Pet.* **2018**, *27*, 105–110. [CrossRef]
14. Wellington, S.; Richardson, E. Low surfactant concentration enhanced waterflooding. *SPE J.* **1997**, *2*, 389–405. [CrossRef]
15. Gao, T. Injection huff mechanism and application of tight reservoir in Yanchang Oilfield. *Spec. Oil Gas Res.* **2018**, *25*, 134–137.
16. Liu, L. Numerical Simulation of Imbibition Production in Fractured Reservoirs. Master's Thesis, Southwest Petroleum University, Chengdu, China, 2006.
17. Kaul, S.; Putra, E.; Schechter, D. Simulation of spontaneous imbibition using Rayleigh-Ritz finite element method—A discrete fracture approach. In Proceedings of the Canadian International Petroleum Conference, Calgary, AB, Canada, 8–10 June 2004; PETSOC-2004-228.
18. Nguyen, V. A dynamic network model for imbibition. In Proceedings of the SPE Annual Technical Conference and Exhibition, Houston, TX, USA, 26 September 2004; SPE-90365-MS. [CrossRef]
19. Zhu, Y. Study on Oil Recovery by Nano-Active Water Huff and Puff After Fracturing in Tight Reservoirs. Master's Thesis, China University of Petroleum, Beijing, China, 2022.
20. *SY/T 5370-2018*; Test Method for Surface Tension and Interfacial Tension. China Petroleum Industry Publishing House: Beijing, China, 2019.
21. *SY/T 5153-2017*; Test Method of Reservoir Rock Wettability. China Petroleum Industry Publishing House: Beijing, China, 2017.
22. *Q/SY17583-2018*; Technical Specification for Surfactants Used in Binary Composite Flooding. China Petroleum Industry Publishing House: Beijing, China, 2018.
23. *GB/T 28912-2012*; Test Method for Two Phase Relative Permeability in Rock. China Standards Press: Beijing, China, 2012.
24. Green, D.W.; Willhite, G.P. *Enhanced Oil Recovery*; Society of Petroleum Engineers: Richardson, TX, USA, 2018. [CrossRef]
25. Bui, T.; Frampton, H.; Huang, S.; Collins, I.; Striolo, A.; Michaelides, A. Water/oil interfacial tension reduction—An interfacial entropy driven process. *Phys. Chem. Chem. Phys.* **2021**, *44*, 25075–25085. [CrossRef] [PubMed]
26. Najimi, S.; Nowrouzi, I.; Khaksar, A.; Mohammadi, A. Experimental study of the performances of commercial surfactants in reducing interfacial tension and wettability alteration in the process of chemical water injection into carbonate reservoirs. *J. Pet. Explor. Prod. Technol.* **2020**, *10*, 1551–1563. [CrossRef]
27. Liang, T.; Hou, J.; Qu, M.; Xi, J.; Raj, I. Application of nanomaterial for enhanced oil recovery. *Petrol. Sci.* **2022**, *19*, 882–899. [CrossRef]
28. Rosen, M.; Kunjappu, J. *Surfactants and Interfacial Phenomena*, 4th ed.; Wiley: Hoboken, NJ, USA, 2012.
29. Dang, H.; Wang, X.; Duan, W. Research on imbibition displacement of fractured low permeability reservoir in Ordos Basin. *Fault Block Oil Gas Field* **2017**, *24*, 687–690.
30. Li, Y.; Li, M.; Xu, H.; Yang, F.; Yan, M.; Tao, T. Surfactant flooding technology in ultra-low permeability reservoir. *Liaoning Chem. Ind.* **2020**, *49*, 804–806.
31. Zhang, J.; Wang, M.; Wu, J. Study on synthesis and pressure flooding technology of a new nano imbibition agent. *Appl. Chem. Ind.* **2021**, *50*, 1239–1244.
32. Liu, C.; Wang, Y.; Yu, H. Laboratory study on surfactant flooding system in low permeability reservoir. *Oil Gas Chem. Ind.* **2011**, *40*, 486–489.
33. Cui, Z. Research on the Influence of Wettability of Low Permeability Reservoir on oil Recovery. Master's Thesis, Daqing Petrol. Inst., Daqing, China, 2009.
34. Phukan, R.; Saha, R.; Mazumdar, P. Adsorption behavior of surfactants on sandstone reservoir rocks with carbonate cements and its influence on wettability alteration. *J. Surfactants Deterg.* **2024**, *27*, 392–408. [CrossRef]
35. Li, N.; Wang, H.; Li, J. Research status and development trend of surfactants for oil displacement. *Guangdong Chem. Ind.* **2012**, *39*, 98–99+127.
36. Ratanpara, A.; Kim, M. Wettability alteration mechanisms in enhanced oil recovery with surfactants and nanofluids: A review with microfluidic applications. *Energies* **2023**, *16*, 8003. [CrossRef]

37. Malode, G.; Ande, S.; Chavhan, S.; Bartare, S.; Malode, L.; Manwar, J.; Bakal, R. A critical review on nanoemulsion: Advantages, techniques and characterization. *World J. Adv. Res. Rev.* **2021**, *11*, 462–473. [CrossRef]
38. Jalilian, M.; Tabzar, A.; Ghasemi, V.; Mohammadzadeh, O.; Pourafshary, P.; Rezaei, N.; Zendehboudi, S. An experimental investigation of nanoemulsion enhanced oil recovery: Use of unconsolidated porous systems. *Fuel* **2019**, *251*, 754–762. [CrossRef]
39. Xu, H.; Li, Y.; Wu, H.; Ding, Z.; Yuan, S.; Bai, H.; Yao, E.; Zhou, F. Development and performance evaluation of nonionic surfactant-stabilized nanoemulsion for enhanced oil recovery applications in tight reservoir. *Phys. Fluids* **2024**, *36*, 032015. [CrossRef]
40. Ding, X.; Zhou, D.; Wu, K.; Li, S.; He, Y.; Yu, B.; Chen, L. Performance evaluation and application of nanoemulsion imbibition oil-displacing agent. *Oilfield Chem.* **2022**, *39*, 651–657.
41. Morrow, N.R.; Mason, G. Recovery of oil by spontaneous imbibition. *Curr. Opin. Colloid Interface Sci.* **2001**, *6*, 321–337. [CrossRef]
42. Zhang, L.; Wang, J.; Liu, H. Optimization of cyclic water injection parameters in low-permeability reservoirs: Understanding the marginal returns of injection volume. *J. Pet. Sci. Eng.* **2021**, *198*, 108209.
43. Kukharova, T.; Maltsev, P.; Novozhilov, I. Development of a control system for pressure distribution during gas production in a structurally complex field. *Appl. Syst. Innov.* **2025**, *8*, 51. [CrossRef]

*processes*

MDPI

*Article*

# Optimization of Fracturing Sweet Spot in Deep Carbonate Reservoirs by Combining TOPSIS and AHP Algorithm

**Yong Liu [1], Guiqi Xie [1], Honglin Zheng [2,3,*], Xinfang Ma [2], Guangcong Ren [2,4], Xinyuan Feng [1], Wenkai Zhao [1], He Ma [2] and Fengyu Lei [1]**

[1] Drilling and Production Technology Research Institute, PetroChina Qinghai Oilfield Company, Dunhuang 736200, China; liuyzcqh@petrochina.com.cn (Y.L.); zhaowenkaijxqh@petrochina.com.cn (W.Z.)
[2] College of Petroleum Engineering, China University of Petroleum (Beijing), Beijing 102200, China
[3] SINOPEC Research Institute of Petroleum Engineering Co., Ltd., Beijing 102206, China
[4] CNPC Engineering Technology R&D Company Limited, Beijing 102206, China
* Correspondence: zhhl1214.sripe@sinopec.com or zhenghl12144869@163.com; Tel.: +86-18302405560

**Abstract:** The deep carbonate reservoirs in the Yingzhong Block of the Qaidam Basin exhibit strong vertical heterogeneity and complex natural fracture development. Conventional fracability evaluation methods struggle to accurately characterize formation features, thereby affecting the stimulation effectiveness. To enhance the evaluation accuracy of fracturing sweet spot intervals, automatic mineral scanning equipment is employed to obtain formation micro-physical property parameters at continuous depths. Considering the temperature-pressure coupling effect under deep conditions, a rock mechanics computational model based on mineral composition was established to derive macroscopic mechanical parameters such as brittleness index and in situ stress. Based on a combined algorithm of the improved Technique for Order Preference by Similarity to Ideal Solution (TOPSIS) and Analytic Hierarchy Process (AHP), a fracturing sweet spot prediction model integrating micro- and macro-multi-factors is established, and sweet spot index levels are classified. The research results indicate that the rock mechanics computational model demonstrates high accuracy, the calculated macroscopic parameters are reliable, and the fracturing sweet spot index model can fracability and meticulously evaluate the characteristics of deep carbonate formations. The fracturing sweet spots can be classified into three levels: Level I with an index higher than 0.50, Level II with an index between 0.35 and 0.50, and Level III with an index lower than 0.35. After using this method for layer selection, the fracture pressure decreases by 11.6%, and the sand addition success rate increases by 24%. Applying this method to guide the optimization of fracturing intervals demonstrates good on-site practical value, providing an important reference for identifying fracturing sweet spots in deep carbonate reservoirs.

**Keywords:** deep carbonate reservoir; cuttings electron microscopy scanning; combined TOPSIS-AHP algorithm; multi-factors integration; fracturing sweet spot index

## 1. Introduction

With the expansion of global oil and gas exploration and development into deep and ultra-deep fields, carbonate rock reservoirs have become important alternative resources. Both the Tarim Basin and the Qaidam Basin have large areas of carbonate rock reservoirs, which are superimposed and interbedded with high-quality source rocks. They are rich in oil and gas resources and are the focus of oil and gas exploration and development, with huge development potential [1,2]. However, such reservoirs generally have characteristics such as deep burial, strong heterogeneity, and complex development of natural fractures, which leads to low identification accuracy of sweet spots in fracturing and modification projects. In the actual stimulation process, problems such as single well rupture/excessive stimulation pressure, difficulty in sand addition, and poor modification efficiency are often encountered, seriously restricting the economic and effective development of deep carbonate reservoirs. Therefore, it is urgently necessary to explore fine evaluation methods for deep carbonate reservoirs and carry out research on the identification of fracturing sweet spots, in order to improve the success rate of fracturing in such reservoirs.

The core of fracturing layer section optimization is the fine characterization of the reservoir. In recent years, scholars at home and abroad have conducted relevant research. In 2012, Huang Jinliang et al. [3] considered geological engineering conditions such as shale thickness, gas content, burial depth, permeability, and fracture distribution, and selected the favorable area for shale gas exploration and development in the Longmaxi Formation. In 2014, Li Huayang et al. [4] established a calculation model for the inversion of the brittleness index of tight sandstone using conventional logging by applying rock mechanics experiments and whole-rock X-ray diffraction analysis experiments, providing technical support for the selection of oil and gas "sweet spots" in tight sandstone. In 2017, Yang Hongwei et al. [5] established a continuous compressible mathematical model for long horizontal sections of shale based on logging data by using the Analytic Hierarchy process (AHP) and fuzzy mathematics methods, guiding the design of perforation parameters. In 2018, Zhai Wenbao et al. [6] introduced the mutation theory to develop a new method for evaluating the fractionability of shale reservoirs, providing a theoretical basis for pre-pressure evaluation and the effect of stimulation. In 2020, Lin Yu et al. [7] conducted research on OVT domain migration processing and sensitive attribute optimization, and developed a set of fine seismic characterization techniques for small-scale fracture-cave reservoirs in deep carbonate rocks. In 2023, Feng Xinyuan et al. [8] developed a sweet spot evaluation technology for shale oil development projects by constructing a digital model of the spatial structure of mineral composition and calculating the fracability index of shale oil in the Qaidam Basin.

With scholars' further understanding of the complexity of strata, the optimization technology of fracturing sweet spots has gradually shown a development trend of diversified influencing factors and diversified evaluation methods. The evaluation criteria for fracturing sweet spots have evolved from initial reliance solely on brittleness index to a comprehensive quantitative characterization system that incorporates multiple factors including rock mechanical properties, in-situ stress, and natural fractures.. The acquisition of rock mechanical properties and in situ stress parameters is typically based on well logging data such as acoustic wave, density, gamma ray, and caliper logs, with inverse calculations performed using Huang's model. However, for deep reservoir conditions, this method has not undergone block-specific corrections, and both the logging data and the inversion results exhibit limitations in terms of accuracy. Therefore, in existing research methodologies, there remains a problem of insufficient availability of formation data and inadequate adaptability of model representation. Deep carbonate rock formations have the characteristics of ultra-high temperature and ultra-high pressure. Key parameters such as acoustic waves and density in logging data are significantly affected by wellbore conditions, and logging quality degradation may occur; secondly, core sampling is costly and has limited representativeness, making it difficult to fully reflect the heterogeneity of the reservoir. In terms of sweet spot prediction models, current research predominantly focuses on sandstone and shale formations, commonly employing the Analytic Hierarchy Process (AHP) to establish linear fracability index calculation models. However, these models often neglect the influence of reservoir burial depth and fail to account for the lateral heterogeneity of the reservoir, which limits the effective application of existing achievements in the fracturing of deep carbonate reservoirs.

This paper fully considers the characteristics of deep carbonate reservoirs, introduces automatic mineral identification scanning equipment, and conducts electron microscopy scanning experiments on logging cuttings to obtain microscopic physical parameters such as formation mineral content, the number of micro-fractures and porosity under continuous well sections. Considering the temperature–pressure coupling effect under deep conditions, a rock mechanics calculation model based on mineral components was established to obtain macroscopic mechanical parameters such as formation brittleness index and in situ stress. Based on the TOPSIS-AHP joint algorithm, fracability considering four micro-macro data such as porosity, the number of micro-fractures, brittleness index, and bidirectional stress difference coefficient, a multi-factor fusion fracturing sweet spot prediction model was established. The fracturing sweet spot index ($I_{FSS}$) was calculated and the sweet spot levels were classified, achieving a comprehensive characterization of deep carbonate rock formations. It provides a new idea for the selection of sweet spot layers for fracturing in this type of reservoir.

## 2. Methods for Obtaining Microscopic-Macroscopic Parameters

### 2.1. Methods for Obtaining Microscopic Parameters

Clarifying the mineral composition and micro-pore structure of reservoirs is a prerequisite for understanding their fundamental characteristics. To precisely characterize the features of deep carbonate formations, formation mineral data, pore structure data, and the number of micro-fractures are obtained through cutting scanning electron microscopy (SEM) experiments. The equipment used for cutting scanning is the RoqSCAN automated mineral analysis SEM (CGG Company, Houston, TX, USA), a novel quantitative analysis technology for reservoir rock mineral composition and microstructure. The scanning device is equipped with a high-resolution SEM (Carl Zeiss, New York, NY, USA), a backscattered electron detector, a high-performance energy-dispersive spectrometer (Bruker Corporation, Billerica, MA, USA), and automated analysis software, providing essential parameters for reservoir evaluation through its scanning results.

The automated mineral analysis SEM employs electron imaging, with its working principle illustrated in Figure 1. The electron gun at the top of the column generates electrons in a vacuum environment, and the released electrons accelerate towards the positively charged anode. After passing through the condenser lens, they form a high-energy electron beam that bombards the surface of the scanned sample. When electrons collide with the sample surface, various types of photons, electrons, and rays are produced. The equipment identifies different elements by capturing the characteristic X-rays of the scanned points and then classifies these elements into different mineral types [9,10]. Pores and micro-fractures are identified through the grayscale values of backscattered images, enabling the derivation of scanning data at different formation depths [11,12].

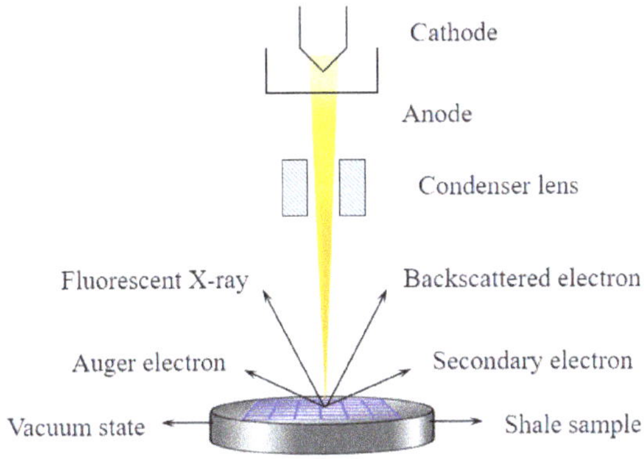


**Figure 1.** Schematic diagram of the working principle of a scanning electron microscope.

The flowchart of the RoqSCAN rock cuttings scanning experiment is shown in Figure 2. The preparation of the experimental samples mainly consists of three steps: rock cuttings sample curing, sample surface grinding, and sample surface carbon plating. The rock cuttings samples are cured using acrylic resin powder and liquid curing agent. After curing, the sample size is Φ 30 mm × H13 mm (±2 mm). The surface of the rock cuttings samples was roughly ground, finely ground and polished using a geological grinding instrument to expose the fresh side of the rock cuttings, eliminate the shadow effect in backscattered imaging. A carbon film was deposited onto the surface of the sample using a Leica EM ACE200 carbon coater to mitigate the charging effect during electron microscopy scanning and enhance the signals of secondary electrons and backscattered electrons. After the sample preparation is completed, it is placed in the vacuum chamber stage of the RoqSCAN equipment. The instrument is evacuated and then demagnetized, filament correction, focusing, contrast adjustment and other operations are carried out in sequence. After debugging, the sample is scanned point by point. After the scanning is completed, high-resolution scanning result images, mineral composition and content data, pore size distribution and porosity data, and the number of micro-fractures are generated through SmartPI software (Version 5.06), providing data support for calculating macroscopic rock mechanical parameters.

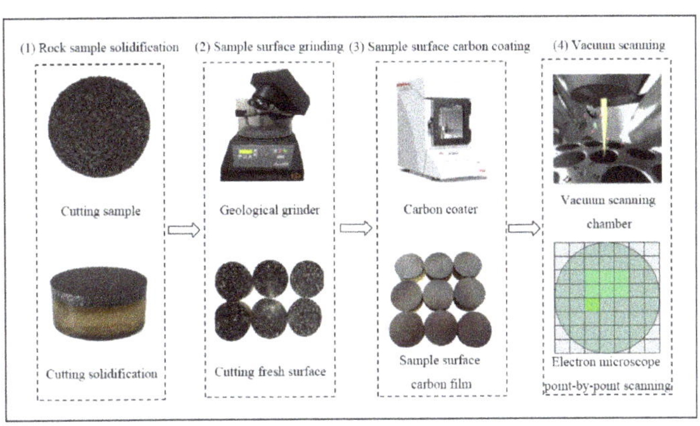


**Figure 2.** Physical diagram of production process of rock cutting scanning sample.

### 2.2. Macroscopic Rock Mechanics Parameter Calculation Model

#### 2.2.1. Rock Mechanics Model

The construction of the macroscopic rock mechanical model primarily encompasses three core components: a dry rock mechanical model, a saturated rock mechanical model [13–15], and an in situ stress calculation model. The calculation workflow for macroscopic mechanical parameters based on mineral composition is illustrated in Figure 3. The model first calculates the initial elastic parameters of the rock based on the mineral components and their contents. Combining factors such as pore structure data, effective formation pressure, rock stiffness tensor, and coordination number, it then employs a model based on Hertz-Mindlin theory to derive the mechanical properties of the dry rock formation. The Hertz-Mindlin model is applicable to unconsolidated or weakly consolidated granular media. Its fundamental principle involves simulating the elastic contact behavior of spherical particles under effective stress, calculating the effective modulus based on Hertzian contact theory.

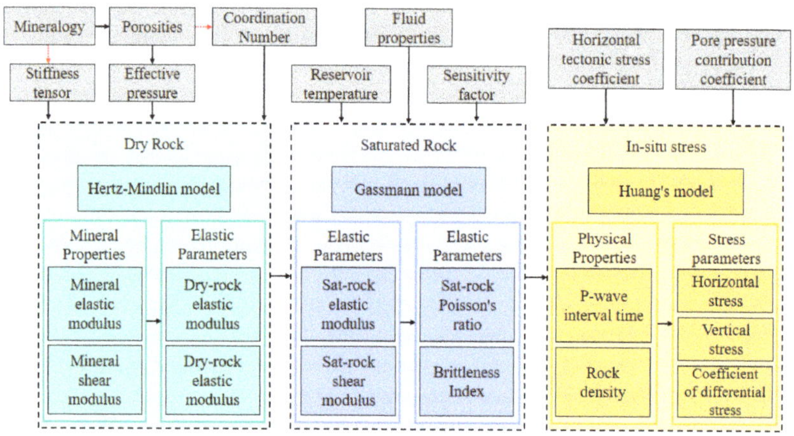


**Figure 3.** Calculation procedure for the macroscopic rock mechanics model.

Building upon the dry rock modulus, the model further assumes that the pore fluid has a zero shear modulus (i.e., the fluid's influence on shear stiffness is neglected). Simultaneously, considering the properties of the formation fluid and the impact of reservoir temperature changes on fluid properties, a temperature sensitivity factor is introduced. The Gassmann model is then used for fluid substitution to compute the mechanical characteristics of the saturated rock. The Gassmann model is suitable for low-frequency saturation conditions and requires the rock to be macroscopically homogeneous, with connected pores and no chemical interaction between the fluid and the rock frame. Its basic principle relies on the contribution of the pore fluid to the bulk modulus, calculating the saturated rock modulus through the relationship between the fluid and the dry rock frame modulus.

Finally, based on the calculated saturated rock mechanical parameters (such as dynamic elastic modulus) and incorporating formation temperature and pressure conditions, Huang's model is adopted to compute macroscopic mechanical parameters like the brittleness index and in situ stress. This model comprehensively accounts for the complex influence of formation temperature and pressure on rock mechanical properties and is suitable for the quantitative assessment of stress parameters in tight reservoirs. Through the synergistic application of the aforementioned models, cross-scale prediction from mineral composition, pore structure, and temperature-pressure environment to macroscopic mechanical responses is achieved, providing reliable theoretical support for reservoir mechanical characterization and engineering decision-making.

Regression calculation of rock properties based on weighted average values of different mineral inherent property parameters:

$$E_{\mathrm{m}} = \sum_i f_i E_i \tag{1}$$

$$\mu_{\mathrm{m}} = \sum_i f_i \mu_i \tag{2}$$

In Equations (1) and (2), $E_{\mathrm{m}}$ represents the initial elastic modulus of the rock, GPa; $\mu_{\mathrm{m}}$ is the initial rock shear modulus, GPa; $E_i$ is the elastic modulus of the $i$-th mineral itself, GPa. $\mu_i$ is the shear modulus of the $i$-th mineral itself, GPa. $f_i$ represents the percentage of the $i$-th mineral content, %.

In deep carbonate reservoirs, formation pressure has a significant impact on the internal pore structure of rocks. In dry rock mechanics calculation models, the influence of confining pressure needs to be considered. The dry rock mechanics model based on the contact conditions of rock component particles and pore structure is:

$$E_{\mathrm{dry}} = \left[\frac{C^2(1-\phi)^2 E_{\mathrm{m}}^2}{18\pi^2(1-\mu_{\mathrm{m}})^2} P_{\mathrm{eff}}\right]^{1/3} \tag{3}$$

$$\mu_{\mathrm{dry}} = \frac{5-4\mu_{\mathrm{m}}}{5(2-\mu_{\mathrm{m}})}\left[\frac{3C^2(1-\phi)^2 \mu_{\mathrm{m}}^2}{2\pi^2(1-\mu_{\mathrm{m}})^2 M} P_{\mathrm{eff}}\right]^{1/3} \tag{4}$$

In Equations (3) and (4), $E_{\mathrm{dry}}$ represents the elastic modulus of the dry rock sample, GPa; $C$ is the particle contact coefficient of the rock component, dimensionless. $E_{\mathrm{m}}$ is the initial elastic modulus of the rock, GPa; $\mu_{\mathrm{dry}}$ is the shear modulus of the dry rock sample, GPa; $\mu_{\mathrm{m}}$ is the initial rock shear modulus, GPa; $\mu_{\mathrm{m}}$ is the shear modulus, GPa; $\varphi$ represents total porosity, %; $P_{\mathrm{eff}}$ stands for effective formation pressure, in MPa. $M$ is the rock stiffness tensor, GPa.

In deep carbonate reservoirs, minerals exhibit temperature sensitivity. Under high-temperature conditions, carbonate minerals may undergo thermal pyrolysis. By introducing a temperature sensitivity coefficient, we can characterize the rate of change in elastic parameters under deep high-temperature conditions. When the rock pores are given a fluid saturated state, the calculation model for the mechanical property parameters of saturated rocks considering the temperature sensitivity coefficient is:

$$E_{\mathrm{sat}} = \left[E_{\mathrm{dry}} + \frac{\left(1 - E_{\mathrm{dry}}/E_{\mathrm{m}}\right)^2}{\phi/E_{\mathrm{fl}} + (1-\phi)/E_{\mathrm{m}} - E_{\mathrm{dry}}/E_{\mathrm{m}}^2}\right] \cdot [1 - \alpha_{\mathrm{E}}(T - T_0)] \tag{5}$$

$$\mu_{\mathrm{sat}} = \mu_{\mathrm{dry}} \cdot [1 - \alpha_{\mathrm{v}}(T - T_0)] \tag{6}$$

$$v_{\mathrm{sat}} = \frac{3E_{\mathrm{sat}} - 2\mu_{\mathrm{sat}}}{2(3E_{\mathrm{sat}} + \mu_{\mathrm{sat}})} \tag{7}$$

In Equations (5)–(7), $E_{\mathrm{sat}}$ represents the elastic modulus of saturated rock, GPa; $E_{\mathrm{fl}}$ is the elastic modulus of the fluid, GPa. $\alpha_{\mathrm{E}}$ is the temperature sensitivity coefficient of the elastic modulus, with a value of $1.49 \times 10^{-3}$, 1/°C. $\alpha_{\mathrm{v}}$ is the temperature sensitivity coefficient of Poisson's ratio, with a value of $1.53 \times 10^{-3}$, 1/°C [16]. $T$ represents the formation temperature, in °C. $T_0$ is room temperature, °C. $\mu_{\mathrm{sat}}$ is the shear modulus of saturated rock samples, GPa; $v_{\mathrm{sat}}$ is the Poisson's ratio of saturated rocks and is dimensionless.

Based on the calculation models of rock elastic modulus and Poisson's ratio, the calculation models of formation rock density, longitudinal wave time difference, brittleness index and in situ stress can be further constructed:

$$\rho_R = (1-\phi)\rho_{\mathrm{m}} + (\phi - \phi_{\mathrm{s}})\rho_{\mathrm{f}} + \phi_{\mathrm{s}}\rho_{\mathrm{w}} \tag{8}$$

$$V_{\mathrm{p}} = \sqrt{\left(K_{\mathrm{sat}} + \frac{4}{3}v_{\mathrm{sat}}\right)/\rho_R} \tag{9}$$

$$BI = \frac{K_{\mathrm{sat}} - K_{\min}}{2(K_{\max} - K_{\min})} + \frac{v_{\max} - v_{\mathrm{sat}}}{2(v_{\max} - v_{\min})} \tag{10}$$

$$\sigma_{\mathrm{H}} = \left(\frac{v}{1-v} + \beta_{\mathrm{H}}\right)(\sigma_{\mathrm{v}} - VP_{\mathrm{eff}}) + VP_{\mathrm{eff}} \tag{11}$$

$$\sigma_{\mathrm{h}} = \left(\frac{v}{1-v} + \beta_{\mathrm{h}}\right)(\sigma_{\mathrm{v}} - VP_{\mathrm{eff}}) + VP_{\mathrm{eff}} \tag{12}$$

$$\sigma_{\mathrm{v}} = \int_0^H \rho_{\mathrm{R}} dh \tag{13}$$

$$C_{\mathrm{h}} = (\sigma_{\mathrm{H}} - \sigma_{\mathrm{h}})/\sigma_{\mathrm{h}} \tag{14}$$

In Equations (8)–(14), $\rho_{\mathrm{R}}$ represents the density of the formation, in $\mathrm{g/cm^3}$. $\rho_{\mathrm{m}}$ is the density of pure rock, $\mathrm{g/cm^3}$. $\rho_{\mathrm{f}}$ represents the density of the reservoir fluid, $\mathrm{g/cm^3}$. $\rho_{\mathrm{w}}$ is the density of bound water, $\mathrm{g/cm^3}$. $\varphi$ represents total porosity, %; $\varphi_{\mathrm{s}}$ is the porosity occupied by the imflowable fluid, %; $V_{\mathrm{p}}$ represents the velocity of the longitudinal wave, in m/s. $v_{\mathrm{sat}}$ is the Poisson's ratio of saturated rocks and is dimensionless. $BI$ is the brittleness index and is dimensionless. $E_{\max}$ is the upper limit of the elastic modulus of saturated rocks, GPa; $E_{\min}$ is the lower limit of the elastic modulus of saturated rock, GPa; $v_{\mathrm{sat}}$ is the Poisson's ratio of saturated rock, dimensionless. $v_{\max}$ is the upper limit of the Poisson's ratio of saturated rocks and is dimensionless. $v_{\min}$ is the lower limit of the Poisson's ratio of saturated rocks and is dimensionless. $\sigma_{\mathrm{H}}$ represents the maximum horizontal principal stress, in MPa; $\sigma_{\mathrm{v}}$ is the vertical stress, MPa; $\sigma_{\mathrm{h}}$ is the minimum horizontal principal stress, MPa; $\beta_{\mathrm{H}}$ and $\beta_{\mathrm{h}}$ are horizontal structural stress coefficients, dimensionless. $V$ is the contribution coefficient of formation pore pressure, dimensionless; $C_{\mathrm{h}}$ is the coefficient of stress difference in both directions, dimensionless.

### 2.2.2. Model Verification

A typical carbonate reservoir fracturing vertical well in the Anglo-Chinese block was selected as the verification object. The distribution of carbonate rocks in the controlled reservoir of this well is stable and the data is complete. The reservoir of the example well was buried at a depth of 5600 to 5700 m. Electron microscopy scans of 20 groups of rock cuttings samples were conducted for the test oil layer. Based on the above rock mechanics calculation model, the reservoir density, longitudinal wave time difference and other rock physical property parameters were calculated. The key input parameter values of the model are shown in Table 1. The accuracy of the model was verified using the actual logging data of this well. The measured data were fitted through the calculation results of the model. The fitting results are shown in Figure 4. The range of the actual logging acoustic time difference data is 174.9–201.5 μs/m, and the density range is 2.645–2.785 $\mathrm{g/cm^3}$. According to the Pearson correlation calculation, the correlation coefficient between the longitudinal wave time difference calculated by the model and the logging acoustic data is 0.631, and the correlation coefficient of the density data is 0.817. The main reason for the difference is that the immersion of the drilling fluid or formation fluid on the wellbore will affect the acoustic characteristics of the rock, thereby affecting the acoustic time difference. Secondly, the accuracy of electrical measuring instruments and operation methods, among other factors, can all lead to data errors. The Pearson correlation coefficients between the electrical measured values and the calculated values are all greater than 0.6, which proves that the two sets of data have a high similarity, verifies the agreement between the model's calculated results and the measured data, and indicates that the model in this paper can accurately calculate the mechanical parameters of reservoir rocks.

**Table 1.** Key input parameters for rock mechanics calculation model.

| Parameter Name | Parameter Value | Unit |
|---|---|---|
| Elastic modulus of quartz minerals $E_q$ | $7.85 \times 10^4$ | MPa |
| Elastic modulus of calcite mineral $E_c$ | $5.80 \times 10^4$ | MPa |
| Elastic modulus of feldspar minerals $E_a$ | $4.5 \times 10^4$ | MPa |
| Percentage of quartz mineral content $f_q$ | 5.7~13.1 | % |
| Percentage of calcite mineral content $f_c$ | 16.0~25.3 | % |
| Percentage of plagioclase mineral content $f_a$ | 8.6~17.8 | % |
| Particle contact coefficient C | 9.0 | Dimensionless |
| Total porosity $\varphi$ | 1.82 | % |
| Porosity of non-flowable fluids $\varphi_s$ | 0.58 | % |
| Effective formation pressure $p_{eff}$ | 54.9 | MPa |
| Rock stiffness tensor $M$ | 0.82 | MPa |
| Density of pure rock $\rho_m$ | 2.65 | g/cm$^3$ |

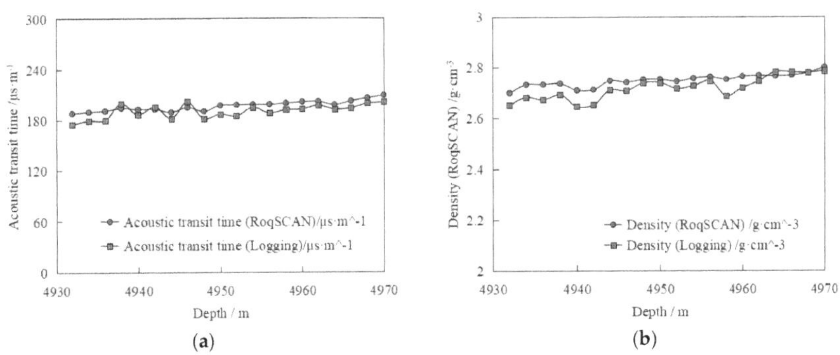


**Figure 4.** Calculation results of fitting correlation between acoustic time difference and density curve. (**a**) Correlation coefficient of acoustic transit time = 0.631. (**b**) Correlation coefficient of density = 0.817.

### 2.3. Fracturing Sweet Spot Index Model

In view of the characteristics of deep carbonate reservoirs, when optimizing the fracturing layer sections, both geological and engineering sweet spots should be fracability considered [17]. Perforation fracturing modification should be carried out in the layer sections with good pore connectivity, well-developed micro-fractures, high brittleness index, and small bidirectional stress difference [18]. To quantitatively evaluate the influence of various indicators on the optimization of fracturing layer sections, a combination of the improved superior and inferior solution distance method (TOPSIS) and the analytic hierarchy process method [19] was adopted to establish the calculation model of the fracturing sweet spot index ($I_{FSS}$), and to precisely and quantitatively evaluate the reservoir properties.

#### 2.3.1. Model Establishment

(1) Positive indicator transformation: Unify the indicator types and convert all indicators into extremely large ones. That is, the larger the indicator value, the more favorable it

is for the evaluation result. In the optimization of fracturing layers, four indicators obtained by RoqSCAN technology, namely the hole-to-surface ratio, the number of micro-fractures, the brittleness index, and the bidirectional stress difference coefficient, are mainly selected as evaluation indicators. Among them, the smaller the bidirectional stress difference coefficient, the more likely the artificial fractures are to turn, and the higher the complexity of the fracture network. Therefore, the bidirectional stress difference coefficient needs to be positively processed.

(2) Standardization of positive indicators: Eliminate the influence of the dimensions of various indicators and standardize the positive indicators. Suppose there are $n$ evaluation schemes and 4 evaluation indicators, then the forward matrix is:

$$X = \begin{bmatrix} x_{11} & x_{12} & \cdots & x_{14} \\ x_{21} & x_{22} & \cdots & x_{24} \\ \cdots & \cdots & \cdots & \cdots \\ x_{n1} & x_{n2} & \cdots & x_{n4} \end{bmatrix},$$

Its normalization matrix is denoted as Z, and the zij calculation formula for each element in the Z matrix is:

$$z_{ij} = x_{ij} / \sqrt{\sum_{i=1}^{n} x_{ij}^2} \tag{15}$$

(3) Construct the calculation formula for the fracturing sweet spot index: assuming there are $n$ evaluation schemes and 4 evaluation indicators, then the normalized matrix Z is:

$$Z = \begin{bmatrix} z_{11} & z_{12} & \cdots & z_{14} \\ z_{21} & z_{22} & \cdots & z_{24} \\ \cdots & \cdots & \cdots & \cdots \\ z_{n1} & z_{n2} & \cdots & z_{n4} \end{bmatrix},$$

Define the maximum value:

$$Z^+ = (Z_1^+, Z_2^+, \ldots, Z_4^+) = (\max\{z_{11}, z_{21}, \ldots, z_{n1}\}, \max\{z_{12}, z_{22}, \ldots, z_{n2}\}, \ldots, \max\{z_{14}, z_{24}, \ldots, z_{n4}\})$$

Define the minimum value:

$$Z^- = (Z_1^-, Z_2^-, \ldots, Z_4^-) = (\min\{z_{11}, z_{21}, \ldots, z_{n1}\}, \min\{z_{12}, z_{22}, \ldots, z_{n2}\}, \ldots, \min\{z_{14}, z_{24}, \ldots, z_{n4}\})$$

Define the *i*-th ($i$ = 1, 2, ...) The calculation formulas for the distances $D_i^+$ from the maximum value and $D_i^-$ from the minimum value of $n$ evaluation objects are:

$$z_{ij} = x_{ij} / \sqrt{\sum_{i=1}^{n} x_{ij}^2} \tag{16}$$

$$z_{ij} = x_{ij} / \sqrt{\sum_{i=1}^{n} x_{ij}^2} \tag{17}$$

In the formula,

$D_i^+$—represents the distance between the evaluated object and the maximum value;

$D_i^-$—The distance between the evaluated object and the minimum value;

$\omega_j$—The weight of the evaluation index, calculated by the Analytic Hierarchy Process [20].

The $i$ ($i$ = 1, 2, ...) The calculation formula for the fracking sweet spot index of $n$ evaluation objects is:

$$I_{FSS} = \frac{D_i^-}{D_i^+ + D_i^-} \tag{18}$$

In the formula, $I_{FSS}$ represents the fracking sweet spot index.

In the fracturing engineering sweet spot evaluation of deep carbonate reservoir, considering the convenience of obtaining parameters and field treatment experience, the influencing indicators in the sweet spot index model are ranked in descending order of importance as follows: brittleness index, pore-to-surface ratio, number of micro-fractures, and difference in horizontal biaxial stresses [21,22]. An Analytic Hierarchy Process (AHP) judgment matrix $A$ is constructed (Equation (15)):

$$A = \begin{pmatrix} 1.00 & 1.50 & 2.00 & 2.50 \\ 0.67 & 1.00 & 1.50 & 2.00 \\ 0.50 & 0.67 & 1.00 & 1.50 \\ 0.40 & 0.50 & 0.67 & 1.00 \end{pmatrix} \tag{19}$$

After calculation, the consistency index ($CI$) of judgment matrix $A$ is 0.0031, and the consistency ratio ($CR$) is 0.0035. To enhance the robustness of the weight calculation results, the arithmetic mean method, geometric mean method, and eigenvalue method are employed respectively to calculate the weight vector of judgment matrix $A$. The calculation results are shown in Table 2.

**Table 2.** Weight scale for fracturing sweet spot influencing factors.

| Influencing Factors | Arithmetic Mean Method | Geometric Mean Method | Eigenvalue Method | Comprehensive Weight |
|---|---|---|---|---|
| Brittleness Index | 0.3855 | 0.3855 | 0.3856 | 0.3855 |
| Pore-to-Surface Ratio | 0.2773 | 0.2774 | 0.2773 | 0.2773 |
| Number of Micro-fractures | 0.1962 | 0.1961 | 0.1961 | 0.1961 |
| Difference in Horizontal Biaxial Stresses | 0.1411 | 0.1409 | 0.1410 | 0.1410 |

#### 2.3.2. Fracking Sweet Spot Index Classification

Classifying the fracking sweet spot index into different levels is conducive to quantitatively evaluating the quality of layers at different depths. Due to the different evaluation models, there is currently no unified classification standard for fracking desserts in the industry. Based on the existing research foundation of brittiness classification and compressible classification of different blocks, and in accordance with the mineral composition characteristics and rock mechanical parameter characteristics of the deep carbonate reservoirs in the $E_3{}^2$ layer of the Yingzhong area, the fracturing sweet spot index classification standard suitable for this block is determined: $I_{FSS} \geq 0.50$, which is a Class I fracturing sweet spot layer section, representing the best fracturing sweet spot of reservoir quality. The $I_{FSS}$ is between 0.35 and 0.50, which is a type II fractured sweet spot layer section, representing a fractured sweet spot with medium reservoir quality. If $I_{FSS} \leq 0.35$, it is classified as a sweet spot section of Class III fracturing, indicating poor reservoir quality. It is not recommended to be used as a modified section.

## 3. On-Site Practical Application

The application workflow for optimizing fracture engineering sweet spots based on the integrated TOPSIS-AHP algorithm is shown in Figure 5. First, scanning electron microscope (SEM) experiments are conducted on cuttings from a continuous carbonate reservoir interval to obtain microphysical properties such as mineral composition, pore structure data, and micro-fracture count. Next, parameters including mineral composition and content, effective pressure, fluid properties, and reservoir temperature are used as model inputs. A rock mechanics model is then applied to calculate macro-mechanical parameters such as Young's modulus, Poisson's ratio, and in situ stress parameters, from which the brittleness index and horizontal stress difference coefficient are further derived. The obtained pore data, micro-fracture count, brittleness index, and stress difference coefficient are used as factors influencing fracture sweet spot identification and incorporated into the $I_{FSS}$ calculation model based on the integrated TOPSIS-AHP algorithm, resulting in a continuous $I_{FSS}$ profile along the sampled well section. Finally, based on the variation characteristics of the $I_{FSS}$ curve, intervals with high $I_{FSS}$ values are selected as target zones for fracture stimulation, thereby completing sweet spot optimization.

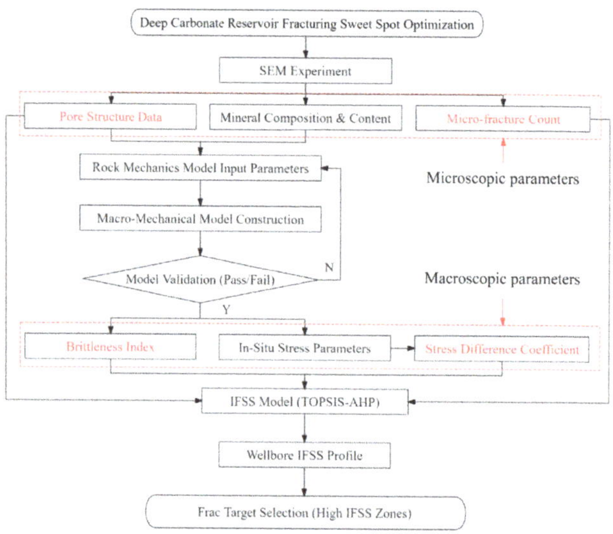


**Figure 5.** Fracture sweet spot optimization workflow based on the integrated TOPSIS-AHP algorithm.

The field application of the layer section optimization method was carried out by taking the typical carbonate reservoir pre-exploration well S1 as the example well. This well is located in the Yingzhong No. 2 structural Zone of the Yingzhong area in the Qaidam Basin, with a completion depth of 5500 m and the completion layer position being the upper section $E_3^2$ of the Lower Ganchaigou Formation. The effective permeability of the formation is 0.08 mD, the formation pressure coefficient is 2.09, and the formation temperature gradient is 3.12 °C/100 m.

### 3.1. Cuttings Scanning Results and Analysis

The RoqSCAN equipment was applied to conduct rock cuttings scanning tests on the 4932–4970 m well section of the $E_3^2$-IV oil group in the S1 well's test oil layer. The sampling interval was 2 m, and a total of 20 rock cuttings samples were prepared. The scanning experiment results were used to analyze the formation mineral composition characteristics and pore development degree characteristics of the sampled well section.

#### 3.1.1. Characteristics of Mineral Composition in Strata

The coloring diagrams of mineral composition at different depths in the sampling section of Well S1 are shown in Figure 6. According to the scanned mineral images, it can be seen that the composition and content of rock minerals in strata at different depths are complex and variable. Within the depth range of 4932–4940 m, the content of potassium feldspar minerals in the strata is relatively high. Within the range of 4942–4964 m, the mineral content of carbonate rocks mainly composed of calcite is relatively high. Within the range of 4966–4970 m, the content of clay minerals mainly composed of Immon mixed layers is relatively high.

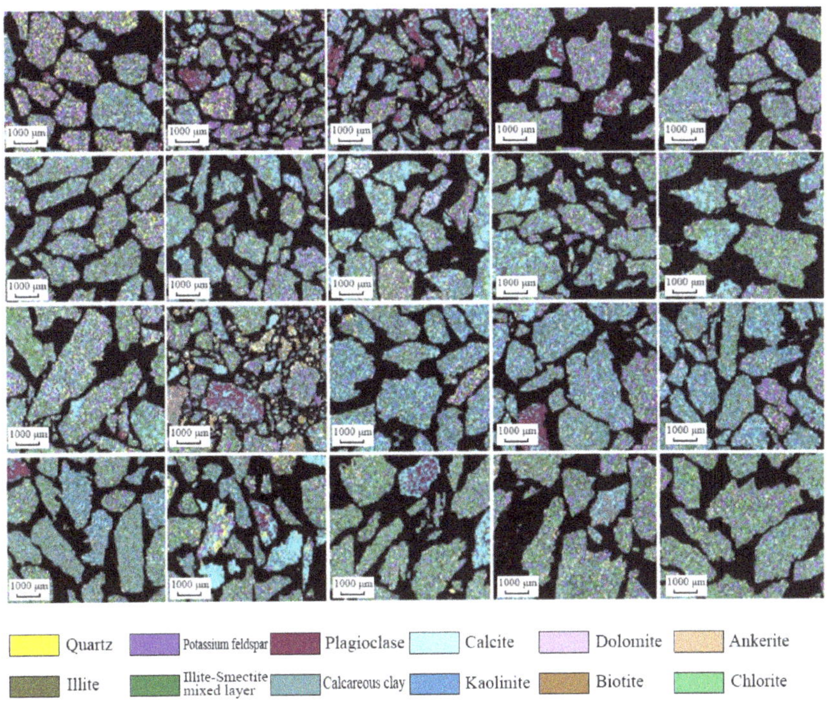


**Figure 6.** Coloring diagram of mineral composition at different depths at 4932–4970 m (sampling interval of 2 m) of well S1.

The main mineral components and content data are shown in Table 3. The stratum is dominated by clay minerals, with an average proportion of 44.7%. The content of carbonate minerals ranks second, with an average proportion of 30.2%, mainly consisting of calcite, dolomite and iron dolomite minerals. Silicate minerals have the lowest content, averaging 19.8%, mainly consisting of quartz, plagioclase and potassium feldspar minerals.

**Table 3.** Data of various main mineral compositions and contents in the sampling well section of well S1.

| Mineral Classification | Main Mineral Names | Main Mineral Content Range (%) | Average Value of Major Minerals (%) | Total Average Value (%) |
|---|---|---|---|---|
| Silicates | Quartz | 5.2–18.7 | 8.90 | 19.8 |
| | Potassium feldspar | 0.25–1.81 | 0.61 | |
| | Plagioclase | 3.36–21.40 | 10.31 | |
| Carbonates | Calcite | 16.8–37.6 | 26.80 | 30.2 |
| | Dolomite | 0.05–3.92 | 1.00 | |
| | Iron dolomite | 1.18–3.76 | 2.40 | |
| Clay minerals | Imon mixed floor | 4.99–20.42 | 11.66 | 44.7 |
| | Illite | 3.99–17.22 | 10.17 | |
| | Calcareous clay | 14.33–33.81 | 22.82 | |
| | Kaolinite | 0–0.16 | 0.05 | |
| Accessory minerals | Anhydrite | 0.08–13.32 | 2.74 | 5.3 |
| | Hematite | 0–0.07 | 0.02 | |
| | Pyrite | 0.37–1.03 | 0.63 | |
| | Siderite | 0–0.11 | 0.03 | |

3.1.2. Characteristics of Formation Pore Development

The pore size distributions at different sampling points below the sampling location in Well S1 are shown in Figure 7. Based on the statistical analysis of pore sizes obtained from scanning, pores with sizes ranging from 0 to 100 μm are classified as small pores, those from 100 to 200 μm as mesopores, and pores larger than 200 μm as macropores. By examining the proportions of small, meso-, and macropores, along with the total pore area ratio data, the degree of formation connectivity and pore development at different depths can be evaluated. At the depths of the sampled interval (4932~4970 m) in Well S1, the pore area ratio ranges from 0.19% to 3.30%, with an average of 0.98%. Among them, the proportion of small holes (0~100 μm) was 21.14–68.54%, with an average of 40.49%. The proportion of mesopores (100~200 μm) was 5.43–15.78%, with an average of 9.55%. The proportion of macropores (>200 μm) ranged from 14.30% to 62.21%, with an average of 40.17%. Overall, the proportion of small pores and macropores was comparable, and the connectivity was relatively good.

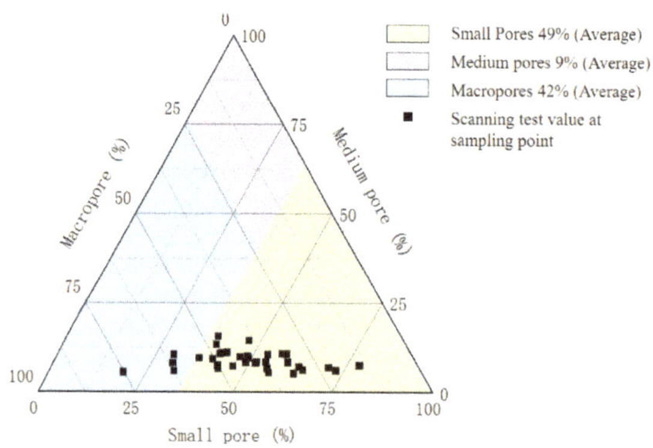


**Figure 7.** Variation of pore structure data of well S1 with depth.

The RoqSCAN electron microscope scanning technique is capable of identifying pore structures with apertures as small as 0.1 μm. Among these, nanoscale microcracks, with widths less than 1 μm, are typically intergranular microcracks, while microscale microcracks have apertures ranging from 1 to 100 μm and are usually interparticle microcracks. To differentiate microscale microcracks from intergranular pores or other minute structures, the focus is placed on examining the development extent of microscale microcracks with apertures > 1 μm and an aspect ratio (length-to-width ratio) of 30 or more. Within the sampled interval, the number of identified micro-fractures per cutting sample ranged from 1 to 7, with an average of 3.1 micro-fractures per sample. The vertical distribution of micro-fractures exhibited strong heterogeneity; micro-fractures were poorly developed at depths of 4932~4936 m, with only 1~2 micro-fractures detected per sample, whereas they were relatively well-developed at depths of 4958~4962 m, with 5~7 micro-fractures detected per sample, as shown in Figure 8. The precise identification of micro-fracture counts provides robust support for the fracability evaluation of deep carbonate reservoirs.

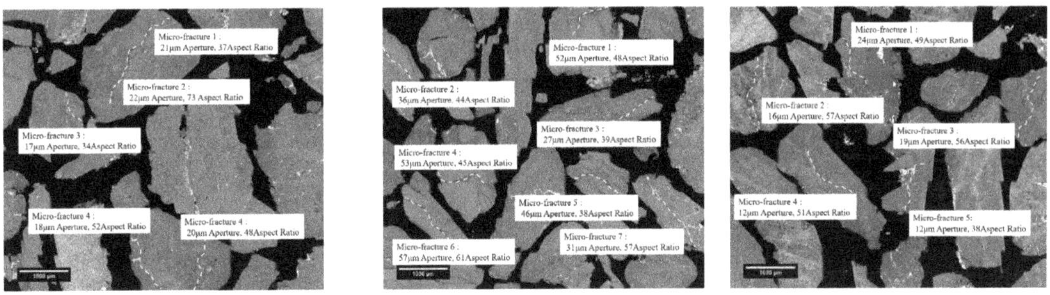


(**a**) 4958 m (**b**) 4960 m (**c**) 4962 m

**Figure 8.** Scan results for micro-fracture identification at 4958~4962 m.

### 3.2. Analysis of Rock Mechanics Characteristics

Based on the macroscopic rock mechanics model considering the temperature-pressure coupling effect, the elastic modulus, Poisson's ratio, brittleness index and in situ stress parameters under the scanning well section were calculated. The calculation results are shown in Table 4. The elastic modulus ($E$) of the deep carbonate rock strata in the Anglo-Chinese block ranges from 43.15 GPa to 54.89 GPa. The Poisson's ratio ($v$) ranges from 0.261 to 0.330. The brittleness index ($BI$) ranges from 25.06 to 48.89, with an average of 33.03. The brittleness of the reservoir is at a moderately weak level. The range of the minimum horizontal principal stress ($\sigma_h$) is 108.5 to 109.4 MPa, the stress difference in both horizontal directions is 8.7 MPa, and the range of the stress difference coefficient ($C_h$) is 0.067 to 0.090.

**Table 4.** Summary of calculation results of rock mechanical parameters in sampling well section.

| Value Range | $E$ (GPa) | $v$ | $BI$ (%) | $\sigma_h$ (MPa) | $C_h$ |
|---|---|---|---|---|---|
| Max | 54.89 | 0.330 | 48.89 | 109.5 | 0.067 |
| Min | 43.15 | 0.261 | 25.06 | 108.3 | 0.090 |
| Average | 49.51 | 0.311 | 33.03 | 108.9 | 0.079 |

### 3.3. Interval Optimization Based on Fracturing Sweet Spot Index

The fracture sweet spot index ($I_{FSS}$) model was utilized to calculate the sweet spot index at the sampling depths of Well S1. The calculation results of the $I_{FSS}$ are shown in Table 5. A continuous profile of the $I_{FSS}$ was generated using the cubic spline interpolation method, and the resulting index profile is illustrated in Figure 9. According to the calculation results, the range of the fracturing sweet spot index of Well S1 is 0.045 to 0.712. According to the classification standard of the fracturing sweet spot index of deep carbonate reservoirs in the Yingzhong area, the $I_{FSS}$ corresponding to the depth of 4954 m~4964 m is relatively high, with an average value of 0.622, which belongs to the type I fracturing sweet spot layer section. Therefore, this interval is selected for perforation fracturing modification. The fracturing operation flow rate was 7 $m^3/min$, with a total fluid inflow of 605 $m^3$ and a total sand inflow of 45 $m^3$, meeting the fracturing design requirements and the stimulation process went smoothly.

**Table 5.** Calculation results of dominant perforation index for S1 Well.

| Depth (m) | BI (%) | Microcrack / | Porosity (%) | $C_h$ / | $I_{FSS}$ | Level |
|---|---|---|---|---|---|---|
| 4932 | 48.89 | 2 | 3.2996 | 0.088 | 0.590 | I |
| 4934 | 39.63 | 1 | 1.4956 | 0.090 | 0.301 | II |
| 4936 | 37.42 | 2 | 1.7216 | 0.079 | 0.401 | II |
| 4938 | 33.45 | 4 | 0.5802 | 0.091 | 0.270 | III |
| 4940 | 29.22 | 3 | 0.3551 | 0.086 | 0.188 | III |
| 4942 | 30.01 | 2 | 0.3569 | 0.084 | 0.133 | III |
| 4944 | 29.10 | 3 | 0.3039 | 0.079 | 0.225 | III |
| 4946 | 33.44 | 3 | 0.2709 | 0.071 | 0.285 | III |
| 4948 | 29.68 | 2 | 0.3376 | 0.082 | 0.154 | III |
| 4950 | 29.36 | 2 | 0.1854 | 0.082 | 0.145 | III |
| 4952 | 32.64 | 4 | 0.2089 | 0.084 | 0.259 | III |
| 4954 | 57.49 | 4 | 2.1675 | 0.068 | 0.636 | I |
| 4956 | 56.42 | 4 | 2.6979 | 0.068 | 0.712 | I |
| 4958 | 31.79 | 5 | 1.8582 | 0.075 | 0.568 | I |
| 4960 | 30.94 | 7 | 1.6563 | 0.069 | 0.619 | I |
| 4962 | 25.06 | 5 | 1.4142 | 0.067 | 0.492 | II |
| 4964 | 47.17 | 4 | 2.594 | 0.069 | 0.687 | I |
| 4966 | 29.67 | 3 | 1.4265 | 0.083 | 0.361 | II |
| 4968 | 33.46 | 1 | 1.3609 | 0.087 | 0.273 | III |
| 4970 | 28.83 | 1 | 0.2899 | 0.089 | 0.047 | III |

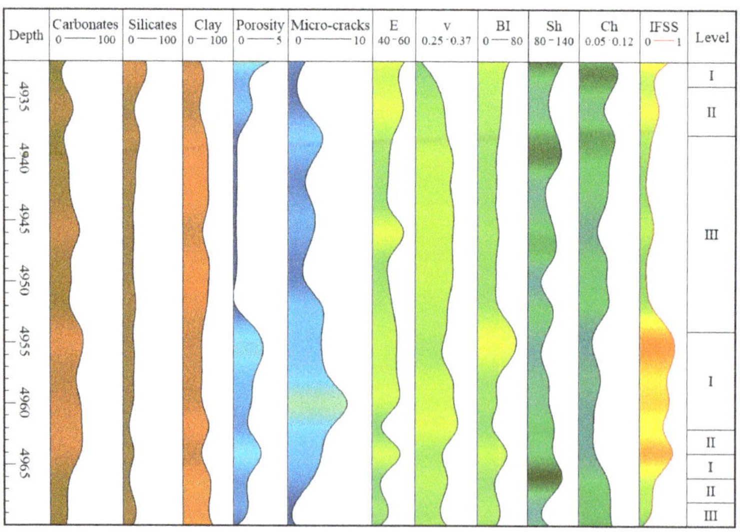


**Figure 9.** The continuous profile of the $I_{FSS}$ obtained through cubic spline interpolation.

The adjacent well S2 at the same layer was selected as the control well. Well S2 had a total depth of 5330 m and was also in the $E_3{}^2$ formation, with geostress gradients and pressure coefficients comparable to those of Well S1. Well S2 employed conventional methods for selecting fracturing intervals, i.e., choosing locations with high brittleness indices in well-interpreted logging sections as perforation zones. The brittleness index was calculated using a general logging regression empirical formula, which lacked reservoir specificity. Well S2 targeted the 5034–5046 m interval for fracturing, with a stimulation scale similar to that of Well S1. A comparison of the stimulation pressure curves for the two wells is shown in Figure 10. It can be observed that during the main proppant addition phase in Well S1, the stimulation pressure ranged from 95.4 to 103.5 MPa, averaging 97.5 MPa. Small fluctuations in the pressure curve might be attributed to the activation of natural fractures in the formation, while the pressure curve remained relatively stable during proppant addition, with the actual proppant volume reaching 100% of the designed amount. In Well S2, during proppant addition, the stimulation pressure ranged from 82.9 to 123.9 MPa, averaging 108.8 MPa, with frequent pressure fluctuations. Between 76 and 107 min of stimulation, the pressure showed an overall upward trend with intensified fluctuations, indicating difficulties in proppant addition and a risk of sand plugging in the formation [23,24]. The actual proppant volume added in Well S2 only reached 76% of the designed amount.

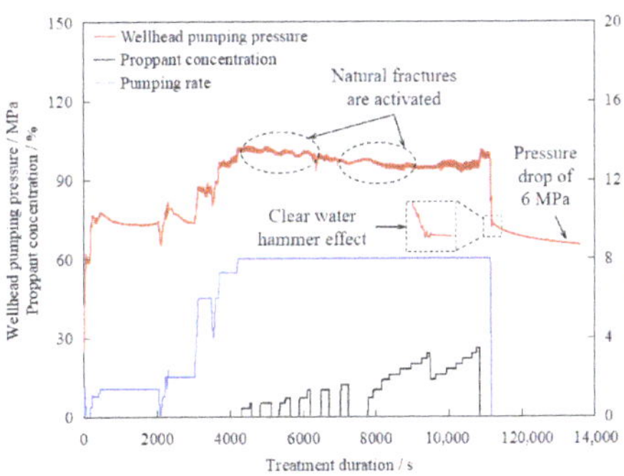


(**a**) Well S1

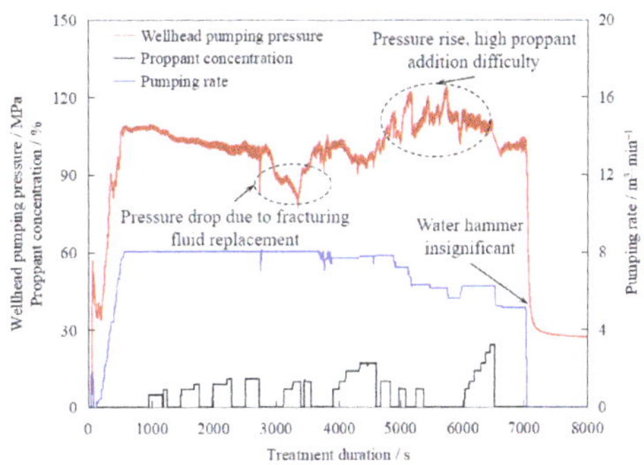


(**b**) Well S2

**Figure 10.** Comparison of stimulation curves between S1 and S2 wells.

Comparing the pump shutdown phases of the two wells, Well S1 exhibited a significant water hammer effect upon instantaneous pump shutdown, with a pressure drop of 6 MPa after 30 min, indicating obvious pressure diffusion within the formation and suggesting high fracture network connectivity. In contrast, Well S2 showed no obvious water hammer effect upon instantaneous pump shutdown, with the pressure drop remaining essentially unchanged after shutdown, indicating poor conductivity of artificial fractures in the formation and a failure to form an effective reserve production zone. Compared to adjacent Well S2, Well S1 achieved an approximately 11.6% reduction in average stimulation pressure and a smooth stimulation process, effectively addressing issues such as high treatment pressure and difficulties in proppant addition in deep carbonate reservoirs.

## 4. Conclusions

(1) In light of the characteristics of deep carbonate reservoirs in the Yingzhong Block of the Qaidam Basin, combined with automatic mineral identification scanning experiments, and considering the deep temperature-pressure coupling effect, a rock mechanics model based on mineral components is established. Upon verification, the correlation coefficient between the acoustic transit time calculated by the model and the actual logging acoustic transit time is 0.631, while the correlation coefficient

for density is 0.817. The verification results demonstrate that this model can accurately characterize the mechanical and in situ stress properties of carbonate reservoirs under high-temperature and high-stress conditions, providing data support for the evaluation of formation fractionability.

(2) The porosity and the number of micro-fractures in the scanning results are selected as the microscopic physical property indicators affecting the fracturing sweet spot, and the brittleness index and the bidirectional stress difference coefficient calculated by the rock mechanics model are selected as the macroscopic mechanical indicators. A multi-factor fusion fracturing sweet spot prediction model fracability considering reservoir physical properties, mechanical properties and stress states was established by using the TOPSIS-AHP joint algorithm, and the fracturing sweet spots of deep carbonate reservoirs were divided into three levels: $I_{FSS} \geq 0.50$ belongs to Class I desserts, $0.35 \leq I_{FSS} < 0.50$ belongs to Class II desserts, and $I_{FSS} < 0.35$ belongs to Class III desserts.

(3) The formation evaluation technology based on the fracturing sweet spot index was successfully applied to the test oil Wells in the study work area. By using this method for fracturing section selection, the stimulation operation pressure was reduced by 11.6%, and the sand addition success rate was increased by 24%, effectively improving the fracturing effect of deep oil Wells. This research has positive guiding significance for the benefit development of deep carbonate reservoirs.

**Author Contributions:** Conceptualization, Y.L. and X.M.; methodology, H.Z.; software, G.R.; validation, H.M.; formal analysis, G.X.; investigation, X.F.; resources, W.Z.; data curation, F.L.; writing—original draft preparation, H.Z.; writing—review and editing, Y.L.; visualization, G.X.; supervision, X.M.; project administration, Y.L.; funding acquisition, Y.L. All authors have read and agreed to the published version of the manuscript.

**Funding:** This research received no external funding.

**Data Availability Statement:** The original contributions presented in this study are included in the article. Further inquiries can be directed to the corresponding author(s).

**Conflicts of Interest:** Authors Yong Liu, Guiqi Xie, Xinyuan Feng, Wenkai Zhao and Fengyu Lei were employed by the company PetroChina Qinghai Oilfield Company. Author Honglin Zheng was employed by the company SINOPEC Research Institute of Petroleum Engineering Co., Ltd. Author Guangcong Ren was employed by the company CNPC Engineering Technology R&D Company Limited. The remaining authors declare that the research was conducted in the absence of any commercial or financial relationships that could be construed as a potential conflict of interest.

## References

1. He, Z.L.; Ma, Y.S.; Zhu, D.Y.; Duan, T.Z.; Geng, J.H.; Zhang, J.T.; Ding, Q.; Qian, Y.X.; Wo, Y.J.; Gao, Z.Q. Theoretical and technological progress and research direction of deep and ultra-deep carbonate reservoirs. *Oil Gas Geol.* **2021**, *42*, 533–546. [CrossRef]
2. Shao, X.P.; Guan, S.W.; Jin, J.Q.; Shen, Y.; Meng, Q.Y.; Wang, L.N.; Wang, P. Differences in structural characteristics and control factors of Yingxiong Range structural belt in the western Qaidam Basin. *Pet. Geol. Recovery Effic.* **2018**, *25*, 67–72. [CrossRef]
3. Huang, J.L. Shale gas accumulation conditions and favorable zones of Silurian Longmaxi Formation in south Sichuan Basin, China. *J. China Coal Soc.* **2012**, *37*, 782–787.
4. Li, H.Y.; Zhou, C.C.; Li, C.X.; Li, C.L.; Shi, Y.J.; Wang, C.S. Logging Evaluation and Application of Brittleness Index in Tight Sandstone Reservoir—A case study of Chang -7 tight sandstone reservoir in Longdong area of Ordos basin. *Xinjiang Pet. Geol.* **2014**, *35*, 593–597.
5. Yang, H.W. Quantitative evaluation of shale fracability based on logging data. *Fault-Block OilGas Field* **2017**, *24*, 382–386.
6. Zhai, W.B.; Li, J.; Zhou, Y.C.; Huang, T.; Song, X. New evaluation method of shale reservoir fracability based on logging data. *Lithol. Reserv.* **2018**, *30*, 112–123.

7. Lin, Y.; Li, C.M.; Gu, W.; Luo, W.J.; Wang, Z.Y.; Yu, Z.; Zhang, J.; Bie, J.; Li, W.Q. Seismic fine characterization of deep carbonate fractured-vuggy reservoir: Case study of the 4th Member of Sinian Denying Formation in Anyue Gas Field, Sichuan Basin. *Nat. Gas Geosci.* **2020**, *31*, 1792–1801. [CrossRef]
8. Feng, X.Y.; Zhang, C.J.; Xie, G.Q.; Guo, D.L.; Liu, Y. Evaluation techniques and applications of sweet spots in shale oil engineering in the Qaidam Basin. *Nat. Gas Oil* **2023**, *41*, 63–69.
9. Zhu, J.Y.; Tan, Z.J.; Zhang, G.Q.; Li, D. A comprehensive lithology identification technology for the buried-hill reservoirs with complex lithology in the Bohai Sea. *Oil Drill. Prod. Technol.* **2018**, *40*, 62–65.
10. Ashton, T.; Vinh Ly, C. Portable Technology Puts Real-time Automated Mineralogy On The Well Site. In Proceedings of the SPE Unconventional Resources Conference and Exhibition-Asia Pacific, Brisbane, Australia, 11–13 November 2013. [CrossRef]
11. Oliver, G.; Spence, G.; Davis, A.; Stolyarov, S. Advanced cuttings analysis provides improved completion design, efficiency and well production. *Unconv. Carbon Capture Storage* **2016**, *34*, 69–76. [CrossRef]
12. El Sherbeny, W.; Nuic, I.; Madkour, A. Significance of advanced cuttings evaluation (ACE) technologies for Chemostratigraphy purposes while drilling; technology overview & applications. In Proceedings of the SPE North Africa Technical Conference and Exhibition, Cairo, Egypt, 14–16 September 2015.
13. Yang, G.; Zhang, S.; Zheng, H.; Tian, G. Analysis of Cuttings' Minerals for Multistage Hydraulic Fracturing Optimization in Volcanic Rocks. *ACS Omega* **2020**, *5*, 14868–14878. [CrossRef]
14. Duan, Y.T.; Feng, X.T.; Li, X. Study on the influence of mesomineral bands of shale on its macro failure Mode. *Chin. J. Rock Mech. Eng.* **2021**, *40*, 43–52. [CrossRef]
15. Ly, C.V.; Oliver, G.M.; Spence, G.; Centurion, S.; Jackson, C.E.; Hearn, F.P.; Palomarez, V. Cross Correlation of Logging Data with SEM-Based Mineralogical and Textural Well Data: A New Tool for Optimized Completion Design. In Proceedings of the SPE Russian Oil and Gas Exploration & Production Technical Conference and Exhibition, Moscow, Russia, 14–16 October 2014. [CrossRef]
16. Yang, W.; Wang, B.; Yao, J.; Jing, W.; Zhang, X. Experimental study on mechanical properties of carbonate rocks under real-time high temperature and heat treatment under triaxial compression. *Chin. J. Rock Mech. Eng.* **2024**, *43*, 1347–1358.
17. Wang, J.B.; Feng, M.G.; Yan, W.; Liu, S. Influence factors and evaluation methods for shale reservoir fracability in Jiaoshiba Area. *Fault-Block Oil Gas Field* **2016**, *2*, 216–220.
18. Zhang, M.L.; Zhang, J.P.; Tang, Z.Y.; Li, M.Q.; Xu, Y.M.; Luo, W.J.; Peng, Y.; Jiang, R. Preferential selection of favorable development areas of tight sandstone gas reservoir in the second Member of Xujiahe Formation in Anyue gas field of central Sichuan Basin. *Fault-Block Oil Gas Field* **2022**, *29*, 171–177.
19. Zhang, J.; Liang, C. Application of TOPSIS Model Based on Gray Correlation Coefficient Matrix in the Evaluation of Water Environmental Quality. *J. Sichuan Univ.* **2009**, *41*, 97–101.
20. Liu, Y.T.; Sun, B.L.; Yu, Y.S. Fuzzy recognition and quantitative calculation method for large pore channels. *Oil Drill. Prod. Technol.* **2003**, *25*, 054–059.
21. Jin, X.; Shah, S.N.; Roegiers, J.C.; Zhang, B. An integrated petrophysics and geomechanics approach for fracability evaluation in shale reservoirs. *SPE J.* **2015**, *20*, 518–526. [CrossRef]
22. Li, Z.H.; Deng, P.; Yang, C.H.; Guo, Y.T.; Hou, L.F. Study on mechanical properties and compressibility evaluation of carbonate reservoir. *J. Guangxi Univ.* **2019**, *44*, 1450–1460.
23. Xiao, Z.H.; Liu, J.S.; Chen, Y.G. Analysis of Fracturing Operation Curve Characteristics and Their Applications. *Pet. Geol. Eng.* **2008**, *22*, 99–102.
24. Huang, J.; Gong, Y.G.; Xu, Y.T.; Guo, B.M.; Zhang, B.K. The Application of G Function in Bohai Oilfield Frac-pack Treatment. *Dffshore Oil* **2019**, *39*, 57–61.

processes

MDPI

*Article*

# Analysis of the Effectiveness Mechanism and Research on Key Influencing Factors of High-Pressure Water Injection in Low-Permeability Reservoirs

**Yang Li [1], Hualei Xu [2,3,*], Shanshan Fu [2,3], Hongtao Zhao [1], Ziqi Chen [2,3], Xuejing Bai [1], Jianyu Li [2,3], Chunhong Xiu [1], Lianshe Zhang [1] and Jie Wang [2,3,*]**

1 Luming Oil and Gas Exploration and Development Co., Ltd., Shengli Oilfield, Sinopec, Dongying 257064, China; liyang986.slyt@sinopec.com (Y.L.); zhaohongtao966.slyt@sinopec.com (H.Z.); baixuejing.slyt@sinopec.com (X.B.); xiuchunhong.slyt@sinopec.com (C.X.); zhanglsh.slyt@sinopec.com (L.Z.)

2 Cooperative Innovation Center of Unconventional Oil and Gas, Yangtze University, Wuhan 430100, China; 15110202382@163.com (S.F.); 2024710333@yangtzeu.edu.cn (Z.C.); 2024710353@yangtzeu.edu.cn (J.L.)

3 State Key Laboratory of Low Carbon Catalysis and Carbon Dioxide Utilization, Yangtze University, Wuhan 430100, China

* Correspondence: hualeixu@yangtzeu.edu.cn (H.X.); wangjie@yangtzeu.edu.cn (J.W.)

**Abstract:** Low-permeability oil reservoirs, due to their weak seepage capacity and high start-up pressure, have limited yield-increasing effects through conventional water injection development methods. High-pressure water injection can significantly change the seepage environment around the well and within the reservoir, expand the effective swept volume of injected water, and thereby greatly enhance the oil recovery rate of water flooding. However, there is still a relative lack of research on the mechanism of high-pressure water injection stimulation and its influencing factors. This paper systematically analyzes the effectiveness mechanism of high-pressure water injection technology in the exploitation of low-permeability reservoirs. The internal mechanism of high-pressure water injection for effective fluid drive and production increase is explained from the aspects of low-permeability reservoir seepage characteristics, capacity expansion and permeability enhancement by high-pressure water injection, and the dynamic induction of micro-fractures. Based on geological and engineering factors, the main factors affecting the efficiency enhancement of high-pressure water injection are studied, including formation deficit, reservoir heterogeneity, dominant channel development and fracturing stimulation measures, injection displacement and micro-fractures, etc. The results of numerical simulation showed the following: (1) formation depletion, reservoir heterogeneity, and the formation of dominant channels significantly affected the effect of water flooding development and (2) engineering factors such as the fracture direction of hydraulic fracturing, water injection rate, and the development of micro-fractures under high-pressure water injection directly determined the propagation path of reservoir pressure, the breakthrough speed of the water drive front, and the ultimate recovery factor. Therefore, during the actual development process, the construction design parameters of high-pressure water injection should be reasonably determined based on the geological reservoir conditions to maximize the oil production increase effect of high-pressure water injection. This study can successfully provide theoretical guidance and practical support for the development of low-permeability oil reservoirs.

**Keywords:** high-pressure water injection; micro-fractures; dominant channels; formation depletion; control water and oil increase

## 1. Introduction

Statistics show that about 38% of global oil reserves are located in low-permeability reservoirs [1]. The development of low-permeability oil reservoirs has long faced problems such as poor injection capacity of water injection wells and difficulty in achieving results from oil wells, resulting in continuous low liquid and oil production, which seriously affects the development effect [2,3]. In response to this problem, major oil fields have developed and implemented technologies such as pressurized water injection and surfactant water injection, which have achieved certain results in reservoirs with relatively high permeability and weak heterogeneity [4,5]. However, for tight oil reservoirs with extremely low permeability, these traditional technologies are still difficult to meet the development requirements. In order to break through this technical bottleneck, Shengli Oilfield in China has innovatively carried out research on the pressurized water injection development technology above the fracture pressure in recent years. Subsequently, field tests of high-pressure water injection have been successively carried out in multiple blocks [6–8]. This technology combines hydraulic fracturing equipment with water injection development. Through high pressure (wellhead injection pressure is generally greater than 35 MPa) and high speed (daily water injection volume of a single well is generally greater than 1000 $m^3/d$), it rapidly supplements reservoir energy and increases formation pressure in a short period of time, thereby significantly increasing the production pressure difference in oil wells, effectively enhancing the liquid and oil production [9,10]. The high-pressure water injection mechanism and influencing factors of this technology in low-permeability reservoirs have received extensive attention, and the related achievements provide important technical support for improving the water injection development effect of low-permeability reservoirs [11].

Regarding the mechanism of pressure-increasing water injection above the fracture pressure for low-permeability reservoirs, Wang et al. [12] divided the mechanism of this technology into increasing water injection volume, establishing effective displacement, and expanding sweep efficiency. Wang et al. [13], based on the analysis of oilfield production data, believed that waterflood-induced fracture dynamic propagation is the main factor causing changes in reservoir properties of low-permeability reservoirs, and in combination with the dynamic formation mechanism of fractures and its influence on water injection development, found that the existence of dynamic fractures exacerbates reservoir heterogeneity. Zhao et al. [14,15] took the Chang 6 oil layer group in the Ansai Oilfield of the Ordos Basin as an example to specifically analyze the characteristics, mechanism, identification methods, and key influencing factors of water injection dynamic fractures, providing reference data for the research on water injection-induced dynamic fractures. Wang et al. [16] applied various methods to study the dynamic opening mechanism of fractures during water injection development. They believed that the dynamic change in fractures is related to water injection pressure and the in situ stress in the direction of the connection line between injection and production wells. Meanwhile, they studied the opening pressure limit and extension mechanism of fractures and proposed a guidance scheme for the later well pattern infill adjustment. Liang et al. [17] established a characterization method for water injection-induced dynamic fracture formation and conducted corresponding reservoir numerical simulation research, simulating the dynamic changes in bottom hole pressure of injection wells in ultra-low-permeability reservoirs.

The factors influencing the effect of high-pressure water injection mainly include geological and engineering factors [18–22]. Comparison of the effective range zones of high-pressure water injection is shown in Figure 1. During the high-pressure water injection development process, the deformation mechanism of the enhanced permeability zone (the zone of preferential flow channels) is influenced by the coupling of multiple factors, mainly manifested as morphological expansion, dynamic evolution of permeability, and migration of water flow paths. During the pressurized fracture initiation stage, the fractures continue to expand. The front of the expanding fluid approaches the front of the fracture network. The injected water is mainly concentrated within the fractured and stimulated zone. The velocity of the pressure front > the velocity of the fluid front > the velocity of the fracture network front. Ma et al. [23] studied the well selection method of high-pressure water injection from aspects of geological reservoir and engineering integration. Cui et al. [24] established a pressure analysis model for injection wells that considers the influence of fractures induced by high-pressure water injection. They divided the fluid flow in high-pressure injection wells into five stages: the initial stage of fracture generation, the stage of fracture propagation, the stage of linear flow, the stage of transitional flow, and the stage of boundary control flow. Zhu et al. [25] studied the injection capacity of high-pressure water injection technology in combination with the background and reservoir physical properties and found that the production pressure difference is the key factor determining the injection capacity.

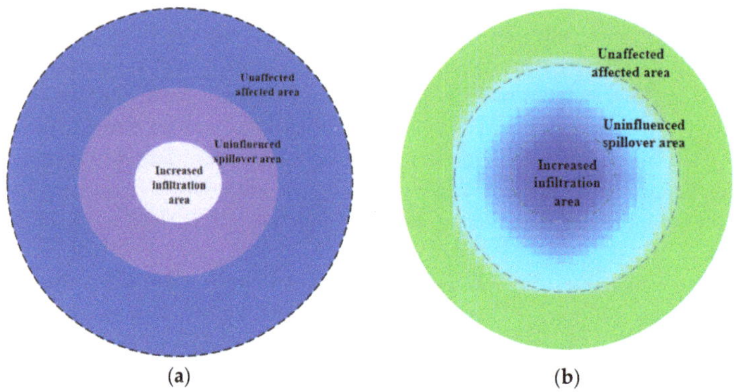


**Figure 1.** Comparison of the effective range zones of high-pressure water injection. (**a**) Conventional water injection; (**b**) high-pressure water injection.

Based on the above literature, it can be seen that in-depth research has been conducted on the effect, influencing factors, and field application of high-pressure water injection technology at home and abroad. However, there has been no systematic study on the influence of laws of various factors on the effect of high-pressure water injection in low-permeability reservoirs from both the geological and engineering perspectives. Based on the above issues, this paper analyzes the mechanism of production enhancement by high-pressure water injection technology in low-permeability reservoirs. From multiple perspectives, such as reservoir seepage characteristics, the effect of high-pressure water injection on expanding capacity and enhancing permeability, and the induced generation of dynamic fractures, it deeply analyzed the internal mechanism of high-pressure water injection for achieving efficient displacement of reservoir fluid and production enhancement. Finally, from two aspects of geological factors (degree of formation energy deficit, reservoir heterogeneity, and strength of water flow dominant channel) and engineering factors (hydraulic fracturing, water injection rate, and existence of dynamic micro-fractures), the influence of these factors on the enhanced oil recovery effect of high-pressure water injection was systematically analyzed. The results of this study provide theoretical innovation and technical guidance for the efficient development of low-permeability reservoirs.

## 2. Effective Mechanism of High-Pressure Water Injection Development in Low-Permeability Reservoirs

Low-permeability reservoirs have weak seepage capacity and high start-up pressure, and the efficiency of conventional water injection development is limited. High-pressure water injection is an important technology to enhance the water flooding development effect of low-permeability reservoirs. It can effectively expand the swept volume of injected water, improve the formation permeability around the injection well, and then enhance the oil recovery efficiency. This section analyzes the mechanism of the enhanced production and efficiency of high-pressure water injection in low-permeability oil reservoirs from multiple perspectives, including fluid dynamics characteristics, the induction mechanism of microfracture propagation, and changes in formation permeability.

### *2.1. Seepage Characteristics of Low-Permeability Reservoirs*

The seepage behavior in low-permeability oil reservoirs exhibits non-Darcy flow characteristics, and an apparent threshold pressure gradient is present. The corresponding seepage equation of is given by Equation (1). When fluid flows in low-permeability reservoirs, the seepage state is closely related to the displacement pressure gradient. Fluid seepage can only occur when the displacement pressure gradient is greater than a certain pressure gradient. The seepage curve can be divided into three stages, as shown in Figure 2. Non-flow section (OA): When the pressure gradient is lower than λ, the fluid does not flow. Nonlinear flow section (AB): The pressure gradient is slightly higher than λ, and the seepage velocity increases slowly. Quasi-linear flow section (BC)—the pressure gradient is much higher than λ, and the seepage approaches Darcy flow.

$$v = \frac{k}{\mu}\left(\frac{\Delta p}{L} - \lambda\right) \tag{1}$$

where $k$ is formation permeability, $10^{-3}$ μm$^2$; $\lambda$ is the minimum starting pressure gradient, MPa/m; $\mu$ is the fluid viscosity, mPa·s; $\frac{\Delta p}{L}$ is the displacement pressure gradient, MPa/m.

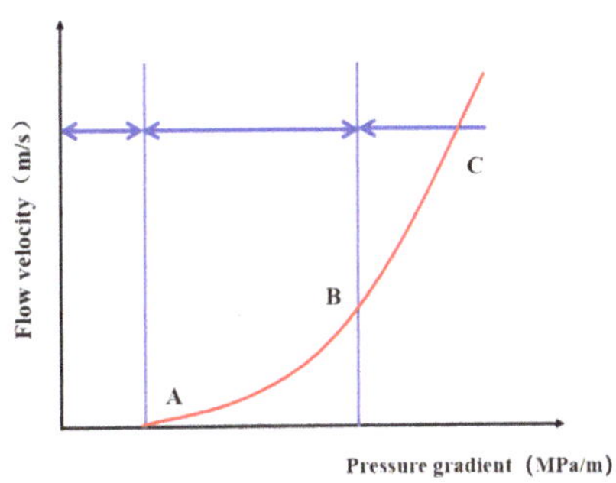


**Figure 2.** Low-permeability reservoir permeability curve.

### *2.2. The Mechanism of High-Pressure Water Injection Expansion and Permeability Enhancement in Low-Permeability Rock Strata*

The mechanism of high-pressure water injection expansion and permeability enhancement in low-permeability rock strata mainly achieves effective fluid driving by changing the pressure field and flow characteristics inside the formation [26,27]. Fractures are usually blocked by materials such as gels, particles, and fibers. As shown in Figure 3, by injecting high-pressure fluid, the rock strata produce certain expansion under water injection pressure, thus changing the microscopic pore structure of the rock strata and enhancing its permeability. The high-pressure water injection expansion and permeability enhancement technology can not only improve the seepage characteristics of the reservoir but also effectively enhance the effect of water injection drive [28]. Specifically, as the injection pressure increases, the micro-fractures and pore network structure within the rock strata change, resulting in the improvement of the fluid flow path, thereby increasing the effective permeability of the fluid. The schematic diagram of high-pressure water injection fluid opening natural fractures is shown in Figure 4. The core of this mechanism lies in that by injecting with a certain external pressure, the natural barrier of the low-permeability reservoir can be overcome, allowing the fluid to enter the target area more smoothly and further enhancing oil and gas recovery.

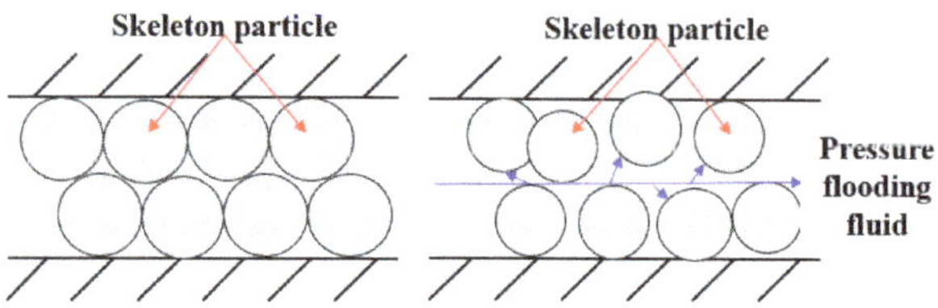


**Figure 3.** High-presser injection fluid induces micro-fractures.

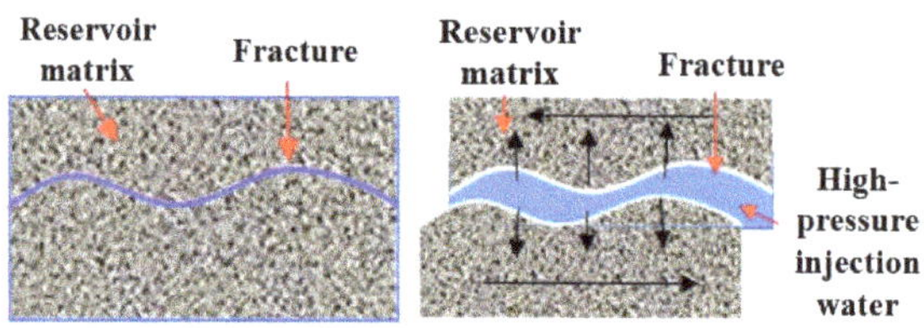


**Figure 4.** High-pressure injection fluid opens natural fractures.

### *2.3. Dynamic Fracture Induction Mechanism*

During the injection process of high-pressure injection fluid, the internal microstructure of the rock strata will undergo certain changes. Especially under high-pressure conditions, the expansion of old fractures or the formation of new fractures will occur inside the formation. This type of fracture is called dynamic fractures [29]. The formation of these fractures will significantly increase the permeability of the reservoir and enable the injected fluid to better penetrate deep into the reservoir, thus effectively driving the flow of crude oil [30,31]. In addition, the effect of high-pressure water injection on rock strata may also lead to the change in the original pores within the rock strata, enhancing their adsorption and conduction capabilities for fluids. Therefore, a reasonable water injection strategy and efficient high-pressure water injection technology can achieve remarkable yield-increasing effects in actual development. When the bottom hole pressure reaches the formation fracture pressure, the reservoir fractures to generate a micro-fracture network, which continues to extend under the water injection condition. When water injection stops, under the action of the formation stress, the micro-fracture network will partially close. When the water injection occurs again, the fractures will reopen, thus forming water injection-induced dynamic fractures.

To distinguish the differences between conventional water injection and water injection for pressure drive, two models are adopted: one is a model based on the elastic expansion of conventional rocks, which is used to simulate the situation of conventional water injection; the other is an expansion–recompaction model to simulate the effect of water injection for pressure drive. When simulating high-pressure water injection, the injection well injects water at a rate of 300 $m^3/d$ for 6 months and then stops injection and is converted into a production well. As shown in Figure 5a,b, traditional hydraulic fracturing mainly promotes fracture extension and forms a single symmetrical long fracture. The characteristics of high-pressure water injection fracturing are short a fracture boundary relatively small fracture volume, which slows down fracture extension during the fracturing process. The key is to form a complex fracture network that diffuses outward with the injection well as the center. As shown in Figure 5c,d, the sweep coefficient of high-pressure water injection is high: a more balanced streamline distribution is generated during the high-pressure water injection process. Compared with the conventional injection mode (sweep coefficient 33%), the sweep coefficient after high-pressure water injection is 81%, an increase of approximately 48%.

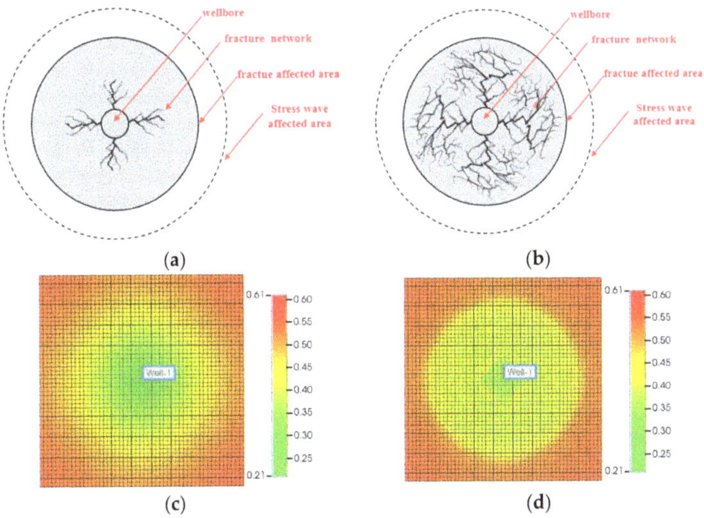


**Figure 5.** Schematic diagram of fracture distribution and oil displacement effect in conventional low-pressure water injection and high-pressure water injection. (**a**) The fracture network formed around the injection well by conventional hydraulic fracturing; (**b**) fractures network in formation around high-pressure water injection wells; (**c**) distribution of remaining oil around conventional water injection wells; (**d**) distribution of remaining oil around high-pressure water injection wells.

In addition, high-pressure water injection has a higher oil displacement efficiency than conventional water injection, as shown in Figure 6. The colors of the bar chart represent the magnitude of permeability, with the unit being mD. High-pressure water injection effectively expands the pore space of the reservoir by promoting the plastic deformation of rocks improves the permeability of the area around the wellbore. Even after the formation pressure drops, this area still maintains a relatively high permeability. The comparison curve of daily oil production from the oil well and the remaining oil in the formation is shown in Figure 7. The daily oil production is high in the initial stage of high-pressure water injection, but the production decreases rapidly. The reason for the analysis is that high-pressure water injection, by forming an area of enhanced permeability around the well, rapidly restores formation energy in a short period of time, thereby driving crude oil to move towards the oil well. As development progresses, the formation energy reaches its peak. However, due to the high seepage resistance of the low-permeability reservoir, the crude oil cannot be effectively extracted, and the remaining oil content in the high-permeability zone gradually decreases, so the oil production of high-pressure injection water will also decrease accordingly.

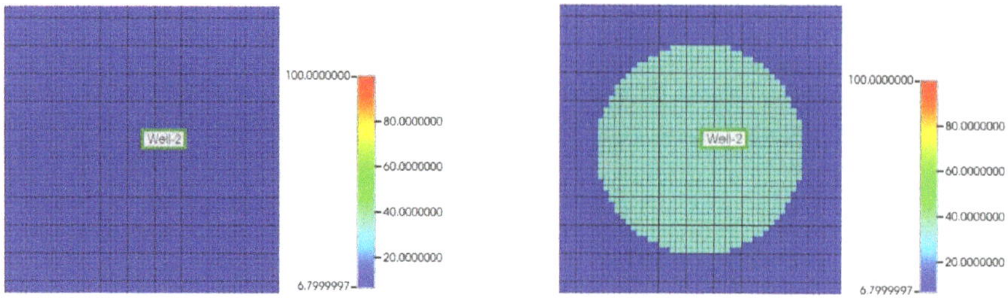


(a) (b)

**Figure 6.** Change in formation permeability around water injection well after stopping water injection. (**a**) Conventional low-pressure water injection; (**b**) high-pressure water injection.

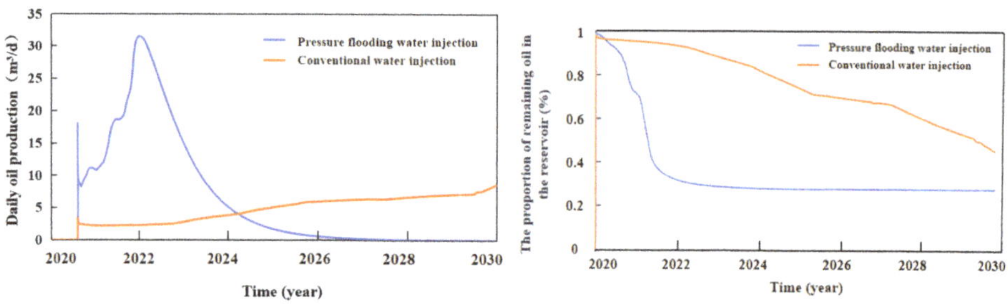


(a) (b)

**Figure 7.** Comparison of daily oil production of oil wells and remaining oil in the formation. (**a**) Daily oil production; (**b**) residual oil.

## 3. Analysis of the Reasons for High-Pressure Water Injection in Research Block

### 3.1. Geological Factors

Geological factors are fundamental factors that affect the water flooding effect during reservoir development, including formation pressure depletion, reservoir heterogeneity, and the formation of dominant flow channels. These factors directly determine the swept volume of injected water, oil recovery, and effective characteristics of oil well.

#### 3.1.1. Formation Pressure Depletion

As shown in Figure 8, in order to simulate the water injection development effect under the condition of formation pressure depletion, a well pattern model of "one injection and two production" is set up. The distances between injection and production wells are both 300 m, and the production method is constant liquid production. The influence of formation pressure depletion on the production effect can be judged by observing the pressure rebound of the production wells during the water injection. In the experimental group ①, it was set that the injection Well-1 started water injection, and the production Well-2 and Well-3 were exploited simultaneously. In the control group ②, it was set that the production Well-2 was exploited for a period of time first, then the injection Well-1 started water injection, and at the same time, the production Well-3 was put into production.

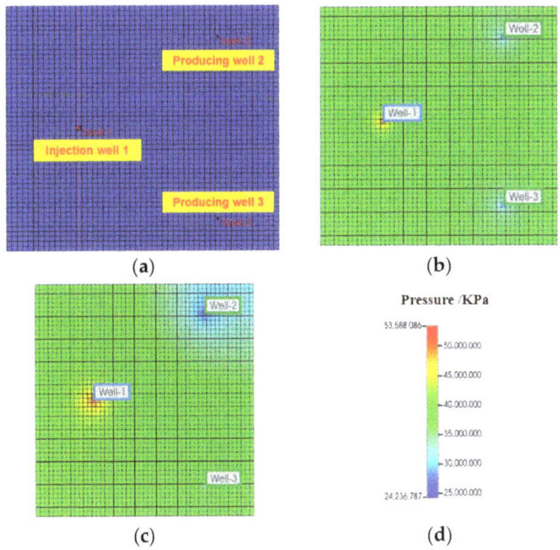


**Figure 8.** Setting of the formation pressure depletion model. (**a**) Well pattern model setup; (**b**) experimental group; (**c**) control group; (**d**) pressure bar chart.

The pressure response results of the production wells are shown in Table 1. The production effect of two production wells in experimental group ① is completely consistent. But in control group ②, under the action of the same water injection well, the production Well-2, which has been exploited in the early stage, has an obvious effect: the bottom hole pressure rises again, and the newly added Well-3 has no obvious pressure response. This indicates that the production well with priority production and formation pressure depletion is more susceptible to the influence of water injection. The effective characteristics of water injection on production wells with formation pressure depletion are shown in Figure 9, and the influence curve of formation pressure depletion on water cut is shown in Figure 10. It is found that production wells with high cumulative production and large formation pressure depletion have a faster increase in water cut after high-pressure water injection to supplement energy. It shows that production wells with large formation pressure deficits, after injecting water to supplement energy, have a faster increase in water cut, and the waterflood front is easier to break through.

**Table 1.** Comparison of pressure responses of production wells.

| Year | Experimental Group, Bottom Hole Flowing Pressure (kPa) | | Control Group, Bottom Hole Flowing Pressure (kPa) | |
|---|---|---|---|---|
| | Well-2 (Deficit Well) | Well-3 (Newly Opened Well) | Well-2 (Deficit Well) | Well-3 (Newly Opened Well) |
| 2020 | 26,471.96 | 26,471.96 | 26,401.14 | 36,430.00 |
| 2021 | 16,885.44 | 16,885.48 | 15,789.76 | 36,256.08 |
| 2022 | 17,040.81 | 17,040.79 | 12,907.02 | 35,563.70 |
| 2023 | 18,581.47 | 18,581.42 | 10,951.85 | 34,639.21 |
| 2024 | 21,986.09 | 21,986.06 | 9274.36 | 33,633.14 |
| 2025 | 25,274.68 | 25,274.66 | 7718.77 | 32,588.81 |
| 2026 | 25,274.60 | 27,878.65 | 6566.71 | 31,530.23 |
| 2027 | 27,878.66 | 29,624.72 | 5882.63 | 30,459.88 |
| 2028 | 29,624.71 | 31,624.32 | 4907.04 | 10,321.08 |
| 2029 | 31,624.30 | 34,015.21 | 7666.73 | 8006.76 |
| 2030 | 34,015.22 | 36,258.50 | 7999.32 | 7718.35 |

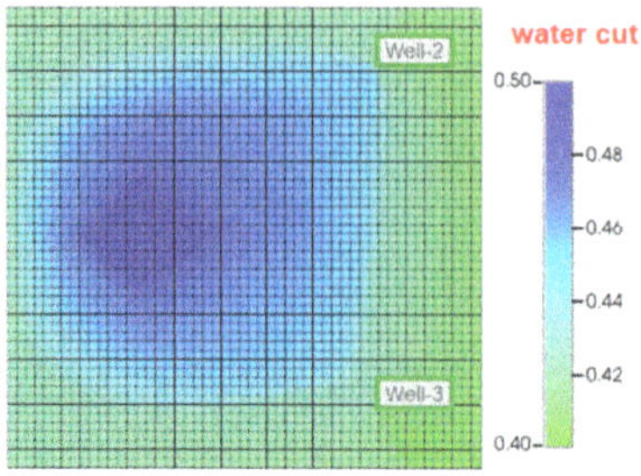


**Figure 9.** Effective characteristics of water injection on production wells with formation pressure depletion.

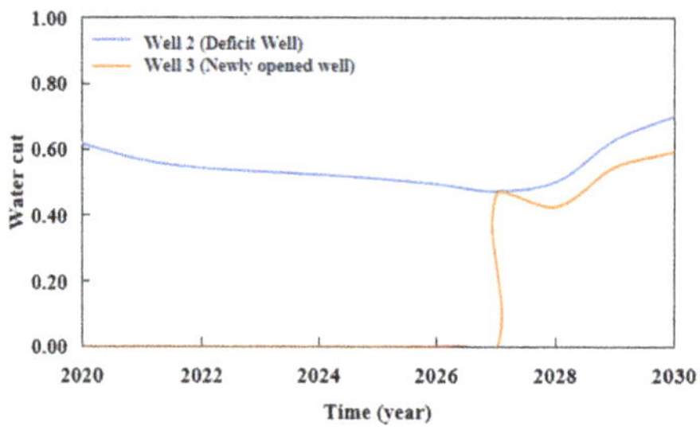


**Figure 10.** Influence curve of formation pressure depletion on water cut.

3.1.2. Reservoir Heterogeneity

A numerical simulation model with a total of six small layers was set up. Among them, layers 1 and 2 are high-permeability layers, and the rest are low-permeability layers. The simulation results are shown in Figure 11. Depletion exploitation was carried out before 2022, and high-pressure water injection was conducted after 2022. High-pressure water injection was carried out when the liquid supply was insufficient and formation water content was high. These injected waters mainly replenish the energy of the high-permeability layers (high-permeability layer 1 and high-permeability layer 2), enabling these layers to restore their liquid production capacity and, thus, increasing the oil production.

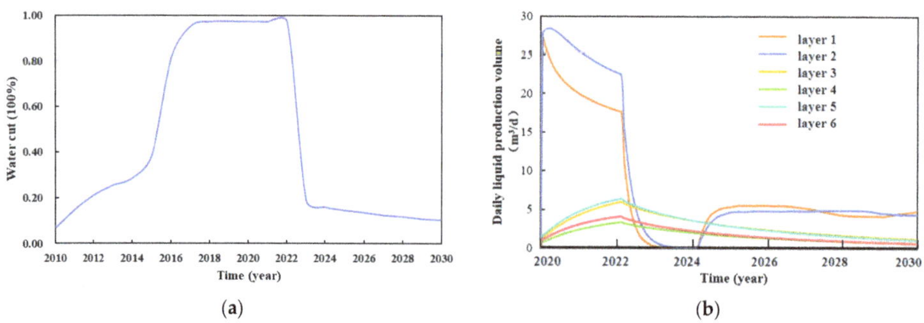


**Figure 11.** Comparison of water injection effects under insufficient liquid supply. (**a**) Change rate of water cut before and after high-pressure water injection; (**b**) liquid production of each small layer after water injection.

On the plane, during the process of water injection development, the injected water will preferentially rush rapidly along high permeability channels (such as fractures and large pores), resulting in a decrease in displacement efficiency. Based on the plane permeability difference in the target well group, an equivalent model is established, as shown in Figure 12a. Observing the pressure propagation during the water injection process, as shown in Figure 12b. The injected water will preferentially flow towards areas with high permeability and low seepage resistance, resulting in faster efficiency, more obvious pressure response, and better effect for production wells in high permeability areas. As shown in Figure 13, the higher the permeability is, the more injected water the high-permeability reservoir absorbs during the process of injecting water to replenish the formation pressure, and the more obvious the pressure recovery is.

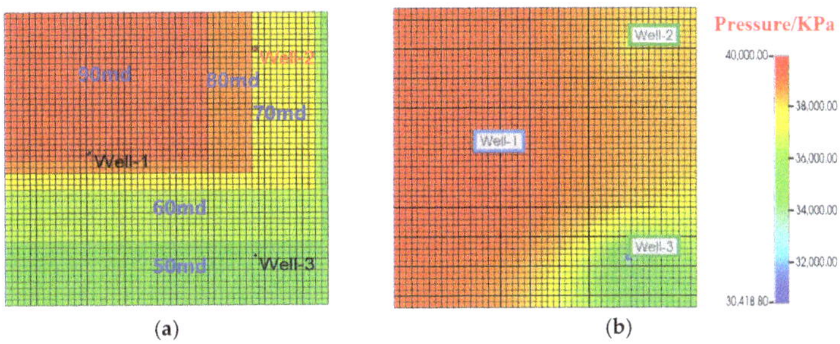


**Figure 12.** Equivalent plane model and pressure response. (**a**) Permeability distribution map; (**b**) pressure propagation map; Well-1 is an injection well, and Well-2 and Well-3 are production wells.

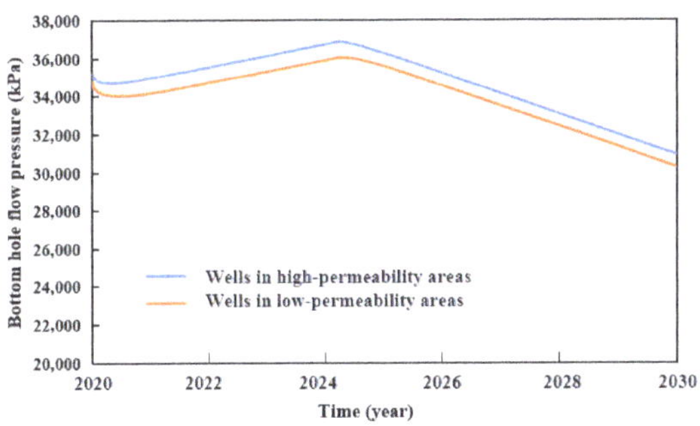


**Figure 13.** Pressure effects of injected water on different permeability zones.

### 3.1.3. Water Flow Dominant Channel

Waterflood-induced fracture refers to the open fractures or water flow dominant channels centered on the water injection well, formed when the water injection pressure exceeds the opening pressure of the natural fractures in the formation during the long-term water injection development process of low-permeability reservoirs. This fracture is usually a naturally formed water flow dominant channel. Based on the "one injection and four extraction" five-point well network, a numerical model including three heterogeneous reservoirs is constructed. By setting different conductivity parameters (medium–high permeability area (conductivity 15 mD) and low permeability area (1 mD)), the development characteristics of dominant channels are quantitatively characterized by setting different conductivity parameters, as shown in Figures 14 and 15. Among them, Well-1 is an injection well, and the remaining four wells are production wells. Based on the equivalent model theory, a simulation study was carried out with the help of numerical simulation software. The model was calibrated through the fitting results. The grid size used in the modeling is a planar step size of 50 m × 50 m. The vertical length is 22 m in total. Each small layer is allocated one layer of grid.

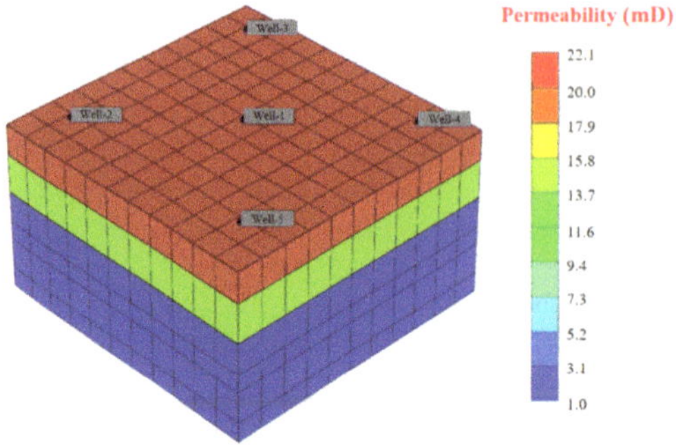


**Figure 14.** "One injection, four production" five-point well pattern model.

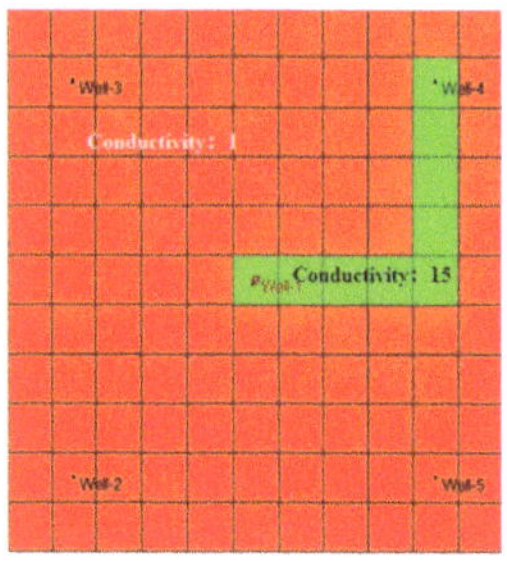


**Figure 15.** Modified conductivity indicates dominant channel.

The simulation results are shown in Figure 16. Due to the high seepage resistance of the injected water in the low-permeability area, the injected water preferentially passes through the water flow dominant channel, resulting in a reduced swept range of injected water. As a result, the low-permeability area around the injection well cannot be fully improved, and the oil well production in the low-permeability area cannot be effectively increased. It indicates that the existence of a dominant seepage channel accelerates the flow velocity of injected water in the reservoir, allowing the waterflood front to reach the production well faster; that is, the production well sees water earlier, and the water cut rises faster.

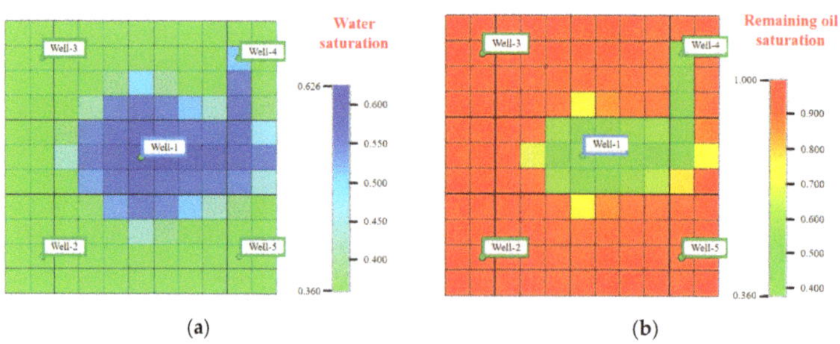


**Figure 16.** Sweep effect of injected water in the presence of water flow dominant channels. (**a**) Water saturation; (**b**) remaining oil saturation.

## 3.2. Engineering Factors

Engineering factors refer to human intervention measures, mainly including fracturing stimulation measures, water injection rate, and micro-fractures produced during high-pressure water injection. These factors have a significant impact on the movement direction of fluids within the oil layer, the sweep range of injected water, and the effect of water flooding to enhance oil recovery. The following will analyze the impact of engineering factors on the development effect of high-pressure water injection in combination with numerical simulation methods.

3.2.1. Fracturing Stimulation Measures

In order to clarify the influence of artificial fracturing on the development effect of high-pressure water injection, a reservoir numerical simulation model was established based on the "one injection and two production" well pattern (Well-1 is the water injection well, and Well-2 and Well-3 are the production wells). Among them, the artificial fracture of production Well-2 is parallel to the injection Well-1, and the artificial fracture of production Well-3 is perpendicular to the injection Well-1. The influence of fracturing dominant channels on the water injection effect is comparatively studied through numerical simulation comparison. The simulation results are shown in Figure 17. ① For the fractured wells parallel to the injection and production direction: Since the dominant channel forms a direct high seepage channel from the injection well to the production well, the pressure response of the production well is rapid, and production takes effect quickly. However, the water cut also rises rapidly; that is, it is easy to cause rapid flooding. ② For fractured wells perpendicular to the injection and production direction, the pressure propagation is relatively slow, and the immediate effect is slightly poor. However, the artificial fractures effectively extend to more oil layer areas. In the long term, obtaining a larger effective water flooding sweep area can enhance the oil displacement effect and the overall development effect. Therefore, when implementing hydraulic fracturing stimulation measures, the stress direction in the reservoir and the direction of injection and production wells should be comprehensively considered, and the direction of the fracturing fractures should be reasonably selected to avoid the premature formation of dominant flow channels between the injection and production wells, which would lead to a rapid increase in water cut of production wells and their scrapping.

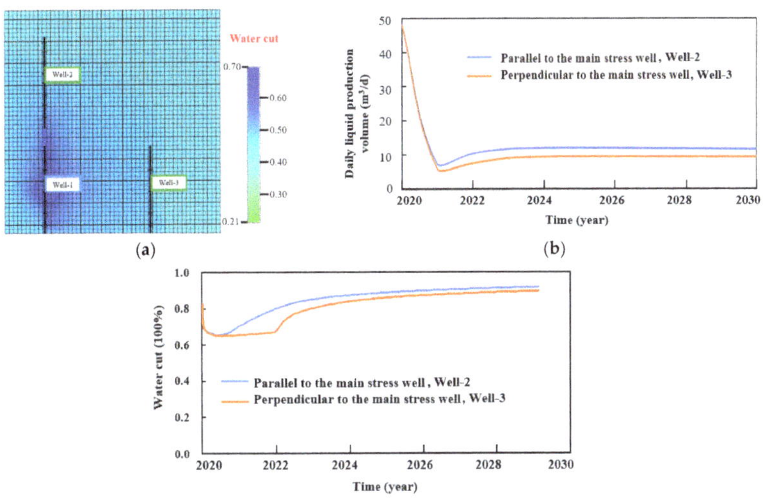


**Figure 17.** Results of the influence of artificial fractures in different directions on water injection development effect. (**a**) The "one injection and two production" well pattern model with artificial fractures; (**b**) daily liquid production rate curve of production well; (**c**) water cut curve of production well.

### 3.2.2. Water Injection Rate

Assuming that the total water injection volume in two groups of injection and production well groups is the same, two groups of simulation schemes (the injection rates of the injection wells c: 50 $m^3$/day and 200 $m^3$/day) are designed to study the influence of water injection rate on effective sweep breakthrough of the waterflood front. The result in Figure 18 shows the following: ① in the short term, the scheme with a high water injection rate has faster water injection pressure propagation, and the dynamic liquid level rises faster; the early oil production effect of production wells is better. But the waterflood front breaks through quickly, and the sweep coefficient decreases. In the long term, the water cut increases significantly and rapidly, and the cumulative recovery rate is relatively low. ② The lower injection rate scheme leads to slower waterflood front breakthrough, and the effective sweep area gradually expands, which is beneficial increasing the water drive effect in long-term development, reducing water cut in the produced fluid, and improving the final cumulative crude oil recovery rate. Therefore, a reasonable water injection rate should be selected in the actual production process, especially for low-permeability reservoirs, and a reasonable and moderate water injection velocity should be maintained to achieve long-term stable and efficient development results.

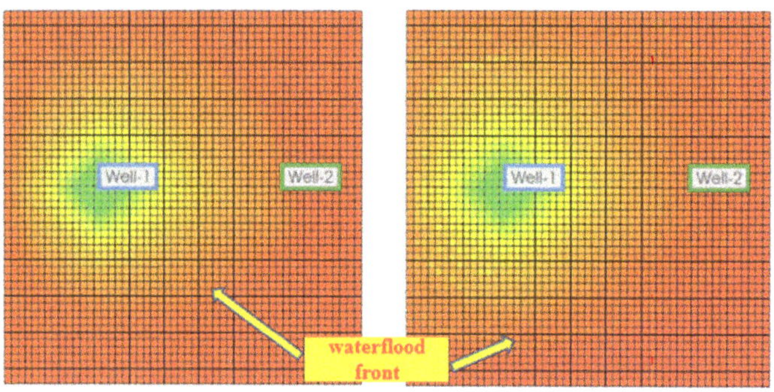


**Figure 18.** Distribution of remaining oil saturation.

### 3.2.3. Microfracture Produced by High-Pressure Water Injection

By numerical simulation method, the influence of microfracture (reservoir permeability enhancement channel) on development effect and water cut change law in high-pressure water injection development is studied [17,32–35]. Oil wells are set in different zones (pure oil zone, swept zone, and permeability enhancement zone), as shown in Figure 19. Four oil wells are set up for production in the reservoir after high-pressure water injection. Among them, Well-2 is located in the enhanced permeability zone, Well-5 is located in the pure oil zone, and Well 3 and Well 4 were located near the injection well and at the edge of the swept zone, respectively. Distance from water injection well: Well-5> Well-4> Well-3> Well-2. The simulation results are shown in Figure 20. The problem of rapid water channeling occurs in production wells in the permeability enhancement area (Well-2), and high water cut appeared earlier. In the swept zone (Well-3, Well-4), the water content at the wellhead increases first and then gradually stabilizes. In the pure oil zone (Well-5), the water cut is relatively low at the beginning, until the front edge of the injection water sweep is reached, and then the water cut begins to increase significantly, showing a sweep law of uniform diffusion outward from the injection well as the center.

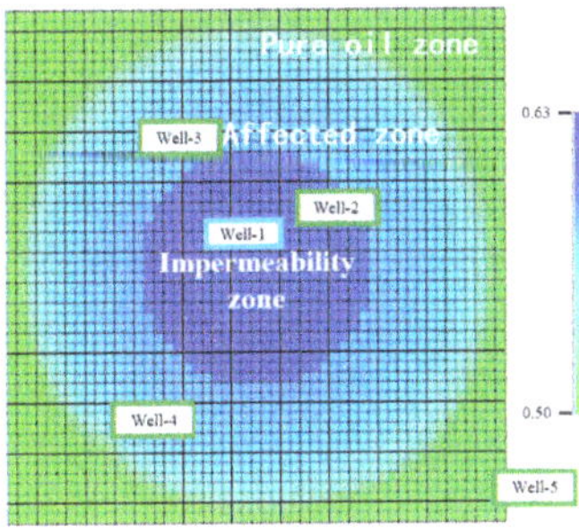


**Figure 19.** Distribution of production well locations.

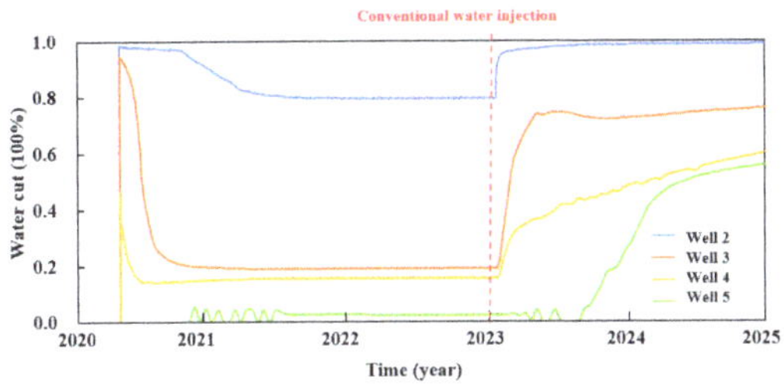


**Figure 20.** Water cut variation curves of production wells in different areas after high-pressure water injection.

As shown in Figures 21 and 22, when a transverse artificial microfracture is added to the model area, the swept area formed by high-pressure water injection again obviously presents a significantly non-uniform expansion distribution: the reservoir permeability enhancement range expands in the direction of the fracture zone, resulting in a significant increase in the reservoir swept area. The existence of micro-fractures in high-pressure water injection leads to the appearance of an obvious directional flow dominant channel for the injected water, and the water flooding sweep direction shows a directional effect. Therefore, in order to give full play to the production increase measures effect of high-pressure water injection, it is necessary to analyze the regional stress distribution field, the extension direction of micro-fractures, and the injection-production well pattern to reasonably optimize the opening and extension of micro-fractures guided by formation stress, further enhance the effect of reservoir water injection and oil displacement, and effectively enhance the oil recovery of the reservoir.

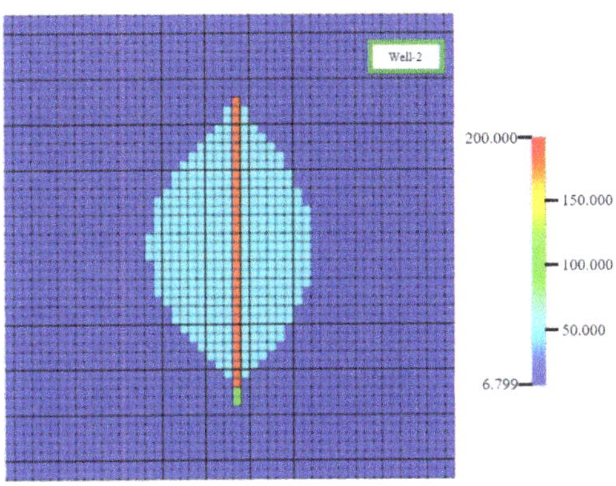


**Figure 21.** Microfracture display and permeability distribution in the enhanced permeability area; color represents permeability size, mD.

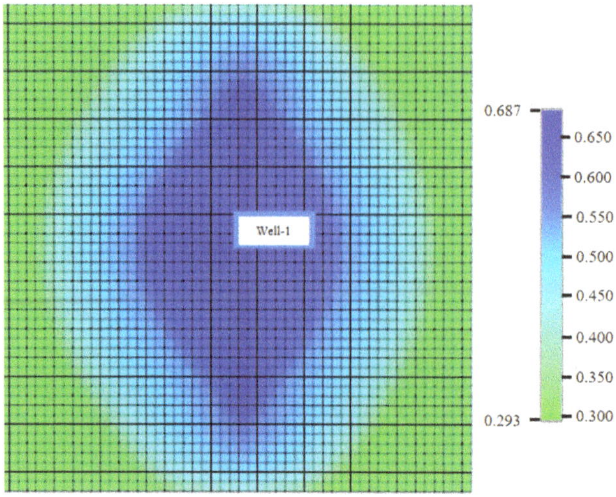


**Figure 22.** Distribution of water saturation in each area after high-pressure water injection; color represents water cut, 100%.

## 4. Conclusions

This article systematically analyzes the mechanism of high-pressure water injection technology in increasing formation water absorption capacity in low-permeability reservoir development and analyzes the inherent relationship between high-pressure water injection to achieve effective fluid drive and production increase from the aspects of seepage characteristics, high-pressure water injection expansion and permeability enhancement, and induced fractures, etc. Finally, by combining geological factors (formation pressure depletion, reservoir heterogeneity, water flow dominant channels) and engineering factors (fracturing stimulation measures, water injection rate, micro-fractures produced by high-pressure water injection), the influence of different factors on water injection development effect was studied. The main conclusions and recommendations are as follows:

(1) Since the mechanism of water flooding during the high-pressure water injection development of low-permeability reservoirs is not yet clear, this study analyzed the influence of geological and engineering factors on the development effect of high-pressure water injection. The research shows that during high-pressure water injection, production wells with a large formation pressure depletion experience a faster increase in water cut after water injection to replenish energy. The existence of preferential water flow channels causes the injected water to flow preferentially along high-permeability channels. High-permeability layers absorb the injected water first, resulting in a relatively rapid increase in the water cut of production wells in these layers, while the water cut of low-permeability layers changes relatively slowly. At the same time, it is necessary to consider the direction of formation stress, combined with a reasonable hydraulic fracture orientation and water injection rate, to increase the low-water-cut oil production time of production wells.

(2) Combining the compaction-expansion mechanism formed in the reservoir during high-pressure water injection, the reservoir after high-pressure water injection was divided into a permeability-increasing zone, a swept zone, and a pure oil zone, and the mining effects of production wells in different zones were numerically simulated. After high-pressure water injection, the change in reservoir water cut shows zonal characteristics: permeability-increasing zone (rapid water flooding) > swept zone (steady increase in water cut) > pure oil zone (delayed increase). It is analyzed that the strong connectivity between

injection and production wells in the permeability-increasing zone is the main reason for the rapid water flooding of production wells in this area.

(3) When the distribution of the water saturation and pressure increase zones formed after high-pressure water injection is affected by various factors in the formation, it usually shows a certain directionality. During the high-pressure water injection process, the micro-fractures formed cause the injected water to exhibit obvious directional flow, which in turn affects the change in water cut. The existence of micro-fractures makes the water flooding sweep direction show a directional effect, further accelerating the rise in water cut. Through numerical simulation of the flow directionality of high-pressure water injection in hydraulic fractures and local high-permeability zones, the results show that the extension direction of high-pressure water injection micro-fractures is comprehensively affected by factors such as high-permeability regions natural or artificial fractures.

**Author Contributions:** Y.L., H.X., H.Z., Z.C., and X.B.: Writing—original draft, methodology, investigation, formal analysis. J.W.: writing—review and editing, conceptualization. J.L., C.X., S.F. and L.Z.: investigation. All authors have read and agreed to the published version of the manuscript.

**Funding:** This work is financially supported by the Key project of scientific research plan of Education Department of Hubei Province, (NO. D20231301) and Open Fund of Hubei Key Laboratory of Oil and Gas Drilling and Production Engineering (Yangtze University) (NO. YQZC202401).

**Data Availability Statement:** The original contributions presented in the study are included in the article, further inquiries can be directed to the corresponding authors.

**Conflicts of Interest:** Authors Yang Li, Hongtao Zhao, Xuejing Bai, Chunhong Xiu, Lianshe Zhang were employed by the company Luming Oil and Gas Exploration and Development Co., Ltd., Shengli Oilfield, Sinopec. The remaining authors declare that the research was conducted in the absence of any commercial or financial relationships that could be construed as a potential conflict of interest. The company had no role in the design of the study; in the collection, analyses, or interpretation of data; in the writing of the manuscript, or in the decision to publish the results.

## References

1. Zeng, F.; Dong, C.; Lin, C.; Tian, S.; Wu, Y.; Lin, J.; Liu, B.; Zhang, X. Pore structure characteristics of reservoirs of Xihu Sag in East China Sea Shelf Basin based on dual resolution X-ray computed tomography and their influence on permeability. *Energy.* **2022**, *239*, 122386. [CrossRef]
2. Xiao, W.; Ren, J.; Pu, W.; Yuan, C.; Meng, L.; Zeng, L.; Zhao, H.; Chen, Q. Laboratory tests and field pilot of foam-assisted deoxidized-air flooding in a low-permeability sandstone reservoir. *Fuel* **2023**, *352*, 129150. [CrossRef]
3. Wang, Z.; Cao, G.; Bai, Y.; Wang, P.; Wang, X. Development Status and Prospect of EOR Technology in Low-Permeability Reservoirs. *Spec. Oil Gas Reserv.* **2023**, *30*, 1–13.
4. Li, X.; Yang, Z.; Li, S.; Huang, W.; Zhan, J.; Lin, W. Reservoir characteristics and effective development technology in typical low-permeability to ultralow-permeability reservoirs of China National Petroleum Corporation. *Energy Explor. Exploit.* **2021**, *39*, 1713–1726. [CrossRef]
5. Park, S.Y.; Ahn, K.I. Evaluation of an accident management strategy of emergency water injection using fire engines in a typical pressurized water reactor. *Nucl. Eng. Technol.* **2015**, *47*, 719–728. [CrossRef]
6. Yang, Y.; Zhang, S.; Cao, X.; Lv, Q.; Wang, J.; Liu, H.; Yu, C.; Sun, H. Practice and understanding of pressure-driven development technology in Shengli Oilfield. *Pet. Geol. Recovery Effic.* **2023**, *30*, 61–71. [CrossRef]
7. Chen, L.; Jia, C.; Zhang, R.; Yue, P.; Jiang, X.; Wang, J.; Su, Z.; Xiao, Y.; Lv, Y. High-pressure capacity expansion and water injection mechanism and indicator curve model for fractured-vuggy carbonate reservoirs. *Petroleum* **2024**, *10*, 511–519. [CrossRef]
8. Worm, J.; Naber, J.; Duncan, J.; Barros, S.; Atkinson, W. Water injection as an enabler for increased efficiency at high-load in a direct injected, boosted, SI engine. *SAE Int. J. Engines* **2017**, *10*, 951–958. [CrossRef]
9. Mahdavi, S.; James, L.A. High pressure and high-temperature study of $CO_2$ saturated-water injection for improving oil displacement; mechanistic and application study. *Fuel* **2020**, *262*, 116442. [CrossRef]
10. Tian, T.; Meng, Y.; Shi, S.; Wang, W. Development technologies and practice of undeveloped reserves in Shengli Oilfield. *Pet. Geol. Recovery Effic.* **2024**, *31*, 122–129. [CrossRef]

11. Ma, X.; Bi, Y.; Jiang, M.; Li, D.; Gu, X. Characteristics of water phase permeability variation in medium-low permeability oil reservoirs during high multiple waterflooding. *Pet. Reserv. Eval. Dev.* **2025**, *15*, 103–109. [CrossRef]
12. Wang, J.; Liu, L.; Wu, Y.; Liu, H.; Jia, Y. Progress and direction of water flooding development technologies in Shengli Oilfield. *Pet. Geol. Recovery Effic.* **2024**, *31*, 99–112. [CrossRef]
13. Wang, Y.; Song, X.; Tian, C.; Shi, C.; Li, J. Dynamic fractures are an emerging new development geological attribute in waterflooding development of ultra-low-permeability reservoirs. *Pet. Explor. Dev.* **2015**, *42*, 222–228. [CrossRef]
14. Zhao, X.; Zeng, L.; Ji, B.; Jiao, J.; Li, P.; Zhang, Y.; Chen, M. Discussion on optimal injection pressure of fractured low-permeability sandstone reservoirs—A case study from Wangyao block in Ansaioilfield, Ordos Basin. *Oil Gas Geol.* **2015**, *36*, 855–861.
15. Zhao, X.; Lv, W.; Wang, C.; Zhu, S.; Fan, J. Major factors controlling waterflooding-induced fracture development in low-permeability reservoirs-A case study of Chang 6 reservoir in W block in Ansai oilfield, Ordos Basin. *Oil Gas Geol.* **2020**, *41*, 586–595.
16. Wang, W.; Peng, Y.; Li, G.; Lei, Z.; Lv, W. Research on Water Flooding Dynamic Fractures to Optimize Infill Drilling Spacing in Ultra-low-permeability reservoirs, Changqing Oilfield. *Pet. Drill. Tech.* **2015**, *43*, 106–110. [CrossRef]
17. Liang, W.; Dang, H.; Liu, B.; Zhang, T.; Wang, X.; Hou, B.; Cui, P.; Wang, Q.; Zhang, F. Experiment and numerical simulation of water injection induced dynamic fractures in ultra-low-permeability reservoirs. *Pet. Geol. Exp.* **2023**, *45*, 566–575.
18. Wang, Y.; Song, X. Comprehensive characterization integrating static and dynamic data for dynamic fractures in ultra-low-permeability reservoirs: A case study of the Chang 6 reservoir of the Triassic Yanchang Formation in the Ordos Basin, China. *Minerals* **2022**, *12*, 1277. [CrossRef]
19. Wang, J.; Zheng, X.; Wang, Z.; Li, B.; Li, Y. Integrated approach to identify and control the water-injection induced fractures in the low-permeability reservoir. In Proceedings of the SPE Annual Caspian Technical Conference and Exhibition, Astana, Kazakhstan, 31 October–2 November 2018. [CrossRef]
20. Zhang, Y.; Yang, Y.; Sun, Z.; Yu, C.; Sun, Q.; Bei, J. Physical simulation of fracturing-flooding and quantitative characterization of fractures in low-permeability oil reservoirs. *Pet. Geol. Recovery Effic.* **2022**, *29*, 143–149. [CrossRef]
21. Lv, Y.; Guo, Z.; Chen, X.; Liu, D.; Zhang, Y.; Wang, Z. Analysis of Key Geological Factors of High Pressure Water Injection in Fractured Reservoir. *Adv. Geosci.* **2019**, *11*, 1050–1055. [CrossRef]
22. Souza, A.L.S.; Fernandes, P.D.; Mendes, R.A.; Rosa, A.J.; Furtado, C.J.A. The impact of injection with fracture propagation during waterflooding process. In Proceedings of the SPE Latin America and Caribbean Petroleum Engineering Conference. Rio de Janeiro, Brazil, 20–23 June 2005. [CrossRef]
23. Ma, X.; Li, H.; Luo, H.; Nie, S.; Gao, S.; Zhang, Q.; Yuan, F.; Ai, W. Research on well selection method for high-pressure water injection in fractured-vuggy carbonate reservoirs in Tahe oilfield. *J. Pet. Sci. Eng.* **2022**, *214*, 110477. [CrossRef]
24. Cui, C.; Li, H.; Wu, Z.; Zhang, C.; Li, H.; Zhang, Y.; Zheng, W. Analysis of pressures in water injection wells considering fracture influence induced by pressure-drive water injection. *Pet. Reserv. Eval. Dev.* **2023**, *13*, 686–694. [CrossRef]
25. Zhu, W.; Gao, Y.; Wang, Y.; Liu, P.; Liu, Y. The injection capacity evaluation of high-pressure water injection in low permeability reservoir: Numerical and case study. *Front. Earth Sci.* **2024**, *12*, 1411451. [CrossRef]
26. Zheng, S.; Yang, M.; Kang, Z.; Liu, Z.; Long, X.; Liu, K.; Li, X.; Zhang, S. Controlling factors of remaining oil distribution after water flooding and enhanced oil recovery methods for fracture-cavity carbonate reservoirs in Tahe Oilfield. *Pet. Explor. Dev.* **2019**, *46*, 786–795. [CrossRef]
27. Wang, X.; Qiao, X.; Mi, N.; Wang, R. Technologies for the benefit development of low-permeability tight sandstone gas reservoirs in the Yan'an Gas Field, Ordos Basin. *Nat. Gas Ind. B* **2019**, *6*, 272–281. [CrossRef]
28. Zeng, L.; Gao, C.; Qi, J.; Wang, Y.; Li, L.; Qu, X. The distribution rule and seepage effect of the fractures in the ultra-low permeability sandstone reservoir in east Gansu Province, Ordos Basin. *Sci. China Ser. D Earth Sci.* **2008**, *51* (Suppl. SII), 44–52. [CrossRef]
29. Prioul, R.; Jocker, J. Fracture characterization at multiple scales using borehole images, sonic logs, and walkaround vertical seismic profile. *AAPG Bull.* **2009**, *93*, 1503–1516. [CrossRef]
30. Wanniarachchi, W.A.M.; Ranjith, P.G.; Perera, M.S.A.; Rathnaweera, T.D.; Zhang, D.C. Investigation of effects of fracturing fluid on hydraulic fracturing and fracture permeability of reservoir rocks: An experimental study using water and foam fracturing. *Eng. Fract. Mech.* **2018**, *194*, 117–135. [CrossRef]
31. Hu, W.; Wei, Y.; Bao, J. Development of the theory and technology for low permeability reservoirs in China. *Pet. Explor. Dev.* **2018**, *45*, 685–697. [CrossRef]
32. Di, S.; Zhao, Y.; Ma, S.; Wei, Y.; Chen, S.; Miu, L. Fracture-to-fracture stimulation method for horizontal wells in fractured tight reservoir. *Pet. Geol. Oilfield Dev. Daqing* **2025**, *44*, 168–174. [CrossRef]
33. Wang, Y.; Song, X.; Meng, F.; Liang, Y.; Jiang, T. Dynamic fracture characteristics and controlling factors of ultra-low permeability reservoirs in Ordos Basin. *Acta Pet. Sin.* **2025**, *46*, 588–598.

34. Yang, Y.; Wang, J.; Bu, Y.; Sun, D. Practice and understanding of pressure-driven development technology in Shengli Oilfield. *Pet. Geol. Recovery Effic.* **2023**, 1–7. Available online: https://link.cnki.net/urlid/37.1359.TE.20230424.1050.002 (accessed on 18 August 2025).
35. Gao, Y.; Li, D.; Chen, H.; Cao, Y.; Yu, J.; Wu, S.; Chen, X.; Yuan, H.; Chen, Z. Mechanism and application of rock dilation-induced injectivity enhancement in injection wells for offshore oil field. *J. Northwest Univ. (Nat. Sci. Ed.)* **2025**, *55*, 613–632. [CrossRef]

processes

MDPI

Article

# A Power-Law-Based Predictive Model for Proppant Settling Velocity in Non-Newtonian Fluid

**Tianbo Liang *, Zilin Deng, Junlin Wu, Fangzhou Xu, Leyi Zheng, Maoqin Yang and Fujian Zhou**

State Key Laboratory of Petroleum Resources and Engineering, China University of Petroleum at Beijing, Beijing 102249, China; 2024211732@student.cup.edu.cn (Z.D.); wjl1242679450@163.com (J.W.); allstarsshining@outlook.com (F.X.); 2024311216@student.cup.edu.cn (L.Z.); 18710390774@163.com (M.Y.); zhoufj@cup.edu.cn (F.Z.)

* Correspondence: btliang@cup.edu.cn

**Abstract:** Effective proppant transport is critical to the success of hydraulic fracturing, particularly when using a non-Newtonian fluid. However, accurately predicting the proppant settling behavior under complex rheological conditions is still a significant challenge. This study proposes a new method for estimating the velocity of proppant settling in the power-law non-Newtonian fluid by accounting for spatial variations in viscosity within the fracture domain. The local shear rate field is first obtained using an analytical expression derived from the velocity gradient, and then used to approximate spatially varying viscosity based on the power-law rheological model. This allows the modification of Stokes' law, which was initially developed for Newtonian fluid, to be used for the power-law non-Newtonian fluid. The results indicate that the model achieved high accuracy in the fracture center region, with an average relative error of 8.2%. The proposed approach bridges the gap between traditional settling models and the non-Newtonian behavior of the fracturing fluid, offering a practical and physically grounded framework for predicting the velocity of proppant settling within a hydraulic fracture. By considering the distribution of the shear rate and viscosity of the fracturing fluid, this method enables an accurate prediction of proppant settling velocity, which further provides theoretical support to the optimization of pumping schedules and operation parameters for hydraulic fracturing.

**Keywords:** hydraulic fracturing; power-law fluid; proppant settling; shear rate; non-Newtonian fluid; sand transport; fracture conductivity

## 1. Introduction

Unconventional resources, including tight sandstone and shale formations, have emerged as key targets for oil and gas extraction, as conventional oil reserves are being depleted [1–3]. Hydraulic fracturing is frequently used to stimulate these formations due to the low permeability and intricate geological features of tight rocks and shales [4–6]. In this process, a high-pressure fluid is injected to generate fractures, and proppants are introduced to keep these fractures open, thereby creating conductive channels for the hydrocarbon flow [7–9].

In order to maximize production and guarantee a long-term fracture conductivity, proppants must be placed within the fracture effectively [10–12]. According to a number of current experimental and numerical investigations, the proppant's effective placement is influenced by its settling velocity [8,13–18]. Stokes' law has been widely applied to predict the settling velocity of the proppant in Newtonian fluid for decades [19], and it is known that the settling velocity is inversely related to the fluid viscosity. However, most fracturing fluids used in field applications exhibit non-Newtonian behavior, particularly a shear-thinning characteristic [20–23]. Experimental results have shown that the proppant settling behavior in non-Newtonian fluid differs significantly from that in Newtonian fluid, especially under varying flow regimes and rheological conditions during hydraulic fracturing [24–31]. This renders Stokes' law unsuitable to predict the particle settling velocity in non-Newtonian fluid.

The existing proppant settling models are mostly based on the modification of Stokes' law or empirical correlations that are obtained use power-law parameters. To study proppant settling behavior, Shah [32] presents a new approach to analyze proppant settling data in power-law non-Newtonian fluid and develops a drag coefficient correlation as a function of rheological parameter $n'$ to investigate the proppant settling velocity. Roodhart [33] introduced a modified model based on Stokes' law, incorporating the concepts of zero-shear viscosity and spatial viscosity anisotropy to enable more accurate calculation of particle settling velocities. In response to the complexity of particle settling in non-Newtonian fluid, researchers have proposed theoretical and empirical models. Ovarlez [34] demonstrated that, within non-Newtonian suspension systems, the velocity of particle settling increases with the applied shear rate, underscoring the critical impact of local shear conditions on the settling velocity. Amaratunga et al. [35] further confirmed that the shear-thinning fluid exhibits a pronounced spatial variation in the shear rate, particularly near boundaries, and these variations correlate with local changes in horizontal and settling velocities. Anyaoku [36] designed a new experimental apparatus capable of measuring localized shear rates; utilizing this measurement, they developed a semi-empirical settling model tailored to viscoelastic fluid to predict its settling velocity. Based on the power-law fluid model and incorporating the fiber volume concentration to modify fluid viscosity and drag coefficient, Bai and Li [37] established a predictive model for the settling velocity of the proppant in fiber-containing fracturing fluid. Despite substantial advances in theory and experiments, the existing models typically ignore the special distribution of the flow rate in the direction of the fracture thickness and are thus unable to calculate the particle settling velocity at different spatial locations within non-Newtonian fluid.

In this study, a predictive method is proposed by combining the spatial distribution of the shear rate with the rheological properties of the non-Newtonian fluid. Specifically, by applying the power-law model and evaluating the local shear conditions within the fracture, the variation in the fluid viscosity and its subsequent impact on the proppant settling velocity can be calculated. This method enables the prediction of proppant settling behavior at different locations within the fracture, thereby offering insights into the transport distance of proppants and facilitating the prediction of the effective length of the propped fracture. The results of this work offer valuable theoretical guidance for optimizing the pumping schedule of hydraulic fracturing and improving the treatment performance in field applications.

## 2. Materials and Methods

### 2.1. Rheology of Fluid Used in This Study

The power-law fluid is used as a non-Newtonian fluid in this study. Its apparent viscosity at different shear rates can be determined as follows.

$$\eta = \frac{\tau}{\gamma} = k\gamma^{n-1} \tag{1}$$

where:

$K$: flow consistency index, Pa·S$^n$;

$n$: fluid fluidity index, unitless;

$\tau$: shear stress, Pa;

$\gamma$: shear rate, s$^{-1}$;

$\eta$: apparent viscosity, Pa·s.

In this study, the fracturing fluid is modeled as a shear-thinning and power-law fluid. Rheometer tests conducted at 25 °C on the fracturing fluid using a rheometer (HAAKE RheoStress 600, Thermo Electron Karlsruhe GmbH, Karlsruhe, Germany) are fitted to the power-law model, which yields a flow behavior index n of 0.46–0.53 and a consistency index $K$ of 138–152 mPa·s. For the simulations, the median values $K$ = 145 mPa·s and $n$ = 0.50 are adopted, and these fall within the typical range for shear-thinning, polymer-based fracturing fluids ($n$ = 0.4–0.9, $K$ = 10–1000 mPa·s). The relationship between apparent viscosity and shear rate is illustrated in Figure 1.

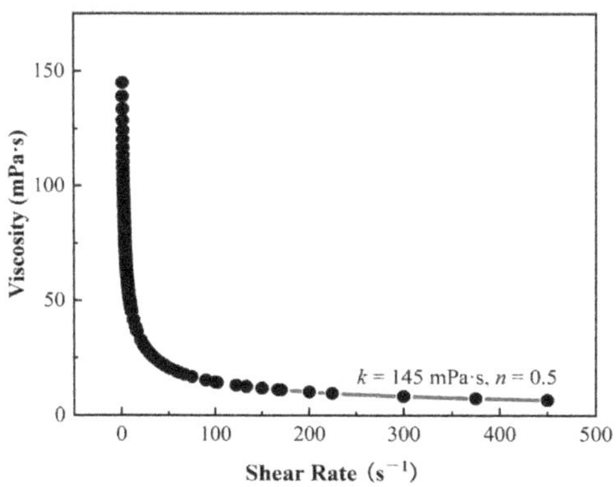


**Figure 1.** The viscosity of the non-Newtonian fluid used in the simulation as a function of the shear rate.

## 2.2. Shear Rate of Non-Newtonian Fluid Within a Hydraulic Fracture

The shear rate of the power-law fluid at various positions within a fracture can be obtained by the formula for calculating the shear rate. Equation (2) is the narrow-fracture model proposed by Hagen and Poiseuille [38], and the fracture is idealized as a rectangular prism with length L, width 2B, and height W (as shown in Figure 2). However, this formulation does not incorporate shear rate effects. Therefore, this study first replaces the pressure P with the flow rate Q by integration, then differentiates to obtain an expression describing the shear rate (Equation (3)).

$$V_Z = \left[\frac{(P_o - P_L)B}{mL}\right]^{\frac{1}{n}} \frac{B}{\left(\frac{1}{n}\right)+1}\left[1 - \left(\frac{x}{B}\right)^{\left(\frac{1}{n}\right)+1}\right] \tag{2}$$

$$\frac{dV_Z}{dx} = \frac{Q}{2W}\frac{1}{B\left(1 - \left(\frac{1}{2+\frac{1}{n}}\right)\right)}\left[-\frac{1}{B}\left(1+\frac{1}{n}\right)\left(\frac{x}{B}\right)^{\left(\frac{1}{n}\right)}\right] \tag{3}$$

where:

$Q$: flow rate, ml/s;

$P_o$: inlet pressure, Pa;

$P_L$: outlet pressure, Pa;

$B$: half fracture width, cm;

$W$: fracture height, cm;

$x$: position from the middle of the fracture, cm;
$n$: fluid fluidity index, dimensionless;
$V_z$: velocity in the z-direction, cm/s.

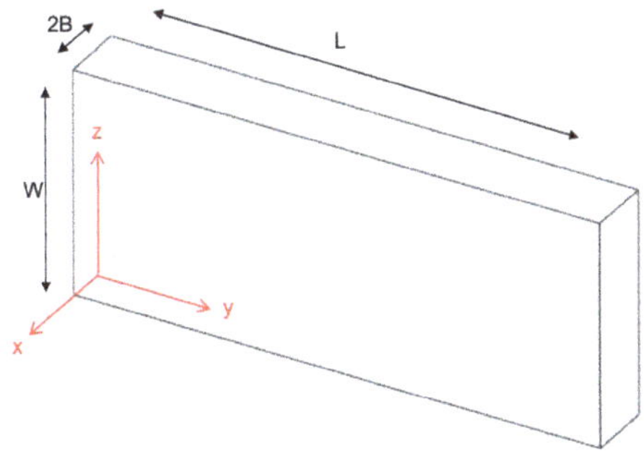


**Figure 2.** Schematic diagram of a hydraulic fracture.

### 2.3. Model of Particle Settling in Newtonian Fluids (Stokes' Law)

In this study, the proppant is treated as a rigid sphere, as in Stokes' law. When the particle motion is at a low Reynolds number (Re < 1), the proppant settles at a constant settling velocity and the force reaches an equilibrium; the settling velocity can be solved by the following equation.

$$V_t = \frac{2}{9} \cdot \frac{\left(\rho_p - \rho_f\right) \cdot g \cdot R^2}{\mu} \tag{4}$$

where:

$R$: particle radius, m;
$\rho_p$: particle density, $kg/m^3$;
$\rho_f$: fluid density, $kg/m^3$;
$g$: gravitational acceleration, 9.81 $m/s^2$;
$\mu$: fluid viscosity, Pa·s or $N{\cdot}s/m^2$.

The shear rate of the power-law fluid at different distances in the direction of the fracture width can be calculated using Equation (2), which then gives the apparent viscosity of the power-law fluid at that location to allow further calculation of the settling velocity of a proppant.

### 2.4. Description of the Simulation Model

We validate the proposed model, which estimates the apparent viscosity of a non-Newtonian fluid at different positions across the fracture width based on the local shear rate and uses it to predict the proppant settling behavior. A computational fluid dynamics-discrete element (CFD-DEM) method is used to simulate the migration and settling of proppants within a vertical fracture. The fluid phase is solved using a CFD approach based on the continuity and momentum conservation equations, while the particle motion is modeled using the DEM method in accordance with Newton's second law. Momentum exchange between the fluid and solid phases is achieved through interphase coupling via pressure and velocity fields. The simulation iteratively updates the flow field and particle trajectories to realize the dynamic coupling between fluid and particle motion, thereby enabling the modeling of the proppant transport and settling processes. The geometry of the vertical fracture is built in the geometry module of ANSYS Fluent (2022R1). Based on field-scale observation, the actual width of the hydraulic fracture typically ranges from 2 mm to 6 mm; therefore, a representative width of 6 mm is selected for this study. The fracture model dimensions are set to 6 mm (width) × 900 mm (height) × 300 mm (depth). In the model, the inlet, outlet, and wall boundaries are defined, as shown in Figure 3.

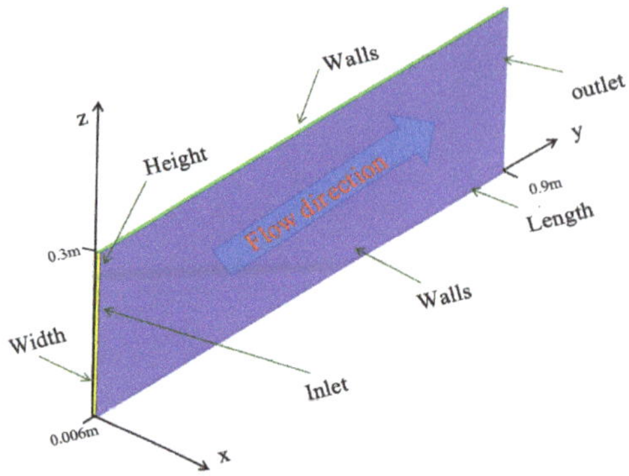


**Figure 3.** Schematic diagram of the hydraulic fracture model.

### 2.5. Validation of the Predictive Model for Proppant Settling Velocity and Design of the Simulation Cases

In order to verify the accuracy of the predictive model for proppant settling velocity, simulations are first conducted in Newtonian fluid. The injection flow rate is selected based on the typical displacement range used in field-scale hydraulic fracturing operations, which generally falls between 2 $m^3/min$ and 23 $m^3/min$; at the laboratory scale, this corresponds to flow velocities ranging from approximately 0.0637 m/s to 0.5537 m/s. To encompass the low-, medium-, and high-displacement-injection conditions that may occur in the actual operating conditions, the design range of the flow velocity is set to 0.06 m/s–0.6 m/s in the numerical simulation. This range not only covers the common displacement variation range in on-site fracturing, but also facilitates the analysis of the settlement behavior of proppants at different flow velocities.

In field applications, proppants with different particle sizes are usually selected according to formation conditions, fracture characteristics, and sand-carrying requirements to achieve effective filling and support for fracturing fractures, so proppants with different particle sizes that are commonly used in the field are selected as the research object, and the particle size settling range is set from 0.00021 m to 0.00084 m (20–70 meshes). Ceramic proppants with a density of 2500 $kg/m^3$ are used in this study. Considering the variation in fracturing fluid viscosity under different reservoir conditions, the simulation scheme also covers a wide viscosity range from 0.005 Pa·s to 1 Pa·s. The numerical results are then compared with theoretical predictions based on Stokes' law, and a high degree of agreement confirmed the validity of the model in the Newtonian fluid. For the non-Newtonian fluid, the simulation scheme included variations in injection location and flow rate along the fracture width (Z-direction). The simulation results are compared with values calculated using the proposed method for predicting the settling velocity of proppant in non-Newtonian fluid.

## 3. Results and Discussion

### 3.1. Influence of Particle Settling Velocity in the Newtonian Fluid

During the hydraulic fracturing operation, particles are typically injected into the fracture fluid at defined concentrations. To better understand the behavior of particles under such conditions, simulations of multi-particle settling at different particle concentrations are conducted in the Newtonian fluid. The same particle is used in the simulation with a diameter of 0.0006 m and a density of 2500 kg/m$^3$; the fluid has a viscosity of 0.001 Pa·s with a horizontal injection velocity of 0.1 m/s.

The figure above presents a comparison of simulation results obtained with different models fitted with experimental data reported by Liu, Richardson, and Daneshy, Maude and Whitmore (Happel, 2012), and Govier and Aziz (Happel, 2012) [39–42]. A good match of simulation and experimental data indicates the accuracy of the model in predicting the particle settling in the Newtonian fluid.

As shown in Figure 4, the average settling velocity of particles decreases significantly with an increase in particle concentrations, which indicates that the settling velocity of multiple particles is slower than that of a single particle. It is likely that particles disturb the surrounding flow field, triggering the back flow of local fluid, which generates a reverse drag force and inhibits the particle settling. Therefore, it is reasonable to focus on a single particle's behavior when evaluating particle settling, and the experimental scheme presented in this study is designed to investigate the settling velocity of a single particle under different experimental conditions.

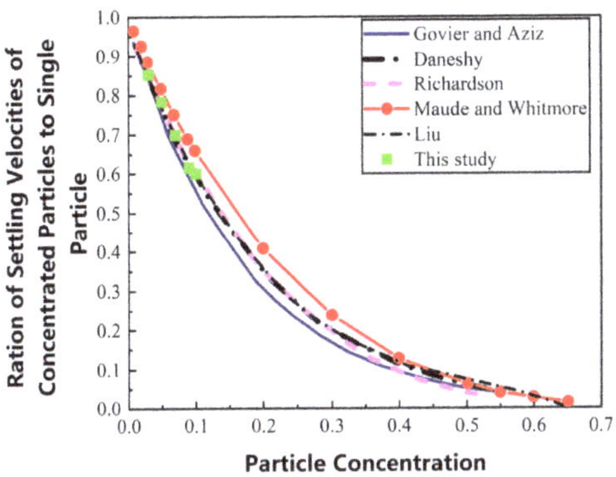


**Figure 4.** Comparison of settling velocities of particles with different particle concentrations using different models.

The impact of different injection flow rates, fluid viscosities, and particle diameters on the settling velocity of particles is systematically investigated in a Newtonian fluid through 17 simulation cases. To investigate the effects of each parameter, a series of controlled simulations are conducted. In cases 1–5, the injection velocity is set at 0.1 m/s and the particle diameter is maintained at 0.0006 m, while the fluid viscosity is varied from 5 to 1000 mPa·s to evaluate its influence. In cases 6–10, with the viscosity fixed at 10 mPa·s and the injection velocity at 0.1 m/s, the particle diameter is varied from 0.00021 to 0.00084 m to examine the effect of the particle size. In cases 11–17, the viscosity and particle diameter are held constant at 10 mPa·s and 0.0006 m, respectively, and the injection velocity is incrementally increased from 0.06 to 0.6 m/s to understand the impact of the flow rate.

Cases 1–5 are used to simulate fracturing fluid systems with different viscosities commonly found in shale reservoirs and conventional reservoirs. This range can effectively represent the differences in fluid characteristics under different sand-carrying capacities in the fracturing process, which can then be used to analyze their influence on the settling rate of particles, providing a theoretical basis for optimizing the fracturing operation parameters. The simulated injection point is set at the midpoint along the width of the fracture inlet, and the injection flow rate is set at 0.1 m/s. The corresponding volumetric flow rate is comparable to that of small-displacement fracturing in the field, which ensures computational feasibility and physical realism, as shown in Figure 5.

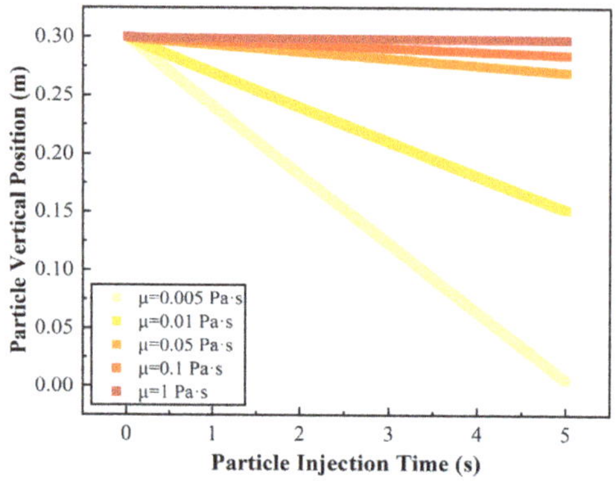


**Figure 5.** Vertical position of particles in Newtonian fluids with various viscosities.

Figure 5 illustrates the settling behavior of particles in Newtonian fluid with various viscosities, all subjected to the same injection velocity of 0.1 m/s. This figure is used to analyze the impact of fluid viscosity on particle settling. The viscosity ranges from 0.005 Pa·s to 1 Pa·s. As shown, the settling velocity is calculated based on the particle trajectory shown in Figure 5, and the vertical settling velocity of the particles significantly decreases as the viscosity increases.

For low-viscosity fluids, such as those with viscosities of $\mu$ = 0.005 Pa·s and $\mu$ = 0.01 Pa·s, the particles rapidly settle to the bottom of the fracture within a short period, which indicates that the fluid's resistance to particle motion is minimal and does not effectively hinder settling under gravitational forces. As the fluid viscosity increases to medium ($\mu$ = 0.05 Pa·s) and high values ($\mu$ = 0.1 Pa·s), the settling velocity of the particles slows down, which suggests that the fluid's resistance to the particles is becoming more significant. Under high viscosity conditions ($\mu$ = 1 Pa·s), the particles remain almost stationary, with negligible settling, which indicates that high-viscosity fracturing fluids exhibit excellent sand-carrying capabilities. These results demonstrate that the fluid viscosity plays a crucial role in determining the settling behavior of particles under a constant injection velocity. High-viscosity fluids significantly enhance the drag force exerted on the particles, prolonging their suspension time in the fracture and thus improving the sand-carrying capacity of the fracturing fluid. Therefore, when choosing the fracturing fluid, it is essential to select a fluid system with an appropriate viscosity based on reservoir conditions and operational parameters to effectively control the settling behavior of particles.

To further validate the accuracy of the simulated particle settling behavior and investigate the influence of key physical parameters on the settling velocity of the particle, cases 6–10 are designed by varying the particle diameter (from 0.00021 m to 0.00084 m) while maintaining a constant horizontal injection velocity of 0.1 m/s. Figure 5 illustrates a comparison between the simulated settling velocities and the theoretical Stokes' settling velocities under these conditions.

The green triangle markers in Figure 6 represent the fitting relationship between the simulated settling velocities and the corresponding theoretical values calculated from Stokes' law under various fluid viscosities, while keeping the particle properties constant. These data points correspond to cases 1–5. The blue star markers depict the fitting results obtained by varying the particle diameters from 0.00021 m to 0.00084 m, corresponding to the 70–20 mesh particles commonly used in field-scale fracturing operations. These data points correspond to cases 6–10. All points are distributed close to the diagonal line y = x, which indicates a strong correlation between the numerical simulation outcomes and theoretical expectations based on Stokes' settling law. This excellent agreement validates the accuracy, reliability, and applicability of the simulation approach adopted in this study for predicting particle settling behavior in Newtonian fluid. As clearly shown in Figure 6, increasing the particle diameter significantly enhances the settling velocity. Larger particles experience stronger gravitational forces relative to the fluid drag, which accelerates their downward motions. Conversely, increasing the fluid viscosity progressively reduces the settling rate. This indicates that viscous drag imposed by the fluid plays a dominant role in resisting the motion of particles, especially under high-viscosity conditions. The observed trends are in excellent agreement with Stokes' equation, which predicts that the settling velocity is directly proportional to the square of the particle diameter and inversely proportional to the fluid viscosity. These findings reinforce the theoretical understanding of settling dynamics and confirm the accuracy of the simulation framework.

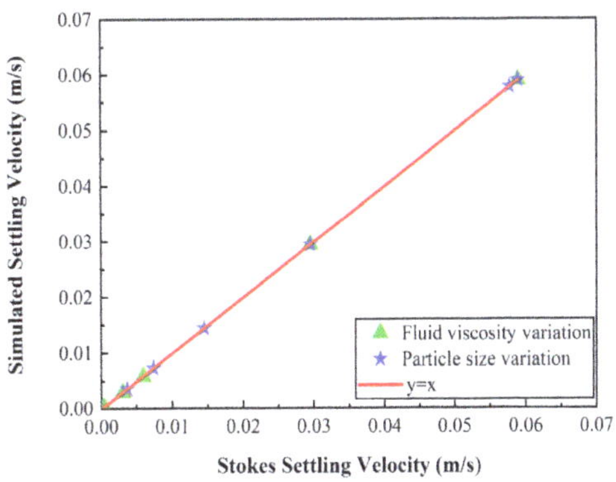


**Figure 6.** Settling velocity of particle in Newtonian fluids with various fluid viscosities and particle diameters.

In summary, in the Newtonian fluid, the settling behavior of the particles is primarily governed by two key parameters: the particle size and fluid viscosity. Adjusting either factor can significantly influence the particle transport performance of the fracturing fluid. Therefore, in practical fracturing fluid design, it is essential to carefully select the particle size and match it with an appropriate fluid viscosity to achieve effective particle suspension and placement. The results of this study not only validate the proposed modeling approach but also provide a theoretical reference for analyzing more complex settling phenomena in non-Newtonian fluids. This contributes to the scientific optimization of proppant design and the enhancement of hydraulic fracturing performance in both conventional and unconventional reservoirs. To investigate the influence of the injection flow rate on particle settling behavior across different fluid types, simulations are performed using non-Newtonian fluids that exhibit the same viscosity as their Newtonian counterparts under low-shear-rate conditions. Figure 7 presents the vertical settling velocities of particles at various injection flow rates in both Newtonian and shear-thinned non-Newtonian fluids. The simulation conditions are kept consistent across both fluid types, with the power-law model being applied to characterize the shear-thinning behavior of the non-Newtonian fluid. Shear thinning refers to the phenomenon where the fluid viscosity gradually decreases with an increasing shear rate, which is a common characteristic of many fracturing fluids used in field operations.

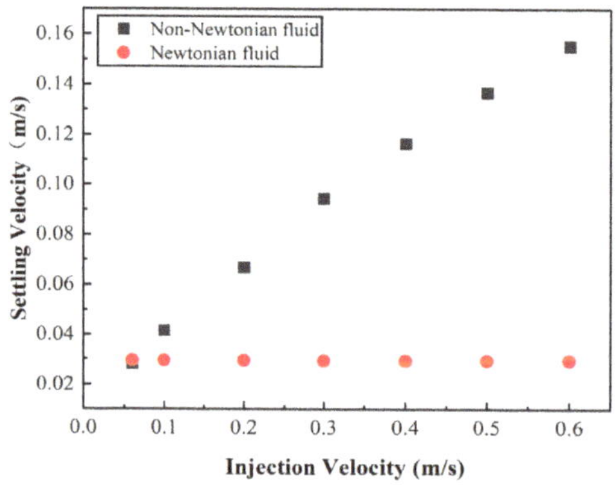


**Figure 7.** Comparison of settling velocity of particles in Newtonian and non-Newtonian fluids under various horizontal flow velocities.

The red dots in Figure 7 represent the settling velocity of particles at different injection rates. The simulation results indicate that, in the Newtonian fluid, the particle settling velocity remains nearly constant across the injection velocity range of 0.06 m/s to 0.6 m/s, with an average value of approximately 0.027 m/s. This is because the viscosity of the Newtonian fluid is constant and unaffected by the shear rate. As a result, variations in the injection velocity influence the fluid's velocity field and horizontal transport path but have minimal impact on the vertical resistance to settling. Under gravity, the settling velocity remains stable and independent of the injection rate.

In contrast, the black dots in Figure 7 represent particle settling velocities in non-Newtonian fluids under identical simulation durations and conditions but with varying injection rates. The results clearly demonstrate that the injection rate significantly affects the particle settling behavior in shear-thinning non-Newtonian fluids, exhibiting a pattern that is distinctly different from that observed in the Newtonian fluid. At lower injection rates, the local shear rate remains low, and the non-Newtonian fluid maintains a high equivalent viscosity. This elevated viscosity generates strong viscous resistance that inhibits particle settling, resulting in a low settling rate and enhanced sand-carrying capacity. However, as the injection rate gradually increases, the local shear rate increases significantly, which leads to a pronounced shear-thinning effect. The resulting decrease in viscosity reduces the viscous drag on the particles, allowing gravitational settling to become more dominant. While higher flow rates may initially enhance particle transport, the rapid viscosity drop weakens the fluid's resistance to settling. This results in a marked increase in settling velocity with an increasing injection rate, causing particles to lose stability and settle more easily within the fractures. Therefore, in the shear-thinning non-Newtonian fluid, higher injection speeds tend to reduce the fluid's sand-carrying capacity, contrary to the behavior observed in the Newtonian fluid.

Shear-thinning non-Newtonian fluids are predominantly employed under actual field fracturing conditions. This phenomenon reveals the existence of a critical sand-carrying velocity in shear-thinning non-Newtonian fracturing fluids. Specifically, when the injection rate exceeds a certain threshold, the fluid's equivalent viscosity decreases significantly, which causes a loss of its effective suspension capability for particles. Consequently, a substantial amount of particles settle prematurely, which adversely affects the effective support of fractures and reduces fracture conductivity. Therefore, during practical fracturing design, excessively high injection rates should be avoided. Instead, an optimal injection rate range must be selected by balancing the fracture propagation capacity with the non-Newtonian sand-carrying characteristics, which thereby ensures uniform particle distribution and stable maintenance of fracture conductivity.

### 3.2. Influence of Particle Settling Velocity in the Non-Newtonian Fluid

To further investigate the settling characteristics of particles in non-Newtonian fluids, this study examines the effects of different injection flow rates and injection positions on the particle settling velocity. The simulations use the non-Newtonian fluid that exhibits pronounced shear-thinning behavior, as detailed in Section 2.1. The same particle is used in the simulation, with a diameter of 0.0006 m and a density of 2500 kg/m$^3$. To investigate the effects of flow parameters on the particle settling behavior in non-Newtonian fluids, 21 numerical simulation cases are designed, with the injection velocity and injection position being the primary control variables. The injection position is set at three levels in ascending order: 0.0006 m, 0.00075 m, and 0.0015 m. For each fixed injection position, the injection velocity is systematically varied across seven levels: 0.06 m/s, 0.1 m/s, 0.2 m/s, 0.3 m/s, 0.4 m/s, 0.5 m/s, and 0.6 m/s. These velocity ranges cover low to moderate flow regimes, allowing for the observation of diverse flow behaviors. In cases 1–7, fixed at an injection position of 0.0015 m, the injection velocities increase incrementally from 0.06 m/s to 0.6 m/s. Cases 8–14, fixed at 0.00075 m, have the same sequential variation in injection velocity as cases 1–7. Cases 15–21, fixed at 0.0006 m, maintain consistency in the injection velocity gradient.

Figure 8 depicts the vertical settling positions of particles over time at various injection rates in a shear-thinning non-Newtonian fluid, corresponding to cases 1–7. Once the particle reaches a steady settling stage, the simulated settling velocity is calculated based on the trajectory. This figure clearly illustrates how the injection rate affects particle settling dynamics in shear-thinning fluids under controlled simulation conditions.

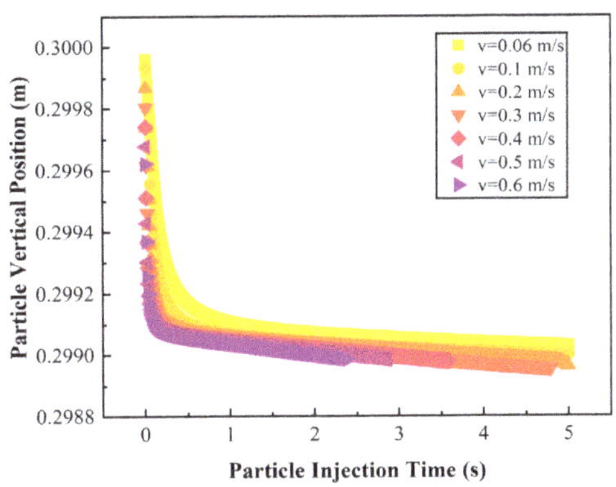


**Figure 8.** Vertical position of particles under various injection velocities of the non-Newtonian fluid.

As shown in Figure 8, the particles exhibit a rapid vertical displacement at the initial stage of injection, which is followed by a significant reduction in settling velocity, eventually approaching a stable rate. This behavior reflects a nonlinear settling trend characterized by an initial rapid descent that transitions into a slower, more stable phase. As the injection velocity increases from 0.06 m/s to 0.6 m/s, the total vertical displacement of the particles within 5 s also increases, which indicates that higher injection velocities facilitate particle settling. This phenomenon is primarily attributed to the shear-thinning behavior of the non-Newtonian fluid. At higher injection rates, the shear rate within the fluid increases substantially, which leads to a decrease in equivalent viscosity. This reduction in viscosity weakens the fluid's viscous drag on the particles, reducing its ability to keep them suspended and allowing gravity to drive their settling more effectively. Additionally, at the early stage of injection, the settling curve exhibits a steep slope, which indicates that particles settle rapidly. This is mainly because the fluid velocity field has not yet fully developed, and the particles are initially in a free-settling state dominated by gravity. At this point, the fluid has not established sufficient drag force to suspend the particles effectively. As time progresses, the flow field gradually stabilizes, the drag force from the fluid increases, and the settling velocity decreases, entering a transitional equilibrium stage. The variation in injection time observed in the graph is caused by the finite length of the computational domain. At higher injection velocities, particles move through the domain more rapidly and reach the outlet boundary earlier, which results in shorter simulation times.

It is noteworthy that higher flow velocities result in significantly greater settling displacement within a short period, as it emphasizes the substantial enhancement of particle settling under high injection rates. Unlike the Newtonian fluid, in the shear-thinning non-Newtonian fracturing fluid, elevated injection rates may significantly increase the risk of particle settling. This is due to the pronounced shear-thinning behavior, which causes a sharp reduction in fluid viscosity and thereby diminishes the fluid's ability to keep particles suspended. As a result, excessively high injection rates may lead to ineffective particle suspension, reduced transport efficiency, and compromised fracture support. In practical fracturing operations, simply increasing the injection rate to enhance fracture propagation should be avoided. Instead, the injection rate should be carefully optimized based on the rheological characteristics of the fracturing fluid to ensure an effective balance between promoting fracture extension and maintaining sufficient proppant-carrying performance. To further investigate the spatial distribution of the particle settling velocity and its relationship with the shear rate, Figure 9 illustrates the correlation between the settling velocity and local shear rate at varying distances from the fracture center, under an injection velocity of 0.1 m/s, in a shear-thinning non-Newtonian fluid.

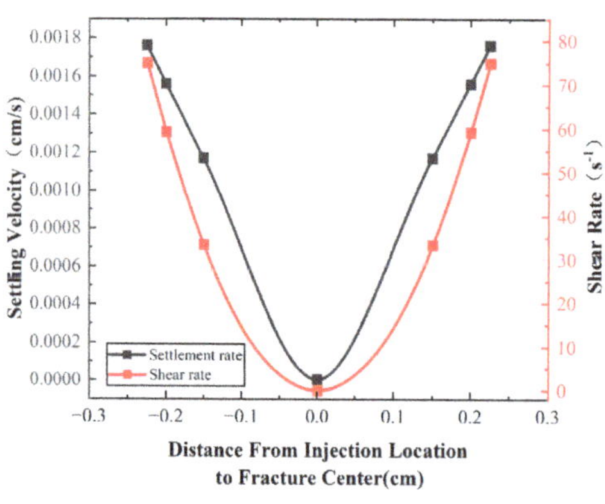


**Figure 9.** Distribution of shear rate and settling velocity at different locations of a hydraulic fracture.

As shown in Figure 9, the shear rate near the center of the fracture is nearly zero, corresponding to the lowest particle settling velocity. In contrast, the shear rate increases significantly near the fracture walls, which results in a marked increase in the particle settling velocity. These results indicate that particle settling in the shear-thinning non-Newtonian fracturing fluid exhibits substantial spatial inhomogeneity. The settling behavior is strongly governed by the local shear field; when the particle is located closer to the fracture wall, it experiences a lower local viscosity, which makes it easier to settle. In the shear-thinning fluid, the equivalent viscosity decreases significantly with an increasing shear rate. At the fracture center, the low shear rate maintains a high viscosity, which increases the drag and strongly suppresses the particle settling. Conversely, near the fracture wall, the high shear rate lowers the local viscosity, reduces fluid resistance, and facilitates particle descent. Therefore, different injection positions correspond to distinct local shear and viscosity conditions, which directly govern the settling dynamics of particles.

Subsequently, to validate the method for determining the settling velocity of particles in the non-Newtonian fluid, the simulated settling velocities obtained by varying the injection position and flow rate in the non-Newtonian fluid are compared with results derived from the settling velocity calculation formula outlined in Section 2.1. The comparison results are presented in Figure 10.

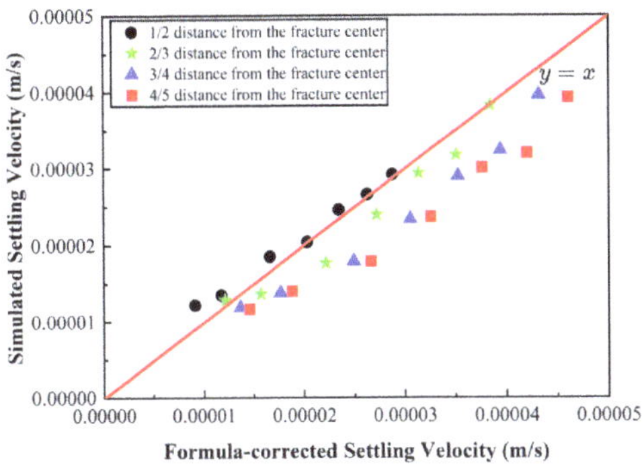


**Figure 10.** Fitting of the calculated settling velocity and the simulated settling velocity at different locations of a hydraulic fracture.

The Reynolds number of the particles in each simulation case ranges from $1.6787 \times 10^{-7}$ to $4.29748 \times 10^{-6}$, with all values remaining well below 0.1. This confirms that the Stokes settling formula is applicable under these conditions. The particle diameter is 0.0006 m, and the particle density is 2500 $kg/m^3$. Figure 10 presents a comparison between the settling velocities predicted by the corrected formula and the simulated values at various positions along the fracture profile. The data points are evenly distributed around the line y = x, confirming the applicability and robustness of the proposed formula for predicting the settling velocity in shear-thinning fluids at low Reynolds numbers.

Although the overall trend shows a good fit, some data points deviate from the ideal line, mainly due to the boundary layer effect. The closer the particle injection location is to the fracture wall, the more susceptible the particles are to wall effects and enhanced shear gradients, which results in a non-uniform distribution of the local fluid viscosity and shear rate. This leads to deviations of the actual settling velocity from the ideal behavior. Quantitative analysis reveals that the average relative error increases from 8.2% near the fracture center to 30.9% close to the wall, as shown in Table 1. This discrepancy arises mainly from the no-slip boundary condition and the finite domain of the fracture, which create velocity gradients and viscous boundary layers that restrict particle motion compared to the infinite medium assumptions underlying the theoretical models. The current models for non-Newtonian power-law fluids are largely based on empirical or semi-empirical correlations derived from experimental data. In contrast, our study proposes an analytical solution that accounts for the spatial variation in shear rate and fluid rheology, thereby overcoming the limitations of traditional approaches. Despite these localized deviations, the proposed method demonstrates strong potential for rapid and accurate prediction of particle settling behavior and the estimation of key engineering parameters under realistic fracturing conditions.

**Table 1.** Average relative error between simulated and formula-corrected settling velocities.

| Distance from Fracture Center | Average Relative Error (%) |
|---|---|
| 1/2 (near center) | 8.2 |
| 2/3 | 10.6 |
| 3/4 | 23.3 |
| 4/5 (near wall) | 30.9 |

When they are suspended in a power-law fluid, the settling velocity of particles varies at different positions from fracture walls, as shown in Figure 11. As previously discussed, this is due to the spatial variation in fluid viscosity. Therefore, the local shear rate at various positions should be calculated using the prediction model proposed in this study to determine the corresponding settling velocities. By incorporating the spatial distribution of the shear rate and viscosity within the fracture, the proposed model enables the estimation of location-dependent settling velocities of particles. This provides a fundamental framework for understanding the influence of rheological properties on particle transport behavior. In particular, the method allows us to predict how particles settle at different locations within the fracture, offering insights into their transport distance and enabling estimation of the effective length of the propped fracture. These findings offer valuable theoretical guidance for optimizing fracturing fluid design and improving treatment performance in field applications.

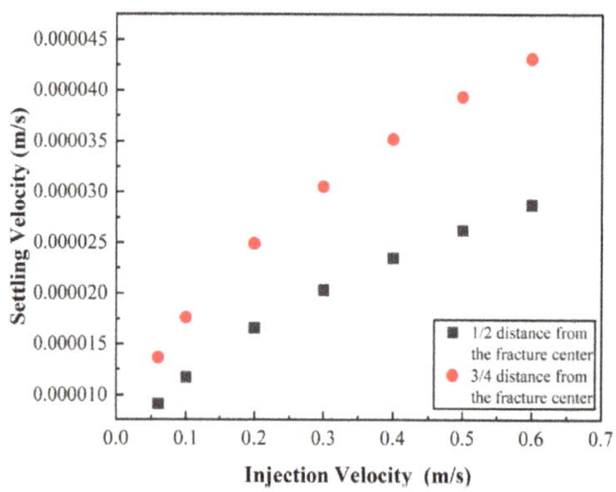


**Figure 11.** Variation in settling velocity at different injection flow velocities at different positions of non-Newtonian fluid.

## 4. Conclusions

This study proposes a novel method for predicting the proppant settling velocity in power-law non-Newtonian fluid, addressing limitations of traditional models in complex rheological environments. The key findings are summarized as follows:

1. Newtonian fluids: The proppant settling velocity is primarily governed by the fluid viscosity and particle diameter. Higher viscosity significantly enhances the drag force, reduces the settling velocity of proppants, and thus improves the proppant suspension capacity. Conversely, larger particle diameters accelerate settling due to the increased gravitational force. These observations align with Stokes' law, validating the foundational principles of particle dynamics in the Newtonian fluid;
2. Non-Newtonian fluids: In power-law shear-thinning fluids, the settling behavior exhibits spatial heterogeneity due to localized variations in shear rate and viscosity. Beyond the viscosity and particle size, the injection position within the fracture

critically influences the settling velocity. Near fracture walls, elevated shear rates reduce fluid viscosity, resulting in a rapid settling. In contrast, low-shear regions near the fracture center maintain a high fluid viscosity, which hinders the proppant settling;

3. Newly proposed model: A practical framework is established by integrating the spatial distribution of the shear rate calculated with the power-law rheological model. The local viscosity is derived from the shear-dependent rheology, and Stokes' law is then adapted to predict the settling velocity of proppant in the non-Newtonian fluid. The results indicate that the model achieved high accuracy in the fracture center region, with an average relative error of 8.2%. Despite localized deviations in the near-wall region, the analytical model proposed in this study can still reliably predict settling velocities, providing theoretical support for the rapid and accurate estimation of particle settling behavior and key engineering parameters in practical fracturing scenarios.

This new model helps bridge theoretical models and field-scale complexities, providing theoretical foundations to optimize fracturing fluid design and operational parameters. By mitigating premature proppant settling, it enhances fracture conductivity and supports efficient hydrocarbon recovery in shale or tight reservoirs. Furthermore, this work underscores the critical role of spatial shear and viscosity heterogeneity in fracture simulations. The proposed approach offers significant engineering value for optimizing proppant transport and ensuring sustainable productivity in hydraulic fracturing operations.

**Author Contributions:** Conceptualization, T.L. and Z.D.; Methodology, Z.D. and L.Z.; Validation, M.Y.; Investigation, Z.D., J.W., L.Z. and M.Y.; Data curation, J.W., F.X. and L.Z.; Writing—original draft, T.L. and Z.D.; Writing—review and editing, T.L.; Visualization, F.X.; Supervision, T.L., M.Y. and F.Z.; Project administration, F.Z.; Funding acquisition, T.L. and F.Z. All authors have read and agreed to the published version of the manuscript.

**Funding:** This work was financially supported by the General Program Grant from the National Natural Science Foundation of China (52274051), and the Key Program of the Joint Fund from the National Natural Science Foundation of China (U23B2084).

**Data Availability Statement:** The original contributions presented in this study are included in the article. Further inquiries can be directed to the corresponding author.

**Acknowledgments:** The authors gratefully acknowledge the technical support from the State Key Laboratory of Petroleum Resources and Engineering at China University of Petroleum, Beijing. Special thanks are extended to the editors and reviewers of Processes for their time and effort in improving the quality of this paper.

**Conflicts of Interest:** The authors declare no conflict of interest.

## References

1. Wu, Z.; Cui, C.; Jia, P.; Wang, Z.; Sui, Y. Advances and challenges in the development of hydraulic fracturing of tight reservoirs: A critical review. *Energy Geosci.* **2021**, *3*, 427–435. [CrossRef]
2. Sun, L.; Zou, C.; Jia, A.; Wei, Y.; Zhu, R.; Wu, S.; Guo, Z. Development characteristics and orientation of tight oil and gas in China. *Pet. Explor. Dev.* **2019**, *46*, 1073–1087. [CrossRef]
3. Zhai, G.M. Speculations on the exploration and development of unconventional hydrocarbon resources. *Nat. Gas Ind.* **2008**, *28*, 1–3.
4. Barati, R.; Liang, J.T. A review of fracturing fluid systems used for hydraulic fracturing of oil and gas wells. *J. Appl. Polym. Sci.* **2014**, *131*, 318–323. [CrossRef]
5. Fredd, C.N.; McConnell, S.B.; Boney, C.L.; England, K.W. Experimental study of fracture conductivity for water-fracturing and conventional fracturing applications. *SPE J.* **2001**, *6*, 288–298. [CrossRef]
6. Li, Q.; Xing, H.; Liu, J.; Liu, X. A review on hydraulic fracturing of unconventional reservoir. *Petroleum* **2015**, *1*, 8–15. [CrossRef]
7. Zhang, Y.; Lu, X.; Zhang, X.; Li, P. Proppant Transportation in Cross Fractures: Some Findings and Suggestions for Field Engineering. *Energies* **2020**, *13*, 4912. [CrossRef]

8. Hou, Z.; Yuan, Y.; Chen, Y.; Feng, J.; Liu, Y.; Zhang, X. Experimental Study on Proppant Migration in Fractures Following Hydraulic Fracturing. *Water* **2024**, *16*, 1941. [CrossRef]
9. Yu, H.; Xu, W.L.; Li, B.; Huang, H.; Micheal, M.; Wang, Q.; Huang, M.; Meng, S.; Liu, H.; Wu, H. Hydraulic fracturing and enhanced recovery in shale reservoirs: Theoretical analysis to engineering applications. *Energy Fuels* **2023**, *37*, 9956–9997. [CrossRef]
10. Kong, C.; Yang, L.; Guo, X.; Tian, F.; Li, Y. Proppant Migration Law Considering Complex Fractures. *Processes* **2023**, *11*, 14. [CrossRef]
11. Huang, H.; Zheng, Y.; Zhang, W. Characteristics of proppant transport and placement within rough hydraulic fractures. *Pet. Explor. Dev.* **2024**, *51*, 453–463. [CrossRef]
12. Yang, S.; Cheng, W.; Hu, Q.; Lei, G. Sensitivity analysis of proppant transportation and settling in a hydraulic fracture. *Phys. Fluids* **2025**, *37*, 013360. [CrossRef]
13. Zeng, J.; Li, H.; Zhang, D. Numerical simulation of proppant transport in hydraulic fracture with the upscaling CFD-DEM method. *J. Nat. Gas Sci. Eng.* **2016**, *33*, 264–277. [CrossRef]
14. Kumar, D.; Ghassemi, A. Modeling and Analysis of Proppant Transport and Deposition in Hydraulic/Natural Fracture Networks. In Proceedings of the 7th Unconventional Resources Technology Conference, Denver, Colorado, 22–24 July 2019. [CrossRef]
15. Zhao, K.; Wang, J.; Xu, H.; Zhang, L.; Jiang, H. Numerical simulation of proppant migration and sedimentation behavior in complex fractures based on computational fluid dynamics. *Phys. Fluids* **2023**, *35*, 093103. [CrossRef]
16. Tang, P.; Zeng, J.; Zhang, D.; Li, H. Constructing sub-scale surrogate model for proppant settling in inclined fractures from simulation data with multi-fidelity neural network. *J. Pet. Sci. Eng.* **2022**, *210*, 110051. [CrossRef]
17. Wu, Z.; Hu, Y.; Lu, B.; Jiang, T. Study on the Law and Influencing Factors of Proppant Settling in Fracturing Fluid. In Proceedings of the International Field Exploration and Development Conference 2020, Chengdu, China, 23–25 September 2020; Springer: Singapore, 2020; pp. 2834–2842.
18. Du, Q.; Yang, H.; He, S.; Deng, P.; Yang, X.; Lin, C.; Sun, Z.; Ren, L.; Yin, H.; He, B.; et al. Numerical Simulation and Application of Coated Proppant Transport in Hydraulic Fracturing Systems. *Processes* **2025**, *13*, 1062. [CrossRef]
19. Stokes, G.G. On the effect of the internal friction of fluids on the motion of pendulums. In *Transactions of the Cambridge Philosophical Society*; Cambridge University Press: Cambridge, UK, 1851.
20. Wrobel, M.; Mishuris, G.; Papanastasiou, P. On the influence of fluid rheology on the hydraulic fracture evolution. *arXiv* **2020**. [CrossRef]
21. Almubarak, T.; Li, L.; Ng, J.H.; Nasr-El-Din, H.; AlKhaldi, M. New Insights into Hydraulic Fracturing Fluids used for High Temperature Wells. *Petroleum* **2020**, *70*, 70–79. [CrossRef]
22. Abdelaal, A.; Aljawad, M.S.; Alyousef, Z.; Almajid, M.M. A review of foam-based fracturing fluids applications: From lab studies to field implementations. *J. Nat. Gas Sci. Eng.* **2021**, *95*, 104236. [CrossRef]
23. Wanniarachchi, W.A.M.; Ranjith, P.G.; Perera, M.S.A. Shale gas fracturing using foam-based fracturing fluid: A review. *Environ. Earth Sci.* **2017**, *76*, 91. [CrossRef]
24. Arabi, A.S.; Sanders, R.S. Particle terminal settling velocities in non-Newtonian viscoplastic fluids. *Can. J. Chem. Eng.* **2016**, *94*, 1092–1101. [CrossRef]
25. Moreira, B.A.; de Oliveira Arouca, F.; Damasceno, J.J.R. Analysis of suspension sedimentation in fluids with rheological shear-thinning properties and thixotropic effects. *Powder Technol.* **2017**, *308*, 290–297. [CrossRef]
26. Alghalibi, D.; Fornari, W.; Rosti, M.E.; Brandt, L. Sedimentation of finite-size particles in quiescent wall-bounded shear-thinning and Newtonian fluids. *Int. J. Multiph. Flow* **2020**, *129*, 103291. [CrossRef]
27. Murch, W.L.; Shaqfeh, E.S.G. Collective effects in the sedimentation of particles in a viscoelastic fluid. *Phys. Rev. Fluids* **2020**, *5*, 073301. [CrossRef]
28. Patankar, N.A.; Huang, P.Y.; Ko, T.; Joseph, D.D. Lift-off of a single particle in Newtonian and viscoelastic fluids by direct numerical simulation. *J. Fluid Mech.* **2001**, *438*, 67–100. [CrossRef]
29. Chhabra, R.P. *Bubbles, Drops, and Particles in Non-Newtonian Fluids*; CRC Press: Boca Raton, FL, USA, 2023.
30. Matveenko, V.N.; Kirsanov, E.A. Rheology of structured liquids. Flow regimes and rheological equations. *Colloid J.* **2025**, *87*, 38–48. [CrossRef]
31. Khataniar, S.; Chukwu, G.A.; Xu, H. Evaluation of rheological models and application to flow regime determination. *J. Pet. Sci. Eng.* **1994**, *11*, 155–164. [CrossRef]
32. Shah, S.N. Proppant-settling correlations for non-Newtonian Fluids. *SPE Prod. Eng.* **1986**, *1*, 446–448. [CrossRef]
33. Roodhart, L.P. Proppant settling in non-Newtonian fracturing fluids. In Proceedings of the SPE Rocky Mountain Petroleum Technology Conference/Low-Permeability Reservoirs Symposium, Denver, Colorado, 19–22 May 1985; SPE: Richardson, TX, USA, 1985. SPE-13905-MS.
34. Ovarlez, G.; Bertrand, F.; Coussot, P.; Chateau, X. Shear-induced sedimentation in yield stress fluids. *J. Non-Newton. Fluid Mech.* **2012**, *177–178*, 19–28. [CrossRef]

35. Amaratunga, M.; Rabenjafimanantsoa, H.A.; Time, R.W. Estimation of shear rate change in vertically oscillating non-Newtonian fluids: Predictions on particle settling. *J. Non-Newton. Fluid Mech.* **2020**, *277*, 104236. [CrossRef]
36. Anyaoku, C.C.; Bhattacharya, S.; Parthasarathy, R. Sheared settling in viscoelastic shear-thinning fluids: Empirical studies and model development. *AIChE J.* **2024**, *70*, e18460. [CrossRef]
37. Bai, Z.; Li, M. Prediction of proppant settling velocity in fiber-containing fracturing fluids. *ACS Omega* **2023**, *8*, 31857–31869. [CrossRef] [PubMed]
38. Liu, Z.; Fan, W. Velocity distribution and scaling properties of wall bounded flow. *J. Zhejiang Univ.-Sci. A* **2010**, *11*, 505–510. [CrossRef]
39. Liu, Y. Settling and Hydrodynamic Retardation of Proppants in Hydraulic Fractures. Ph.D. Thesis, The University of Texas at Austin, Austin, TX, USA, 2006.
40. Richardson, J.F.; Zaki, W.N. Sedimentation and fluidisation: Part I. *Chem. Eng. Res. Des.* **1997**, *75*, S82–S100. [CrossRef]
41. Daneshy, A.A. Numerical solution of sand transport in hydraulic fracturing. *J. Pet. Technol.* **1978**, *30*, 132–140. [CrossRef]
42. Happel, J.; Brenner, H. *Low Reynolds Number Hydrodynamics: With Special Applications to Particulate Media*; Springer Science & Business Media: Berlin, Germany, 2012.

*processes*

MDPI

*Article*

# Study on Evolution of Stress Field and Fracture Propagation Laws for Re-Fracturing of Volcanic Rock

**Honglei Liu [1], Jiangling Hong [1], Wei Shu [1], Xiaolei Wang [1], Xinfang Ma [2,*], Haoqi Li [1] and Yipeng Wang [2]**

1 The First Gas Company of Xinjiang Oilfield, Karamay 834000, China; gcgslhl@petrochina.com.cn (H.L.); hjiangling@petrochina.com.cn (J.H.); cqshuwei@petrochina.com.cn (W.S.); wxlei@petrochina.com.cn (X.W.)

2 State Key Laboratory of Petroleum Resources and Engineering, China University of Petroleum (Beijing), Beijing 102249, China; 2019015104@st.cupk.edu.cn

* Correspondence: maxinfang@cup.edu.cn

**Abstract:** In the Kelameili volcanic gas reservoir, primary hydraulic fracturing treatments in some wells take place on a limited scale, resulting in a rapid decline in production post stimulation and necessitating re-fracturing operations. However, prolonged production has led to a significant evolution in the in situ stress field, which complicates the design of re-fracturing parameters. To address this, this study adopts an integrated geology–engineering approach to develop a formation-specific geomechanical model, using rock mechanical test results and well-log inversion to reconstruct the reservoir's initial stress field. The dynamic stress field simulations and re-fracturing parameter optimization were performed for Block Dixi-14. The results show that stress superposition effects induced by multiple fracturing stages and injection–production cycles have significantly altered the current in situ stress distribution. For Well K6, the optimized re-fracturing parameters comprised a pump rate of 12 $m^3$/min, total fluid volume of 1200 $m^3$, prepad fluid ratio of 50–60%, and proppant volume of 75 $m^3$, and the daily gas production increased by 56% correspondingly, demonstrating the effectiveness of the optimized re-fracturing design. This study not only provides a more realistic simulation framework for fracturing volcanic rock gas reservoirs but also offers a scientific basis for fracture design optimization and enhanced gas recovery. The geology–engineering integrated methodology enables the accurate prediction and assessment of dynamic stress field evolution during fracturing, thereby guiding field operations.

**Keywords:** re-fracturing optimization; in situ stress field evolution; fracture propagation

## 1. Introduction

The Kelameili volcanic rock gas reservoir exhibits dual-medium characteristics (fracture–pore systems), featuring hard lithology, poor reservoir properties, well-developed natural fractures, and low natural productivity. To achieve commercial production, hydraulic fracturing stimulation is essential. However, the initial daily gas production of some gas wells drops rapidly after fracturing, mainly due to the small scale of the initial fracturing and modification, long production time, no contribution of productivity in the near-well area, the reduced conductivity of artificial fractures, and deteriorated supply capacity from the remote reservoir to the near-well area. Re-fracturing has proven an effective approach to unlock remaining development potential in this reservoir. The key challenge for the efficient development of Kelameili gas field lies in optimizing re-fracturing parameters [1]. As a major producing block, Dixi 14 presents complex lithology, poor reservoir quality, and sparsely distributed microfractures that tend to generate complex fracture networks during stimulation [2]. After extended production, the in situ stress distribution has become ambiguous in this block. Current re-fracturing designs lack proper parameter optimization based on stress field evolution, resulting in suboptimal stimulation outcomes [3]. Therefore, comprehensive studies on in situ stress field evolution and re-fracturing parameter optimization are critically needed to enhance recovery efficiency in Kelameili's volcanic gas reservoirs.

In recent years, research on the fracturing technology of volcanic rock reservoirs, as an important means to increase the production and efficiency of this type of unconventional gas reservoir, has received extensive attention and application in various oil fields. Field practice has shown that this technology has significant potential for increasing production. Deng et al. [4] proposed a multi-stage temporary plugging and return fracturing technology approach of "detailed hole filling + pre-pressure energy replenishment + expanded modification volume", which achieved an average daily increase of 4.7 tons in oil per well and a cumulative increase of 832 tons in oil production in the unconventional volcanic rock oil reservoir of Niudong at the Tuha Oilfield. Liu et al. [5] summarized engineering technology and characteristic schemes for the development of volcanic rock gas reservoirs and pointed out that hydraulic fracturing is the key technology for exploiting such gas reservoirs. Liu et al. [6] established a 3D-EPM coupling equation and simulated three-dimensional hydraulic fracture propagation, the interaction with natural fractures, and the influence of in situ stress. Chen et al. [7] conducted laboratory experiments to study the three processes of volcanic rock hydraulic fracturing, acid pressure, and proppant acid pressure and evaluated the average acid erosion depth and proppant embedding depth under acid pressure. Yu et al. [8] compared the fracture pressures of rocks under different experimental conditions and conducted an in-depth study on the fracture propagation behavior of different rocks under different influencing factors. Guo et al. [9] investigated the effect of in situ stress, treatment parameters and different patterns of multi-well fracturing (staggered or tip to tip) on fracture propagation.

During oil and gas development, the in situ stress field undergoes dynamic evolution across different exploration and production stages. The hydraulic fractures generated during primary hydraulic fracturing alter both the magnitude and orientation of the original in situ stress [10]. Furthermore, long-term production post fracturing generates elastic-induced stress and production-induced stress, leading to variations in formation pore pressure and in situ stress redistribution [11]. To address these complexities, researchers have investigated the evolution of in situ stress fields. For example, Li et al. [12] developed a three-dimensional heterogeneous geological model incorporating in situ stress characteristics, reservoir properties, and rock mechanics parameters for a single well. Chen et al. [13] established a highly accurate geomechanical model using compensated neutron density and acoustic logging data. Wang et al. [14] analyzed induced stress near hydraulic fractures in elastic, poroelastic, and dual-porosity media, providing detailed characterizations of low-permeability, sandstone, and fractured formations. While these studies primarily focus on theoretical predictions of in situ stress field distribution, few account for the actual effects of production and fracturing operations on stress field dynamics.

This study investigates the Dixi 14 block, integrating rock mechanics experiments and well logging data to calibrate finite element simulation parameters. An initial geomechanical model was developed, and stress field evolution was analyzed to determine the current in situ stress distribution. Building on the current geomechanical and in situ stress field models, the re-fracturing well construction parameters were optimized, resulting in an optimal parameter combination tailored for the target block.

## 2. Rock Mechanics and In Situ Stress Experiments

The rock mechanical parameters, including Young's modulus, Poisson's ratio, tensile strength, and fracture toughness, serve as the critical link between reservoir geological parameters and geomechanical characteristics [15,16]. These parameters are fundamental to fracture mechanism analysis and reservoir stress field simulation [17,18]. In practice, dynamic rock mechanical parameters (obtained through well log interpretation) are typically converted to static parameters (determined via laboratory rock mechanics tests) to accurately evaluate reservoir rock mechanical properties.

### *2.1. Triaxial Compression Experiment of Volcanic Rock*

We employed a comprehensive rock mechanics testing system to conduct standard triaxial compression tests on core samples, determining the static Young modulus and Poisson ratio of the rock. Ten core samples collected on-site were selected for the experiment. The core samples were from five typical production wells in this block, with a collection depth ranging from 3652 to 3691 m. This depth corresponds to the main development layer system of the Dixi 14 block. Considering that the current formation pressure of Well K6 is 28.7 MPa and the well is in production, the confining pressure was set at 25 MPa. The triaxial compression test procedures and corresponding results are presented in Figure 1 and Table 1, respectively. The experimental results indicate that the volcanic rock samples exhibited an average Poisson ratio of 0.22 and an average Young modulus of 16.97 GPa.

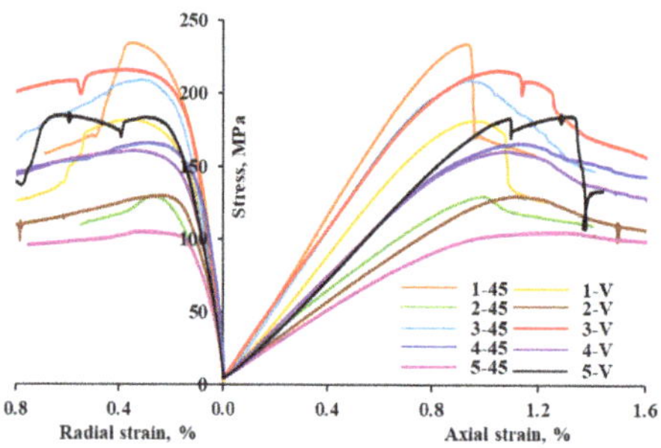


**Figure 1.** Triaxial compression test curve.

**Table 1.** Triaxial compressive strength test scheme and results.

| Core Sample Number | Well Depth (m) | Differential Stress (MPa) | Young's Modulus (GPa) | Poisson's Ratio |
|---|---|---|---|---|
| 1-45 | 3655.86 | 129.82 | 13.51 | 0.247 |
| 1-v | 3655.90 | 129.38 | 12.59 | 0.175 |
| 2-45 | 3696.76 | 209.42 | 26.09 | 0.26 |
| 2-v | 3696.81 | 215.88 | 24.75 | 0.209 |
| 3-45 | 3694.85 | 165.93 | 19.00 | 0.174 |
| 3-v | 3694.90 | 160.64 | 17.58 | 0.217 |
| 4-45 | 3653.81 | 105.16 | 10.67 | 0.172 |
| 4-v | 3653.86 | 186.08 | 19.10 | 0.218 |
| 5-45 | 3840.26 | 131.61 | 16.31 | 0.201 |
| 5-v | 3840.29 | 101.08 | 10.11 | 0.341 |

### *2.2. Volcanic Rock Tensile Strength Test*

Based on the third part of the industry standard recommended by the American Rock Mechanics Association, using the GCTSRTR-1500 high-temperature and high-pressure comprehensive rock testing system, Brazilian splitting tests were conducted on the samples from the Dixi 14 block. The samples were processed into short cylindrical bodies with a diameter of Φ25 mm and a length of 15 mm using a vertical core drilling machine and a face grinding machine. By measuring the failure load in the diameter direction and the size of the sample, the tensile strength of the rock was obtained. The calculation formula for the tensile strength is as follows [19]:

$$\sigma_t = \frac{2P}{\pi D t} \tag{1}$$

where $\sigma_t$ represents the tensile strength, MPa; $P$ is the maximum load when the rock breaks, N; $D$ is the diameter of the rock sample, mm; t is the thickness of the rock sample, mm.

A series of Brazilian splitting tests were conducted on 10 rock specimens to determine their tensile strength. The tested samples are shown in Figure 2, while Table 2 presents the Brazilian tensile strength results for all 10 test groups. The measured tensile strength in the Dixi 14 block ranges from 0.93 to 10.53 MPa.

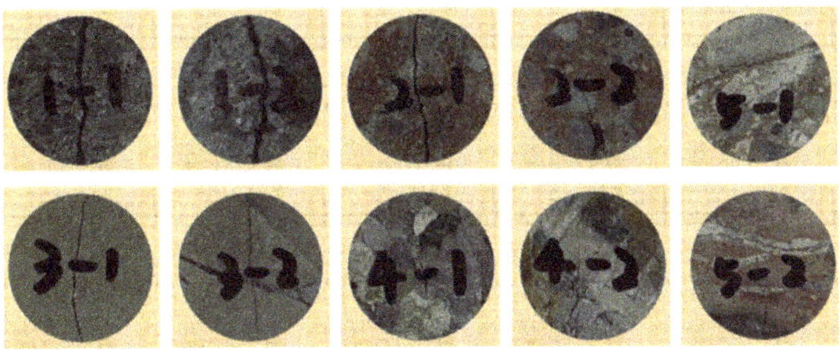

**Figure 2.** Shows the sample after the splitting experiment in Brazil.

**Table 2.** Test results of splitting tensile strength in Brazil.

| Number of Samples | Depth (m) | Diameter (mm) | Thickness (mm) | Load (N) | Tensile Strength (MPa) |
|---|---|---|---|---|---|
| 1-1 | 3691.56 | 25.75 | 15.11 | 4882 | 7.99 |
| 1-2 | | 25.63 | 14.84 | 6320 | 10.58 |
| 2-1 | 3694.85 | 25.57 | 14.52 | 1694 | 2.90 |
| 2-2 | | 25.21 | 14.75 | 1862 | 3.19 |
| 3-1 | 3674.62 | 25.10 | 15.15 | 2066 | 3.46 |
| 3-2 | | 25.39 | 15.10 | 2604 | 4.32 |
| 4-1 | 3653.19 | 25.26 | 14.85 | 1402 | 2.38 |
| 4-2 | | 25.21 | 14.85 | 2670 | 4.54 |
| 5-1 | 3692.42 | 25.05 | 14.84 | 606 | 1.04 |
| 5-2 | | 25.56 | 14.51 | 540 | 0.93 |

### 2.3. Experimental Study on Magnitude of In Situ Stress in Volcanic Rocks

The core samples used in this experiment were obtained from the gas field. Full-diameter field cores were drilled to extract cylindrical specimens (Φ25 × 50 mm) in three orientations: one vertical specimen (Z-axis) and two horizontal specimens (Φ25 × 50 mm each) perpendicular to the core axis, totaling 20 specimens as shown in Figure 3. Triaxial compression tests with acoustic emission monitoring were conducted on specimens from each orientation to determine the Kaiser effect (the Kaiser effect on rocks refers to the phenomenon where under the action of monotonic stress loading, the number of acoustic emissions of rocks suddenly increases when the stress value reaches its in situ stress underground) stress points [20]. These stress values were then substituted into

Equations (2)–(5) [21] to calculate the three principal in situ stresses acting on the rock formation underground.

$$\sigma_v = \sigma_\perp + \alpha P_p - KP_c \tag{2}$$

$$\sigma_H = \frac{\sigma_{0^\circ} + \sigma_{90^\circ}}{2} + \frac{\sigma_{0^\circ} - \sigma_{90^\circ}}{2}(1 + tg^2 2\alpha)^{\frac{1}{2}} + \alpha P_p - KP_c \tag{3}$$

$$\sigma_h = \frac{\sigma_{0^\circ} + \sigma_{90^\circ}}{2} - \frac{\sigma_{0^\circ} - \sigma_{90^\circ}}{2}(1 + tg^2 2\alpha)^{\frac{1}{2}} + \alpha P_p - KP_c \tag{4}$$

$$\tan 2\alpha = \frac{\sigma_{0^\circ} + \sigma_{90^\circ} - 2\sigma_{45^\circ}}{\sigma_{0^\circ} - \sigma_{90^\circ}} \tag{5}$$

where $\sigma_H$, $\sigma_h$, and $\sigma_v$ represent the maximum and minimum horizontal principal stress and vertical stress, MPa; $P_P$ represents the formation pore pressure, MPa; $\alpha$ is the Biot coefficient; $\sigma_\perp$ is the Kaiser point stress of the core in the vertical direction, MPa; and $\sigma_{0^\circ}$, $\sigma_{45^\circ}$, and $\sigma_{90^\circ}$ are the Kaiser point stresses of the core in the three horizontal directions of $0^\circ$, $45^\circ$, and $90^\circ$, respectively, MPa.

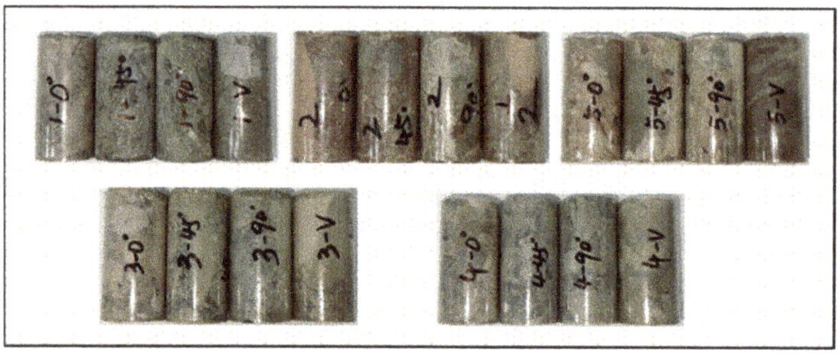

**Figure 3.** Schematic diagram of standard core sampling.

The experiment conducted in situ stress tests on five full-diameter core samples collected from representative wells in the target block. With reference to Well K6′s current formation pressure of 28.7 MPa and considering its production status, the tests were performed under a confining pressure of 25 MPa. Table 3 presents the results of five sets of volcanic rock in situ stress tests. The maximum horizontal principal stress in the Dixi 14 well area ranges from 85.57 to 98.96 MPa, the minimum horizontal principal stress ranges from 51.89 to 64.37 MPa, the vertical stress ranges from 89.50 to 94.06 MPa, and the difference in horizontal stress ranges from 26.41 to 34.95 MPa.

**Table 3.** Results of in situ stress tests.

| Number | Depth (m) | Max Horizontal Stress (MPa) | Min Horizontal Stress (MPa) | Vertical Stress (MPa) | Horizontal Stress Difference (MPa) |
|---|---|---|---|---|---|
| 1 | 3655.86 | 91.19 | 59.74 | 89.50 | 31.45 |
| 2 | 3696.76 | 98.96 | 64.01 | 90.43 | 34.95 |
| 3 | 3694.85 | 85.57 | 51.89 | 90.43 | 33.68 |
| 4 | 3653.81 | 91.06 | 62.64 | 87.10 | 28.43 |
| 5 | 3840.26 | 90.78 | 64.37 | 94.06 | 26.41 |

## 2.4. Dynamic and Static Conversion of Rock Mechanical Parameters

The acquisition of rock mechanical parameters typically relies on laboratory core testing, which provides high-precision mechanical property data through the direct measurement of subsurface core samples. However, this method cannot achieve continuous in situ stress characterization across entire well sections or construct mechanical field models at a regional scale. In contrast, well logging data are more readily available and can provide a one-dimensional distribution of rock mechanical parameters along the borehole with superior vertical continuity. By utilizing array acoustic logging data—including compressional wave (P-wave) and shear wave (S-wave) slowness, bulk density, and shale content—rock elastic parameters and strength properties can be calculated [22–24]. The specific formulas are as follows:

$$E_d = 10^{-3}\rho V_s^2 \frac{3V_p^2 - 4V_s^2}{V_p^2 - V_s^2} \tag{6}$$

$$\mu_d = \frac{V_p^2 - 2V_s^2}{2(V_p^2 - V_s^2)} \tag{7}$$

where $E_d$ represents the dynamic Young modulus, GPa; $\rho$ denotes rock density, g/cm$^3$; $V_p$ is the velocity of the longitudinal wave, m/s; $V_s$ is the velocity of the transverse wave, m/s; and $\mu_d$ is the dynamic Poisson ratio, dimensionless.

Based on the above equations, the dynamic rock mechanical parameters of deep volcanic reservoirs were first calculated using well logging data. The logging-derived results were then calibrated by plotting both dynamic and static rock mechanical parameters at the same depth on a unified coordinate system to establish their correlation and achieve dynamic-to-static parameter conversion. As shown in the cross-plot in Figure 4, the conversion coefficients between dynamic and static rock mechanical parameters exhibit strong positive linear correlations. This analysis further yields dynamic–static conversion Formulas (8) and (9) as follows:

$$E_s = 1.0464 \times E_d - 1.6714 \tag{8}$$

$$\mu_s = 0.8556 \times \mu_d + 0.0126 \tag{9}$$

where $E_s$ represents the static Young modulus, GPa; $E_d$ represents the dynamic Young modulus, GPa; $\mu_s$ represents the static Poisson ratio, dimensionless; and $\mu_d$ represents the dynamic Poisson ratio, dimensionless.

After establishing the calibration relationship between dynamic and static rock mechanical parameters, the well-log-derived profiles can be corrected to obtain triaxial static parameter profiles. This process yields continuous rock mechanical parameter profiles for key well locations, as illustrated in Figure 5.

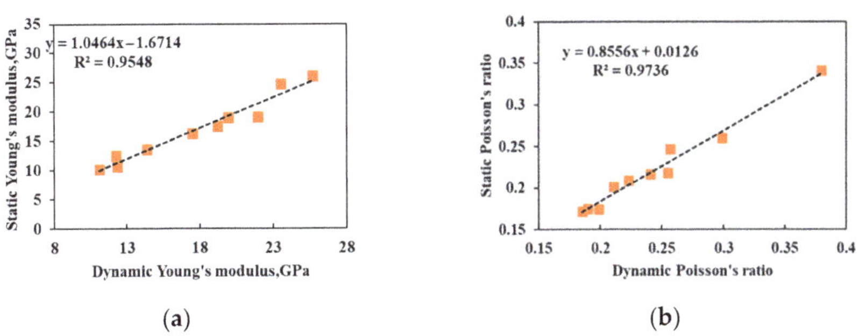


**Figure 4.** (**a**) Young's modulus correlation diagram; (**b**) Poisson's ratio correlation diagram.

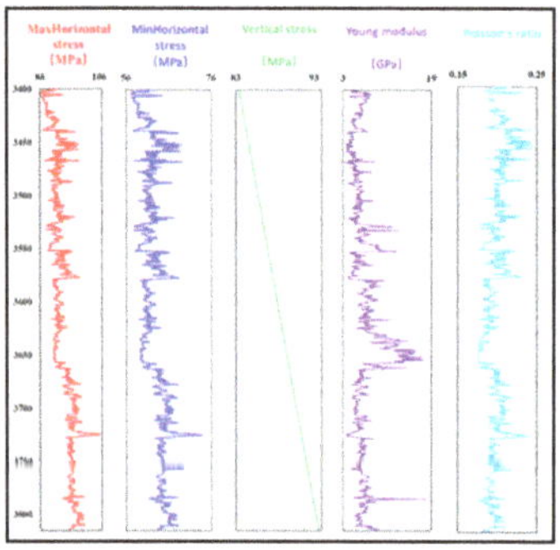

**Figure 5.** The calculation results of static rock mechanical parameters.

## 3. Establishment of Numerical Simulation Models and Applicability Verification

### *3.1. Establishment of Numerical Simulation Model*

Based on the geological data and static rock mechanical parameter calculation results of 13 wells in the Dixi 14 block, a geological mechanics model was established using the integrated geological engineering software. The simulation area (the Dixi 14 rock mass) has a size of 3725 m × 3100 m × 458 m. According to the size of the simulation area, the grid size was determined to be 10 m × 10 m, and the vertical layer was divided into a grid size of 1 m. The total number of grids was 52,816,560 (Figure 6). The initial Young modulus of the Dixi 14 block was 15–32 GPa, with an average of 19 GPa; Poisson's ratio was 0.19–0.24, with an average of 0.20; the tensile strength was 3.08–4.12 MPa, with an average of 3.59 MPa; the initial maximum horizontal principal stress range was 84–104 MPa, with a peak of 95 MPa; the minimum horizontal principal stress range was 57–77 MPa, with a peak of 62 MPa; and the vertical stress range was 82–96 MPa, with a peak of 89 MPa.

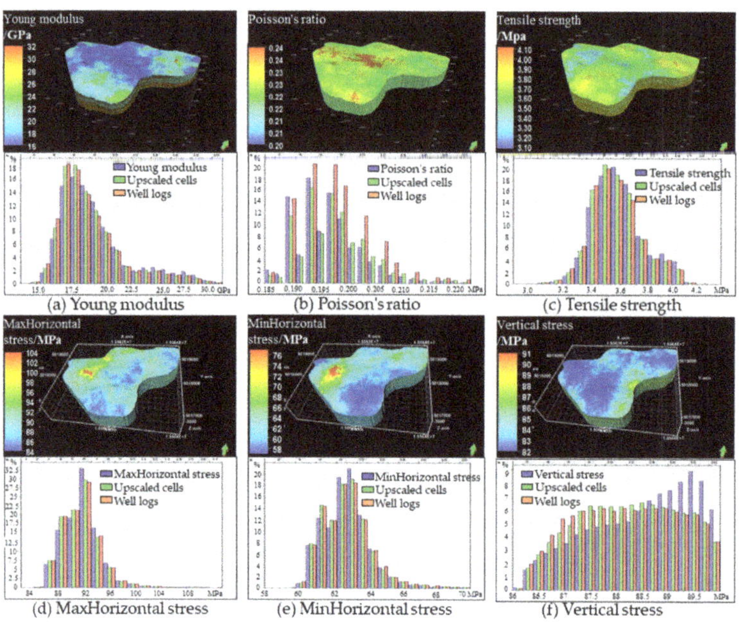


**Figure 6.** Initial geomechanical model.

### 3.2. Model Applicability Verification

Before conducting a series of simulation analyses, the applicability of both fracturing and production simulation processes was first validated through two approaches: (1) Fracture Geometry Characterization: compared to orthogonal fracture [25] or discrete fracture [26] models, fracture propagation simulations based on actual geological data and engineering designs provide the most direct method for characterizing complex fracture morphology. For Well K1 in the Dixi-14 block, net pressure matching was performed using second-by-second fracturing operation data. The inversion fracture parameters (including fracture length and height) are shown in Table 4. Due to the fact that the dynamic fluctuations during construction (such as displacement changes, sand blockage events) and fluid loss differences were not fully considered in the fracturing simulation, there may be slight deviations between the simulated fracture parameters and the inversion fractures, but the error between the two is relatively small. This further confirms that the simulated fracture geometry can effectively approximate the actual fracturing results, indicating that the initial geomechanical parameters are relatively accurate. (2) Production History Matching: constant bottomhole pressure production was simulated using historical pressure data. By comparing the actual production data with the simulated results [27], the final matching outcome (shown in Figure 7) shows that the error between the historical daily gas production and the simulated daily gas production is 5.8–6.9%, and the error between the historical daily oil production and the simulated daily oil production is 4.0–4.7%. Both errors are within a reasonable range, indicating a reliable simulation performance.

**Table 4.** Comparison of hydraulic fracture inversion results and simulated construction crack sizes.

| Fracture Geometry | Fracture Inversion | Simulated Construction | Error Value (%) |
|---|---|---|---|
| Fracture length (m) | 218.6 | 234.8 | 6.9 |
| Fracture height (m) | 45.6 | 47.8 | 4.2 |
| Fracture width (mm) | 11.2 | 10.4 | 7.1 |

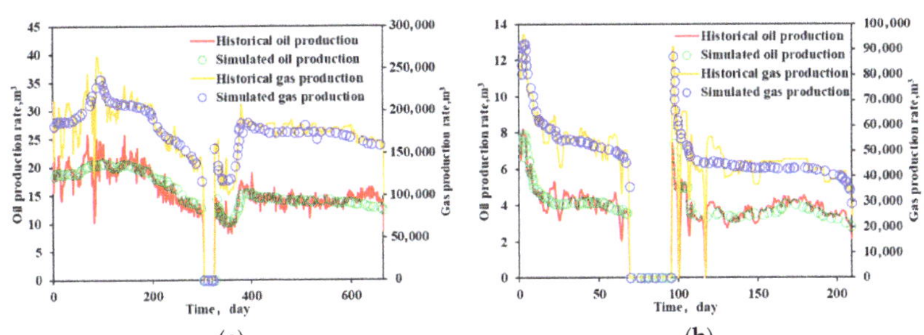


**Figure 7.** (**a**) Fitting results of Well K1; (**b**) fitting results of Well K2.

## 4. Evolution of Geostress Field and Optimization of Fracturing Parameters

The in situ stress distribution prior to re-fracturing shows significant changes compared to the initial state, exhibiting induced stresses from multiple factors—primarily caused by the opening of primary fractures and injection/production activities. Considering fracture dilation, pressure depletion, and other factors, long-term production after fracturing in the development layer with fluid extraction generates both elastic-induced stresses and production-induced stresses. This leads to alterations in formation pore pressure and stress distribution. A finite element geomechanical model was employed to reconstruct the current in situ stress field distribution in the target area, establishing a 3D geomechanical model of the target reservoir. This provides the foundational model for numerical simulations of hydraulic fracture propagation.

### 4.1. The Establishment of the Evolution Model for the Geostress Field

The development process of gas reservoirs involves fluid–solid coupling (fluid flow interacts with rock deformation) interactions between seepage fields and stress fields. The injection of fracturing fluid creates underground seepage effects, altering the stress distribution within the rock matrix and modifying the physical properties of formation rocks. This leads to stress redistribution in the geostress field, ultimately influencing the final geometry of hydraulic fractures in re-fractured wells [28]. Given the significant variation in fracturing times among wells in the Dixi-14 block and the continuous evolution of the geostress field during reservoir development, using initial geostress and fracture data for subsequent fracturing simulations would introduce substantial errors [29]. To accurately characterize the impact of fracturing on geostress field variations, this study adopts a dynamic stress field approach and an integrated geological–engineering workflow. The 3D geomechanical model is updated in real time based on the actual fracturing sequence, incorporating post-fracturing geostress fields and fracture states for each well. This enables a more realistic simulation of subsequent fracturing operations. The fracturing time of each well serves as an evolutionary node for geostress field updates, with wells fractured at similar times grouped under the same node. After compiling the fracturing timeline for all wells in the Dixi-14 block, the resulting stress field evolution sequence was determined and is illustrated in Figure 8.

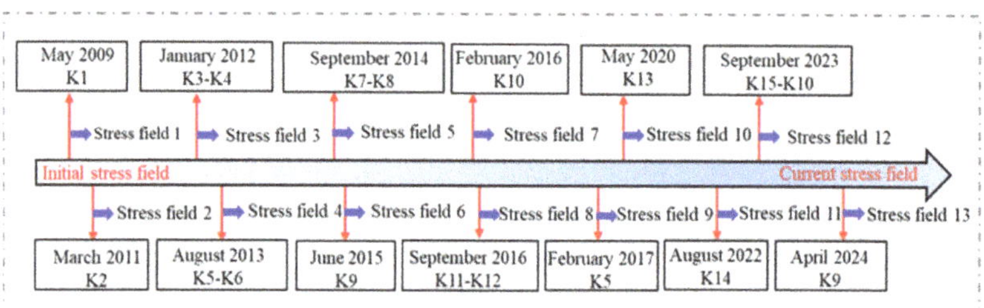


**Figure 8.** The evolution sequence of the geostress field in the Dixi 14 block.

### 4.2. Evolution Analysis of Geostress Field

Figure 9 illustrates the specific simulation workflow for Evolutionary Node Stress Field 1. Well K1, being the first fractured well in the Dixi-14 block, was assumed to have the initial geostress field prior to fracturing due to the short development history of the block at that time. After simulating the fracturing process of Well K1 according to its actual treatment parameters, the post-fracturing stress distribution was obtained. The results demonstrate that fluid injection increased net pressure within fractures, inducing a compressive deformation of the rock matrix through fracture face stresses and altering the stress state of surrounding formations [30,31], with a notable increase in minimum horizontal principal stress. Subsequently, production simulation was conducted for all wells in the block using actual production data, with the simulation period ending at the fracturing time of the next well (K2). Reservoir depletion caused by fluid production led to geostress reduction corresponding to decreasing pore pressure, ultimately establishing Stress Field 1—the first evolutionary node of geostress redistribution in the Dixi-14 block. This Stress Field 1 then served as the baseline geostress condition for Well K2's fracturing simulation. By applying the same workflow, the post-fracturing stress distribution of Well K2 could be accurately derived.

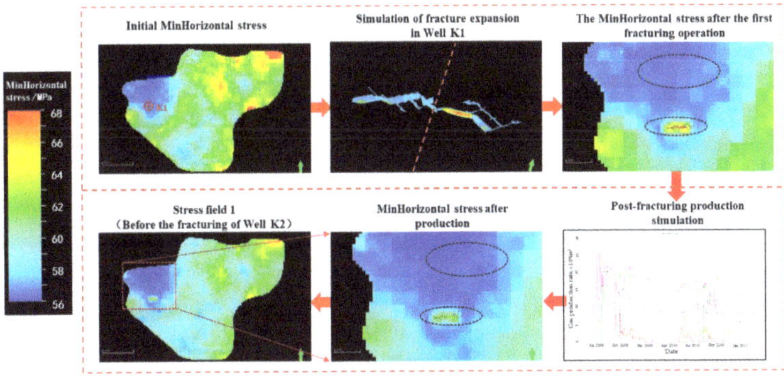


**Figure 9.** Evolution process of Stress Field 1.

Considering factors such as fracture opening and pressure depletion, the geostress field at each fracturing stage was simulated according to the evolutionary sequence of the Dixi-14 block. The current in situ stress distribution of the block is shown in Figure 10 and Table 5. The current geostress fields of the Dixi 14 block are as follows: the minimum horizontal principal stress range is 48–76 MPa, the maximum horizontal principal stress is 78–102 MPa, the vertical pressure is 79–94 MPa, and the formation pressure is 20–35 MPa.

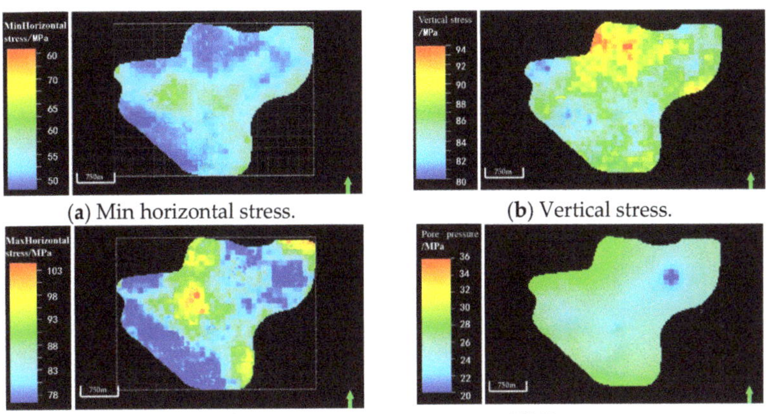


(**a**) Min horizontal stress. (**b**) Vertical stress. (**c**) Max horizontal stress. (**d**) Pore pressure.

**Figure 10.** The distribution of the current ground stress field in block Dixi 14.

**Table 5.** Comparison table of changes in geostress before and after.

| Stage | Max Horizontal Stress (MPa) | Min Horizontal Stress (MPa) | Vertical Stress (MPa) | Pore Pressure (MPa) |
|---|---|---|---|---|
| Initial | 84–104 | 57–77 | 82–96 | 42–51 |
| Current | 78–102 | 48–76 | 79–94 | 20–35 |

## 5. Optimization of Fracturing Construction Parameters

Well K6 is a development well deployed in the Dixi 14 composite lithologic body of the Dixi 14 block in the Kelameili Gas Field. The primary hydraulic fracturing was conducted in August 2013, resulting in an initial daily gas production rate of $6 \times 10^4$ $m^3/d$. After approximately two years of production, the gas rate gradually declined to $3 \times 10^4$ $m^3/d$. Subsequently, the well was operated intermittently, and by September 2024, the daily gas production had declined to $2.5 \times 10^4$ m$^3$/d, indicating an urgent need for re-fracturing and the optimization of treatment parameters. Based on the current geomechanical model of the Dixi 14 block, derived from the evolution of the in situ stress field, a parametric study was conducted to evaluate the effects of the injection rate, fluid volume, prepad ratio, and proppant intensity on fracture geometry.

### *5.1. The Influence of the Injection Rate on Fracture Parameters*

The effectiveness of hydraulic fracturing is closely related to the injection rate. A high injection rate facilitates fracture initiation and promotes the forward propagation of fracture tips, yet excessively high rates impose greater demands on equipment and tubular performance. Conversely, insufficient rates reduce proppant transport capacity. Therefore, determining an optimal injection rate is critical. Simulations were conducted with injection rates of 8, 10, 12, and 14 m$^3$/min, using a pad fluid ratio of 50%, total fluid volume of 725 m$^3$, and proppant volume of 65 m$^3$. The resulting hydraulic fracture geometries are shown in Figure 11.

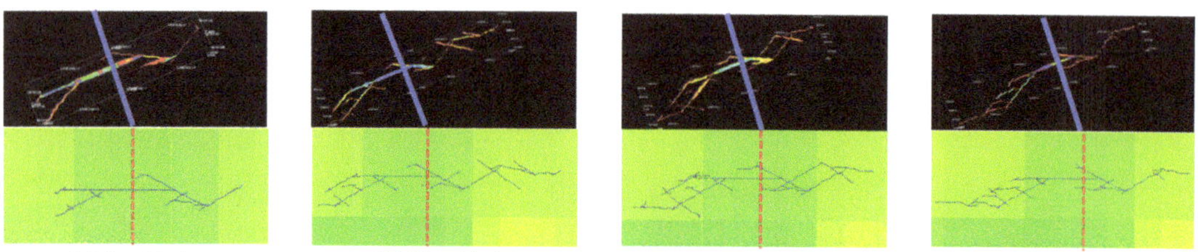

(**a**) 8 m$^3$/min (**b**) 10 m$^3$/min (**c**) 12 m$^3$/min (**d**) 14 m$^3$/min

**Figure 11.** Crack morphology diagrams under different displacements.

As shown in Figure 11, with increasing injection rates, fracture complexity and stimulation volume first increase and then decrease. This is primarily because at excessively high rates, fractures predominantly propagate along the main fracture direction. Figure 12 presents fracture parameter charts under different injection rates. The analysis reveals that fracture length increases linearly with higher injection rates and propped fracture length shows progressively poorer matching with dynamic fracture length. While lower rates achieve the best match between propped and dynamic lengths, the resulting fracture length fails to meet stimulation requirements. Therefore, optimal fracturing performance is achieved when selecting injection rates between 10 and 12 m$^3$/min.

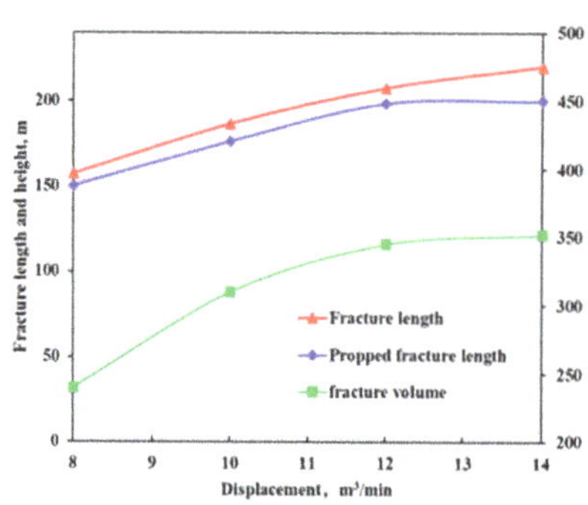


**Figure 12.** Crack parameter plates at different displacements.

### *5.2. The Influence of Injection Liquid Volume on Fracture Parameters*

Optimizing injection fluid volume intensity can enhance fracture complexity and boost well productivity, though some wells exhibit rapid fluid breakthrough with significant wastage [32]. Simulations were performed with fluid volumes of 800, 1000, 1200, and 1400 m$^3$ (the pad fluid ratio: 50%, injection rate: 12 m$^3$/min, and proppant volume: 65 m$^3$). The resulting hydraulic fracture geometries are illustrated in Figure 13.

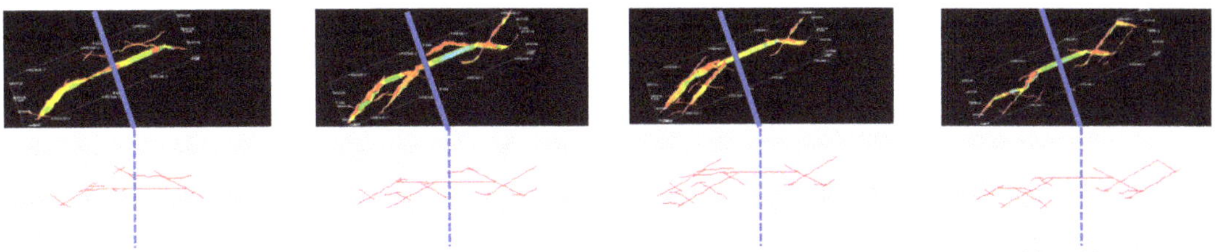

(**a**) 800 m$^3$ (**b**) 1000 m$^3$ (**c**) 1200 m$^3$ (**d**) 1400 m$^3$

**Figure 13.** Crack morphology diagrams under different liquid volumes.

As shown in Figure 13, fracture complexity initially increases and then decreases with higher fluid volumes. The primary reason is that excessive injection may lead to exacerbated fracturing fluid leak-off (into natural fractures or the matrix), reducing the effective energy available for fracture propagation. As a result, branch fractures cannot fully develop. The complexity of the fractures is highest at a liquid volume of 1200 m$^3$. Figure 14 presents fracture parameters under different fluid volumes. It can be observed that dynamic fracture length gradually increases with fluid volume before stabilizing. In addition, the matching degree between propped and dynamic fracture lengths deteriorates progressively, and the stimulated reservoir volume (SRV) first grows linearly and then slows with increasing fluid volume. Therefore, considering both fracturing costs and stimulation requirements, a fluid volume of 1200 m$^3$ is optimal.

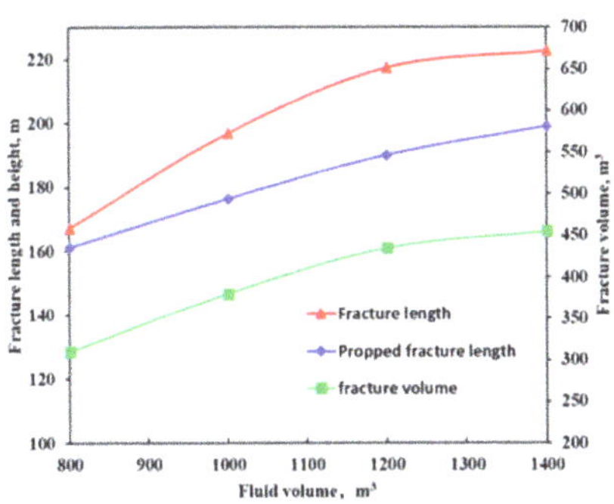


**Figure 14.** The crack parameter plates under different liquid volumes.

### *5.3. The Influence of the Proportion of the Preflush Fluid on the Fracture Parameters*

To minimize formation damage while ensuring effective fracture creation and proppant transport, pad fluid usage should be optimized. Simulations were conducted with pad fluid ratios of 40%, 50%, 60%, and 70% (the total fluid volume: 1200 m$^3$, injection rate: 12 m$^3$/min, and proppant volume: 65 m$^3$). The resulting hydraulic fracture geometries are presented in Figure 15.

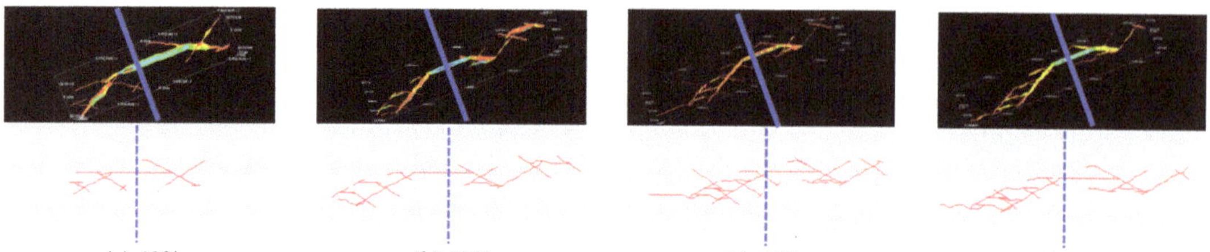

(**a**) 40% (**b**) 50% (**c**) 60% (**d**) 70%

**Figure 15.** Crack morphology diagrams under different proportions of pre-liquid.

As shown in Figure 15, the fracture complexity increases progressively with higher prepad fluid ratios. Figure 16 presents fracture parameter charts under different pad fluid ratios, revealing that while greater pad fluid ratios yield longer fractures, the correlation between dynamic fracture length and propped fracture length deteriorates when the ratio exceeds a certain threshold, resulting in suboptimal stimulation performance. Therefore, a pad fluid ratio between 50% and 60% achieves optimal alignment between propped and dynamic fracture lengths, ensuring effective reservoir stimulation.

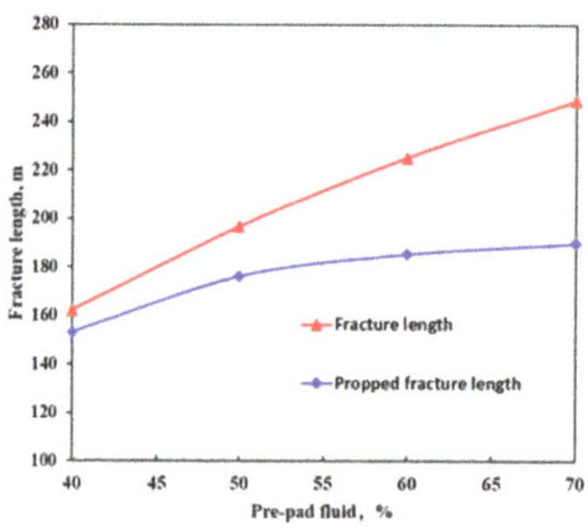


**Figure 16.** The crack parameter plates under different proportions of pre-liquid.

## *5.4. The Influence of Sand Content on Fracture Parameters*

Optimizing fracturing operations requires balancing three key factors: improving proppant transport capacity, maximizing the effective stimulated volume, and controlling costs—a major technical challenge in process enhancement [33,34]. Simulations were conducted with proppant volumes of 45 $m^3$, 60 $m^3$, 75 $m^3$, and 90 $m^3$ (the fluid volume: 1200 $m^3$, injection rate: 12 $m^3$/min, and prepad fluid ratio: 55%). The resulting hydraulic fracture geometries are presented in Figure 17.

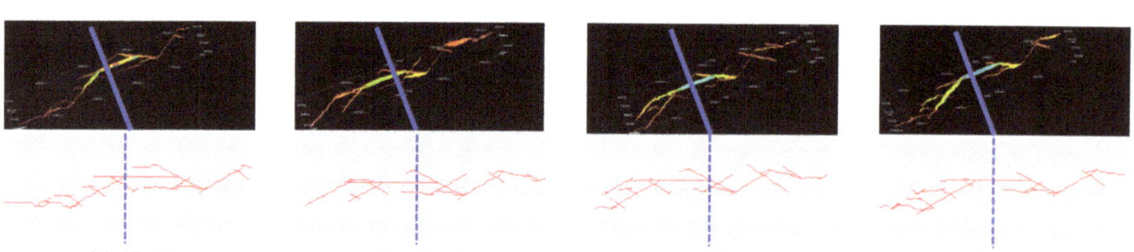

(**a**) 45 $m^3$ (**b**) 60 $m^3$ (**c**) 75 $m^3$ (**d**) 90 $m^3$

**Figure 17.** Crack morphology diagrams under different sand amounts.

As illustrated in Figure 17, fracture complexity exhibits minimal correlation with proppant volume. Figure 18 presents the evolution of fracture parameters (e.g., fracture length and conductivity) under different proppant volumes. It can be observed that proppant volume has limited influence on dynamic fracture length, while propped fracture length initially increases linearly with higher proppant volumes but shows diminishing growth beyond a critical volume. Fracture conductivity improves consistently with increasing proppant volume; however, considering operational costs, a proppant volume of 75 $m^3$ proves sufficient to meet conductivity requirements while maintaining economic efficiency.

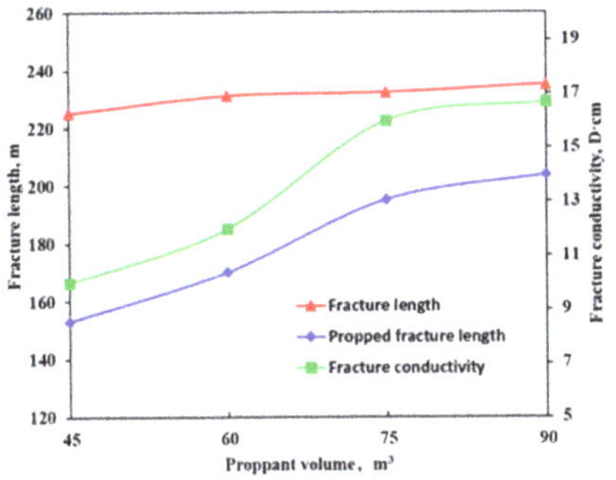


**Figure 18.** Crack parameter plates under different sand contents.

### *5.5. An Effect Evaluation of the Re-Fracturing Treatment*

Based on the optimized re-fracturing design, Well K6 was re-stimulated using a KQ78/65-105 wellhead and 139.7 mm open hole completion. The operation achieved strong alignment with design parameters involving a total fluid volume of 1725 $m^3$, with a prepad ratio of 60%, an average sand ratio of 15.3%, and a main injection rate of 12 $m^3$/min. Overall, the treatment parameters were in good agreement with the optimized design, and the operation proceeded smoothly. The re-fracturing treatment was conducted in two stages. The fracturing treatment curve is shown in Figure 19. The operation was conducted in two stages. At the first stage, initial pad fluid injection showed no pressure buildup, indicating fluid loss into the formation. The main proppant stage proceeded smoothly with stable pressure (24–26 MPa) and a peak concentration of 280 kg/$m^3$. Operations paused after injecting 300 kg of diverting agent due to insufficient fluid volume. The shut-in pressure was 8.8 MPa. At the second stage, an additional 500 kg of diverting agent was injected during the main proppant stage, resulting in a 1–1.5 MPa pressure increase. The operation resumed after fluid replenishment, with extra diverting agent added pre-stimulation.

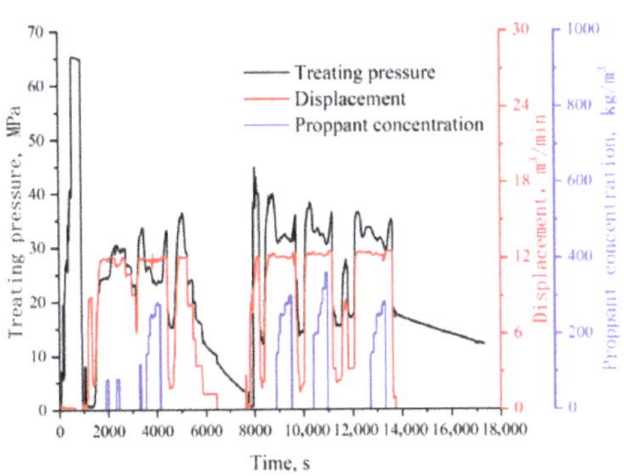


**Figure 19.** The construction curve of repeated fracturing in Well K6.

The production performance of Well K6 before and after re-fracturing is shown in Figure 20. By September 2024, the intermittent production rate had declined to $2.7 \times 10^4$ $m^3/d$ with a wellhead pressure of 4 MPa. After re-fracturing, the daily gas production increased to $4.2 \times 10^4$ $m^3/d$, representing a 56% increase in single-well productivity. The wellhead pressure also rose to 12 MPa, indicating that the re-fracturing treatment achieved a favorable reservoir stimulation effect.

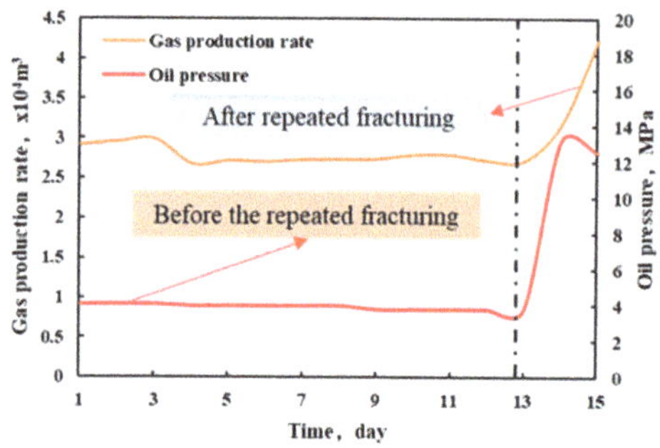


**Figure 20.** Production curves of Well K6 before and after repeated fracturing.

## 6. Conclusions

In this study, the mechanical properties of the rock in Block 14 of Dixi were characterized through indoor mechanical tests and in situ stress tests. A dynamic and static parameter conversion model was constructed using logging data and a continuous static mechanical profile was generated. On this basis, a dynamically updated geomechanical model was established and verified, and the influence of fracturing and production on in situ stress was quantified. Finally, based on the model, the construction parameters of repeated fracturing in Well K6 were optimized and on-site production was guided.

(1) Laboratory triaxial compression tests, Brazilian tensile strength tests, and in situ stress measurements of volcanic rock were conducted to characterize the mechanical properties of the Dixi 14 block. The results indicate that the tensile strength ranges from 0.93 to 10.53 MPa, the average Young modulus is 16.97 GPa, and the average Poisson ratio is 0.22. The maximum horizontal principal stress ranges from 85.57 to 8.96 MPa, the minimum horizontal principal stress from 51.89 to 4.37 MPa, and the vertical stress from 89.50 to 4.06 MPa. The horizontal stress difference ranges from 26.41 to 4.95 MPa.

(2) Dynamic rock mechanical parameters were calculated based on well logging data. Regression analysis was used to establish transformation models between dynamic and static mechanical parameters. The models exhibit high correlation coefficients. Using the transformation models, logging-derived mechanical property profiles were corrected to triaxial static profiles, enabling the construction of continuous static mechanical property profiles at key wells in the Dixi 14 block.

(3) Based on the continuous rock mechanical property profiles at key well locations, a geomechanical model of the Dixi 14 block was established using an integrated geological and engineering software platform. The model was validated through simulations of both hydraulic fracturing and production processes, demonstrating good accuracy. On the basis of the initial geomechanical model, a dynamic stress field concept was introduced to continuously update the stress distribution during fracturing simulations. This approach allows for a more accurate assessment of stress variations caused by fracturing and production activities. The results indicate that hydraulic fracturing increases the stress magnitude in the vicinity of the fractures, while reservoir depletion leads to a reduction in in situ stress. Under the combined effects of fracturing and production, the current stress state of the Dixi

14 block was determined: the minimum horizontal stress ranges from 48 to 76 MPa, the maximum horizontal stress from 78 to 102 MPa, the vertical stress from 79 to 94 MPa, and formation pressure from 20 to 35 MPa.

(4) Based on the current geomechanical model of the Dixi 14 block, a study was conducted on fracture propagation behavior during the re-fracturing of Well K6. The results show that as the injection rate, fluid volume, and prepad ratio increase, the matching between dynamic and propped fracture length tends to deteriorate. Fracture conductivity improves with increasing proppant volume, but this benefit diminishes beyond a certain threshold. The optimal combination of treatment parameters for Well K6 was determined to be an injection rate of 12 $m^3$/min, fluid volume of 1200 $m^3$, prepad ratio of 50–60%, and proppant volume of 75 $m^3$. Field re-fracturing operations were conducted using these optimized parameters. Following re-fracturing, the daily gas production of Well K6 increased by 56%, indicating a significant stimulation effect and successful field application. Other repeated fracturing wells in this block also face the problems of an unreasonable design of construction parameters and a poor repeated fracturing effect. The reservoir physical properties of these wells are relatively close to those of Well K6. The success of Well K6 can provide a reference for the repeated fracturing design of these wells.

**Author Contributions:** Conceptualization, H.L. (Honglei Liu); methodology, J.H.; validation, W.S. and X.W.; formal analysis, H.L. (Haoqi Li); resources, X.M.; data curation, Y.W. All authors have read and agreed to the published version of the manuscript.

**Funding:** This research received no external funding.

**Data Availability Statement:** The original contributions presented in this study are included in the article. Further inquiries can be directed to the corresponding author.

**Conflicts of Interest:** Authors Honglei Liu, Jiangling Hong, Wei Shu, Xiaolei Wang and Haoqi Li are employed by the company The First Gas Company, Xinjiang Oilfield Company, PetroChina. The remaining authors declare that the research was conducted in the absence of any commercial or financial relationships that could be construed as a potential conflict of interest.

## References

1. Tang, P.F.; Wang, Y.Q.; Guo, R.; Qi, S.L.; Yang, C.C. Fracturing Productivity Prediction Technology for Intermediate-Basic Volcanic Gas Reservoirs in Songliao Basin. In Proceedings of the International Field Exploration and Development Conference, Xi'an, China, 12–14 September 2024; Springer: Singapore, 2025; pp. 817–837.
2. Hou, Y.; Wu, C.; Chen, L.; Xie, B.; He, L.; Qin, J. New opinion of development potential of carboniferous volcanic gas reservoir: Taking the Kelameili gas filed as an example. *Front. Energy Res.* **2022**, *10*, 861901. [CrossRef]
3. Li, K.; Ji, G.; Liu, W. A method for calculating current in-situ stress field before refracturing horizontal shale wells. *Nat. Gas Explor. Dev.* **2020**, *43*, 110–118.
4. Deng, Y.; Zheng, B.; Li, Y.; Yue, H.; Li, B. Refracturing technology for horizontal wells in Niudong volcanic oil reservoir. *Well Test* **2024**, *30*, 32–36.
5. Liu, S.; Zhang, Y.; Wu, Q.; Ayers, W.B.; Wang, Y.; Ott, W.K. Crucial Development Technologies for Volcanic Hydrocarbon Reservoirs: Lessons Learned from Asian Operations. *Processes* **2023**, *11*, 3052. [CrossRef]
6. Liu, S.; Zhang, Z.; Huang, Z. Three-Dimensional hydraulic fracture simulation with hydromechanical coupled-element partition method. *Int. J. Geomech.* **2021**, *21*, 04021162. [CrossRef]
7. Chen, X.; Luo, Z.; Zhao, L.; Xiong, X.; Chen, W.; Miao, W.; Chen, X. Hydraulic, acid, and proppant-carrying acid fracturing stimulation of volcanic reservoirs in Sichuan Basin, China: An experimental study. *Pet. Sci. Technol.* **2023**, *41*, 731–749. [CrossRef]
8. Yu, J.; Li, N.; Hui, B.; Zhao, W.; Li, Y.; Kang, J.; Chen, Y. Experimental simulation of fracture propagation and extension in hydraulic fracturing: A state-of-the-art review. *Fuel* **2024**, *363*, 131021. [CrossRef]
9. Guo, T.; Wang, Y.; Chen, M.; Qu, Z.; Tang, S.; Wen, D. Multi-stage and multi-well fracturing and induced stress evaluation: An experiment study. *Geoenergy Sci. Eng.* **2023**, *230*, 212271. [CrossRef]
10. Siebrits, E.; Elbel, J.L.; Detournay, E.; Detournay-Piette, C.; Christianson, M.; Robinson, B.M.; Diyashev, I.R. Parameters affecting azimuth and length of a secondary fracture during a refracture treatment. In Proceedings of the SPE Annual Technical Conference and Exhibition, Paper SPE-48928, New Orleans, LA, USA, 27–30 September 1998.

11. Zhao, J.; Wang, Q.; Hu, Y.; Zhao, C.; Zhao, J. Prediction of pore pressure–induced stress changes during hydraulic fracturing of heterogeneous reservoirs through coupled fluid flow/geomechanics. *J. Eng. Mech.* **2019**, *145*, 05019001. [CrossRef]
12. Li, H.; Lin, C.Y.; Ren, L.H.; Dong, C.M.; Ma, C.F.; Li, S.T.; Liang, S.Y. An Integrated quantitative modeling approach for fault-related fractures in tight sandstone reservoirs. *J. Pet. Sci. Eng.* **2020**, *194*, 107552. [CrossRef]
13. Chen, G.; Wright, C.; Olivas, V. *Geomechanical Study for a Modern Drilling Testing Facility in Grimes County, Texas*; American Rock Mechanics Association: Alexandria, VA, USA, 2018.
14. Wang, Y.L. Coupled fluid flow to geomechanics in fractued reservoir: Governing equation, geomechanics parameters and numerical method. In Proceedings of the SPE Reservoir Characterisation and Simulation Conference and Exhibition, SPE-186041-MS, Abu Dhabi, United Arab Emirates, 8–10 May 2017.
15. Tuzingila, R.M.; Kong, L.; Koy Kasongo, R. A review on experimental techniques and their applications in the effects of mineral content on geomechanical properties of reservoir shale rock. *Rock Mech. Bull.* **2024**, *3*, 100110. [CrossRef]
16. Eyinla, D.S.; Oladunjoye, M.A.; Olayinka, A.I.; Bate, B.B. Rock physics and geomechanical application in the interpretation of rock property trends for overpressure detection. *J. Pet. Explor. Prod.* **2021**, *11*, 75–95. [CrossRef]
17. Lian, P.Q.; Cheng, L.S.; Deng, B. Simulation of ground stress field and fracture anticipation with effect of pore pressure. *Theor. Appl. Fract. Mech.* **2011**, *56*, 34–41. [CrossRef]
18. Wang, J.; Jiang, L.; Cang, T.; Zhou, X.; Wang, B. Simulation of a multi-stage stress field and regional prediction of structural fractures in the tucheng syncline. *Geosciences* **2025**, *15*, 132. [CrossRef]
19. Yun, X.; Mitri, H.S.; Yang, X.; Wang, Y. Experimental investigation into biaxial compressive strength of granite. *Int. J. Rock Mech. Min. Sci.* **2010**, *47*, 334–341. [CrossRef]
20. Bai, X.; Zhang, D.-M.; Wang, H.; Li, S.-J.; Rao, Z. A novel in situ stress measurement method based on acoustic emission Kaiser effect: A theoretical and experimental study. *R. Soc. Open Sci.* **2018**, *5*, 181263. [CrossRef] [PubMed]
21. Lavrov, A. The Kaiser effect in rocks: Principles and stress estimation techniques. *Int. J. Rock Mech. Min. Sci.* **2003**, *40*, 151–171. [CrossRef]
22. Garia, S.; Pal, A.K.; Ravi, K.; Nair, A.M. A comprehensive analysis on the relationships between elastic wave velocities and petrophysical properties of sedimentary rocks based on laboratory measurements. *J. Pet. Explor. Prod. Technol.* **2019**, *9*, 1869–1881. [CrossRef]
23. Zingano, A.C.; Salvadoretti, P.; Rocha, R.U.; Costa, J. Estimating uniaxial compressive strength, density and porosity of rocks from the p-wave velocity measurements in-situ and in the laboratory. *REM-Int. Eng. J.* **2021**, *74*, 521–528. [CrossRef]
24. Shi, W.R.; Wang, X.Z.; Shi, Y.H.; Feng, A.G.; Zou, Y.; Young, S. Application of dipole array acoustic logging in the evaluation of shale gas reservoirs. *Energies* **2019**, *12*, 3882. [CrossRef]
25. Fan, T.G.; Zhang, G.Q. Laboratory investigation of hydraulic fracture networks in formations with continuous orthogonal fractures. *Energy* **2014**, *74*, 164–173. [CrossRef]
26. Sarda, S.; Jeannin, L.; Basquet, R.; Bourbiaux, B. Hydraulic characterization of fractured reservoirs: Simulation on discrete fracture models. *SPE Reserv. Eval. Eng.* **2002**, *5*, 154–162. [CrossRef]
27. Zheng, S.; Xu, W. Calibration of fluid models using multiphase well test data for improved history matching. *J. Pet. Sci. Eng.* **2010**, *75*, 168–177. [CrossRef]
28. Zhao, K.; Huang, P.; He, Y.; Cui, L.; Liu, P.; Feng, Y.; Cao, S. Effect of geostress variation on hydraulic fracturing behavior and stress redistribution in coal seam roofs. *Processes* **2025**, *13*, 1732. [CrossRef]
29. Tang, Z.; Wu, Z.; Jia, D.; Lv, J. The distribution law of ground stress field in yingcheng coal mine based on rhino surface modeling. *Processes* **2024**, *12*, 668. [CrossRef]
30. Berchenko, I.; Detournay, E. Deviation of hydraulic fractures through poroelastic stress changes induced by fluid injection and pumping. *Int. J. Rock Mech. Min. Sci.* **1997**, *34*, 1009–1019. [CrossRef]
31. Sun, C.; Zhang, Y.; Han, L.; Liu, M.; Hu, J. Impacts of fracturing fluid viscosity and injection rate variations on the fracture network propagation in deep coalbed reservoirs. *AIP Adv.* **2025**, *15*, 105–109. [CrossRef]
32. Liu, H.; Huang, Y.; Cai, M.; Meng, S.; Tao, J. Practice and development suggestions of hydraulic fracturing technology in the Gulong shale oil reservoirs of Songliao Basin. *Pet. Explor. Dev.* **2023**, *50*, 603–612. [CrossRef]
33. Wu, Q.; Xu, Y.; Wang, X.Q. Volume fracturing technology of unconventional reservoirs: Connotation, design optimization and implementation. *Pet. Dev.* **2012**, *39*, 352. [CrossRef]
34. Zhou, F.; Su, H.; Liang, X.; Meng, L.; Yuan, L.; Li, X.; Liang, T. Integrated hydraulic fracturing techniques to enhance oil recovery from tight rocks. *Pet. Explor. Dev.* **2019**, *46*, 1065–1072. [CrossRef]

processes

*Article*

# Controls on the Hydrocarbon Production in Shale Gas Condensate Reservoirs of Rift Lake Basins

**Yaohua Li [1,2], Caiqin Bi [2,*], Chao Fu [3], Yinbo Xu [2], Yuan Yuan [2], Lihua Tong [2], Yue Tang [2] and Qianyou Wang [4,*]**

1 State Key Laboratory of Shale Oil and Gas Enrichment Mechanisms and Effective Development, Beijing 100083, China; lyaohua@mail.cgs.gov.cn
2 Oil and Gas Survey, China Geological Survey, Beijing 100083, China
3 Xinjiang Yaxin Coalbed Methane Investment and Development (Group) Co., Ltd., Urumchi 830000, China
4 Department of Earth, Ocean and Ecological Sciences, University of Liverpool, Liverpool L69 3GP, UK
* Correspondence: bicaiqin@mail.cgs.gov.cn (C.B.); qianyou.wang@liverpool.ac.uk or tsianyou@126.com (Q.W.)

**Abstract:** The production of gas and condensate from liquid-rich shale reservoirs, particularly within heterogeneous lacustrine systems, remains a critical challenge in unconventional hydrocarbon exploration due to intricate multiphase hydrocarbon partitioning, including gases ($C_1$–$C_2$), volatile liquids ($C_3$–$C_7$), and heavier liquids ($C_{7+}$). This study investigates a 120-meter-thick interval dominated by lacustrine deposits from the Lower Cretaceous Shahezi Formation ($K_1sh$) in the Songliao Basin. This interval, characterized by high clay mineral content and silicate–pyrite laminations, was examined to identify the factors controlling hybrid shale gas condensate systems. We proposed the Hybrid Shale Condensate Index (*HSCI*), defined as the molar ratios of ($C_1$–$C_7$)/$C_{7+}$, to categorize fluid phases and address shortcomings in traditional *GOR*/*API* ratios. Over 1000 samples were treated by geochemical pyrolysis logging, X-ray fluorescence (XRF) spectrum element logging, SEM-based automated mineralogy, and in situ gas desorption, revealing four primary controls: (1) Thermal maturity thresholds. Mature to highly mature shales exhibit peak condensate production and the highest total gas content (*TGC*), with maximum gaseous and liquid hydrocarbons at $T_{max}$ = 490 °C. (2) Lithofacies assemblage. Argillaceous shales rich in mixed carbonate and clay minerals exhibit an intergranular porosity of 4.8 ± 1.2% and store 83 ± 7% of gas in intercrystalline pore spaces. (3) Paleoenvironmental settings. Conditions such as humid climate, saline water geochemistry, anoxic bottom waters, and significant input of volcanic materials promoted organic carbon accumulation (*TOC* reaching up to 5.2 wt%) and the preservation of organic-rich lamination. (4) Laminae and fracture systems. Silicate laminae account for 78% of total pore space, and pyrite laminations form interconnected pore networks conducive to gas storage. These findings delineate the "sweet spots" for unconventional hydrocarbon reservoirs, thereby enhancing exploration for gas condensate in lacustrine shale systems.

**Keywords:** shale; gas condensate; hydrocarbon; fluid phase; thermal maturity; laminae; lithofacies; lake; rift basins; volcanism

## 1. Introduction

Hybrid condensates require refining processes to eliminate higher-boiling-point hydrocarbon fractions (>$CH_4$). Upon isolation, these residual hydrocarbons undergo liquefaction, yielding natural gas condensate [1]. Such reservoirs exhibit complex thermodynamic responses marked by sub-dew-point pressures that drive condensate precipitation and compositional alteration of reservoir gases [2,3]. Contemporary classification frameworks categorize hybrid shale gas condensate resources into two distinct groups: liquid-lean and liquid-rich systems. Both groups predominantly comprise $C_1$–$C_{33}$ n-alkanes, with the $C_1$–$C_7$ fraction persisting in gaseous states under reservoir conditions [4]. Volatile light hydrocarbons ($C_3$–$C_7$) critically control the generation of liquid-lean condensate through phase partition processes [2,3,5–7].

Recent exploration efforts in continental basins have exposed critical knowledge gaps in characterizing shale gas condensate reservoirs. The interplay between elevated clay mineral content and heterogeneous organic inputs introduces substantial uncertainties in hydrocarbon distribution patterns [8–11]. Hybrid shale gas condensate systems in continental basins represent a critical frontier in unconventional hydrocarbon exploration, given the complexity of multiphase fluid behaviors (gas, volatile liquids, and condensate) with the challenges of heterogeneous lithofacies and diagenetic alterations. While marine shale systems (e.g., Bossier and Haynesville shales) have been extensively studied [12,13], lacustrine systems, particularly those characterized by high clay mineral contents and strong inputs of volcanogenic materials, remain underexplored. Recent advances highlight the role of lamina architectures in enhancing hydrocarbon storage and connectivity [14], the influences of bottom-water redox conditions on organic matter preservation [15], and the development of novel indices to quantify phase partitioning [16]. However, key questions persist regarding the combined effects of volcanic inputs, salinity fluctuations, and laminae-induced fractures in controlling hybrid condensate "sweet spots" within lacustrine shales.

Overall, two principal challenges have been identified in reservoir evaluation:

1. Hydrocarbon phase dynamics. Drilling-induced perturbations of pressure and temperature tend to trigger phase transitions, particularly when reservoir pressures decline below dew-point thresholds near wellbores [17–19]. Differential retention mechanisms further complicate the associated analysis. $C_1$–$C_2$ alkanes demonstrate sluggish diffusion kinetics, while volatile light hydrocarbons ($C_3$–$C_7$) and liquid-phase components ($C_{7+}$) persist in metastable configurations within fracture networks and matrix porosity [6,20,21]. These phenomena limit the efficacy of conventional $C_1$–$C_7$ advanced mud gas logging for quantitative hydrocarbon assessment.
2. Lacustrine reservoir heterogeneity. Compared to marine shales, lacustrine systems exhibit amplified variability of lithofacies due to abrupt facies changes [22,23], multi-scale beddings [24], and lamina-induced fracture systems [25–27] that induce strong anisotropic hydrocarbon distribution and condensate productivity [28–31]. Additional complexity stems from dynamic phase behavior modulated by thermal maturity gradients, in situ P-T conditions, and near-wellbore condensate banking effects [29,31,32]. Consequently, these factors impede the reliable identification of hydrocarbon sweet spots in continental shale reservoirs.

The China Geological Survey implemented a multidisciplinary exploration project of continental shale gas resources based on the Cretaceous Shahezi Formation ($K_1sh$) within the Songliao Basin (JLYY1 borehole, Figure 1). A 120-meter-thick, lacustrine-dominated interval at >3000 m depth was analyzed through an integrated analytical protocol.

- High-resolution geochemical profile via systematic sampling and measurement at a 0.5 to 1 m interval (more than 1000 samples) employing multiple methods such as geochemical logging, in situ gas desorption, and electron microscopic mineralogy.
- Phase-specific hydrocarbon characterization using $S_1$ ($C_7$–$C_{33}$) parameters and total gas content (*TGC*), with speciation of gaseous ($C_1$–$C_2$) versus volatile ($C_3$–$C_7$) components.
- Lithofacies discrimination through the correlation of total organic carbon content (*TOC*) and mineral composition, augmented by elemental proxies for paleoenvironmental reconstruction.

- Nanoscale rock fabric analysis utilizing ultra-high-resolution scanning electron microscopes (SEMs) to detect the occurrence of organic matter and mineral particles, elemental migration pathways, and hydrocarbon–bitumen associations.
- Phase behavior modeling via a proxy termed Hybrid Shale Condensate Index (*HSCI*), incorporating $T_{max}$ values to map maturity-controlled phase distributions across $C_1$–$C_{33}$ hydrocarbons.
- Multivariate statistical integration of *HSCI*, $S_1$, *TGC*, $T_{max}$, *TOC*, and rock compositional datasets to delineate dominant hydrocarbon distribution controls.

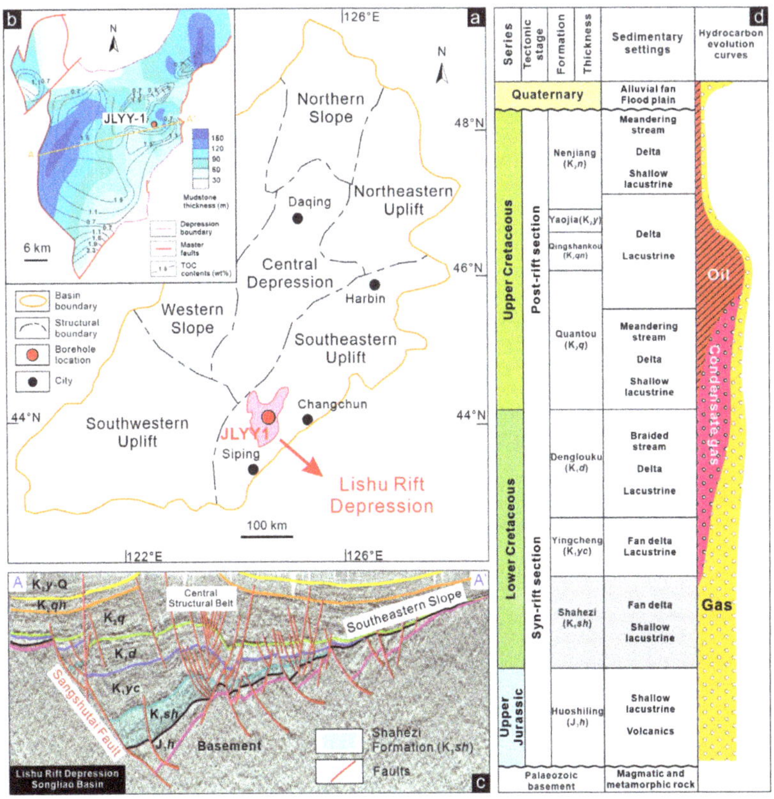


**Figure 1.** (**a**) The tectonic units of the Songliao Basin in northeastern Asia and the location of the Lishu Rift Depression. (**b**) Spatial distribution of mudstones and total organic carbon (*TOC*) concentrations within the Lishu Rift Depression. (**c**) An interpreted seismic reflection profile A–A′ showing the basin geometry, major faults, and depositional filling of the Lishu Rift Depression. The profile location is illustrated in panel (**b**). (**d**) A stratigraphic section measured from drilling boreholes, illustrating the tectonostratigraphic evolution, sedimentary settings, and hydrocarbon phase changes of the Lishu Rift Depression.

Currently available methods for the evaluation of hybrid shale gas condensate systems, e.g., *GOR*/*API* classifications (*GOR*: gas-to-oil production ratio; *API*: American Petroleum Institute gravity), struggle to capture multiphase fluid behavior in lacustrine systems due to their dependence on surface fluid properties. This study addresses this gap by proposing the Hybrid Shale Condensate Index (*HSCI*), which integrates in situ gaseous (*TGC*) and liquid ($S_1$) components to quantify fluid-phase maturity for complex unconventional reservoirs rich in clay minerals. This approach uses $S_1$ and *TGC* as robust proxies for liquid and volatile hydrocarbon phases, respectively, while providing a systematic framework to decipher hydrocarbon occurrence mechanisms in complex lacustrine shale gas condensate systems influenced by volcanic activities, paleoenvironmental settings, water geochemistry, and laminae–fracture synergy.

## 2. Geological Background

The JLYY1 borehole investigated in this study is situated in the eastern slope zone of the Lishu Rift Depression within the Southeast Uplift of the Songliao Basin (Figure 1). The Lishu Rift Depression, covering a total area of 2822 km$^2$, is notable for having the longest rifting period, the most complete stratigraphic development, the thickest sedimentary deposits, and the deepest burial in the southeast uplift area of the Songliao Basin. This depression contains two primary sets of strata: the syn-rift and post-rift sections (Figure 1d). The syn-rift section has a thickness ranging from 0 to 8000 m, with a maximum burial depth exceeding 10,000 m. This section contains the Cretaceous Huoshiling Formation ($K_1h$), Shahezi Formation ($K_1sh$), Yingcheng Formation ($K_1yc$), and Denglouku Formation ($K_1d$), filled with lacustrine and fan delta deposits [22,33]. The post-rift section mainly consists of the Cretaceous Quantou Formation ($K_2q$), Qingshankou Formation ($K_2qn$), Yaojia Formation ($K_2y$), Nenjiang Formation ($K_2n$), and Quaternary strata, characterized by fluvial, deltaic, and lacustrine facies deposition [34]. The Shahezi and Yingcheng Formations are the primary intervals of thick shales developed during the significant rifting subsidence period.

From top to bottom, the JLYY1 borehole drilled through the Quaternary, Upper Cretaceous $K_2qn$ and $K_2q$ Formations, Lower Cretaceous $K_1d$, $K_1yc$, and $K_1sh$ Formations. The measured reservoir pressure of the Shahezi Formation ranges from 19.38 to 26.56 MPa, with a reservoir pressure coefficient ranging between 0.84 and 1.0, averaging approximately 0.98. The reservoir temperature gradient in the Shahezi Formation is between 3.3 °C/100 m and 3.5 °C/100 m, indicating a normal temperature system. The formation temperature at the bottom of the JLYY1 borehole is approximately 120 °C.

## 3. Materials and Methods

### 3.1. Geochemical Pyrolysis-FID Logging and In Situ Gas Desorption

The thermal maturity of source rocks and hydrocarbon fluid properties have been rapidly evaluated by the geochemical pyrolysis-FID (flame ionization detector) logging. The generated hydrocarbons and kerogens can be volatilized and cracked at different temperatures by heating sample powders within an inert gas atmosphere [31,35,36]. The light hydrocarbons ($<C_7$) in the samples are measured as the $S_0$ peak, while the liquid hydrocarbons ($C_8$–$C_{33}$ n-alkanes) are pyrolyzed in the temperature range of the $S_1$ peak. The measured $S_2$ values represent the kerogen and heavy hydrocarbons (greater than $C_{33}$), and the $S_4$ values are contributed by the residual carbon within rocks. The in situ total gas contents (*TGC*, m$^3$/t) of the Shahezi Formation rocks were tested in the well site once the core segments were recovered by pressure coring and sealed in airtight desorption canisters. The coalbed methane and shale gas pulse-type intelligent tester (HTG-Pulse-6, Hengtaishanghe Energy Technology Co., Ltd., Beijing, China) was used to measure the lost gas, desorption gas, and remaining gas. Detailed information on the experimental procedure and equipment can be found in Wang et al. (2023) [31]. In this study, the content of movable liquid hydrocarbons is indicated by $S_1$ values in the geochemical pyrolysis logging, whereas the content of light and gaseous hydrocarbons ($C_1$–$C_7$) is represented by the *TGC* values in the gas desorption tests.

### 3.2. X-Ray Fluorescence (XRF) Spectrum Element Logging

Major and trace element concentrations of the Shahezi Formation were measured by the X-ray fluorescence (XRF) spectrum element logging (CIT-3000SY XRF element analyzer, Sichuan Xinxianda Measurement & Controlling Co., Ltd., Chengdu, China). The XRF elemental capture energy spectrum technology is based on the interaction between high-energy X-rays and the sample. When the X-rays bombard the sample, electrons are released from the outer layers of the atoms, creating electron vacancies. Subsequently, electrons from higher energy levels fall into these vacancies, emitting characteristic X-rays. The variation in energy and wavelength of these emitted X-rays depends on the element involved. Hence, the type and quantity of elements present in the samples can be determined by analyzing the energy and wavelength of the emitted X-rays. The measured elements can be classified into five types: (1) highly migrated elements (Cl, Br, and S); (2) easily migrated elements (Ca, Mg, Na, F, Sr, K, Zn, and P); (3) migrated elements (Cu, N, Co, Mo, V, Mn, P, and S in silicates); (4) stable elements (Fe, Al, Ti, and Se); and (5) hardly migrated element (Si). In this study, twenty elements, including Na, K, Al, Si, Fe, Ca, Mg, Ti, Mn, S, P, Cl, Th, U, Ba, Sr, V, Ni, Cr, and Zr, are measured across the entire shale gas condensate interval to investigate the influence of the paleoenvironmental conditions on the in situ content of shale gas condensate.

### 3.3. SEM-Based Automated Mineralogy and Petrography

The RoqSCAN system is a quantitative and automated SEM-EDS (scanning electron microscopy and energy-dispersive X-ray spectroscopy) developed by Fugro Robertson Ltd. (Leidschendam, The Netherlands) and Carl Zeiss Microscopy Ltd. (Jena, Germany) [37,38]. It can not only be used in the laboratory but can also be prepared as portable, durable core analysis equipment in the field. The RoqSCAN used in this study is equipped with a Carl Zeiss EVO 50 SEM (Carl Zeiss Ltd., Jena, Germany), a Bruker AXS X-ray detector (Bruker Corporation, Billerica, MA, USA), a Bruker Pulse Processor (Bruker Corporation, Billerica, MA, USA), and Zeiss SmartPI$^{TM}$ software (version 1.0, Carl Zeiss Ltd., Jena, Germany) [33], allowing for mineralogical and petrographic analysis of rock cuttings, cores, sidewall coring, and thin sections. It provides data on element concentrations, mineral types, rock fabrics, and petrophysical parameters (e.g., porosity, pore-size distribution, fracture numbers, and fracture density) by analyzing high-resolution SEM-EDS images (pseudo-color mineral images with a maximum resolution of 250 nm) [39–41]. Besides these data, the RoqSCAN system can also give information about bulk-rock density, brittleness, and elasticity [42,43]. In this study, the RoqSCAN is used to quantify the mineral composition and determine the rock fabrics. The images of laminated mudstones, composed of various mineral compositions, help to investigate the influences of lithofacies and laminae-induced fractures on shale gas condensate gas contents.

Besides RoqSCAN technology, an ultra-high-resolution FEG (field emission gun)-SEM (Zeiss GeminiSEM 450, Carl Zeiss Ltd., Jena, Germany) was utilized to detect dispersed organic matter and mineral particles across scales ranging from μm to nm, at the SEM Shared Research Facility, University of Liverpool. Backscattered (BSE, 20 kV accelerating voltage, 1 nA current) and secondary electron (SE, 5 kV, 1 nA) microscopy were performed on the Ar-ion beam-milled rock blocks to produce photomicrographs. Details of sample preparation and experimental procedure can be found in Wang et al. (2023 and 2025) [22,33].

### 3.4. Hybrid Shale Condensate Index (HSCI)

The fluid types on the surface are typically determined based on the gas-to-oil ratio (*GOR*) and oil viscosity (*API*), as illustrated in Table 1 and previous studies [9,29,31]. Both *GOR* and *API* are determined by the light hydrocarbon components ($C_1$–$C_7$) and heavy hydrocarbon components ($C_{7+}$) [44–46]. The content of gaseous hydrocarbons ranges from mixtures of methane and ethane with very few other constituents (dry gas) to mixtures containing a range of hydrocarbons from methane to pentane, and even hexane ($C_6H_{14}$) and heptane ($C_7H_{16}$) (wet gas). In both cases, some carbon dioxide ($CO_2$) and inert gases, including helium (He), are present along with hydrogen sulfide ($H_2S$) and trace amounts of organic sulfur [47–49].

**Table 1.** Measured fluid composition and fraction properties for typical condensate gas regions. Separator gas and oil samples were captured and recombined to the reported *GOR* in the laboratory.

| *GOR* (scf/STB) [1] | 3000–4000 | 4000–5000 | 5000–8000 | 8000–15,000 | 15,000–50,000 |
|---|---|---|---|---|---|
| Specific gravity (°API) | 52.1 | 55.5 | N/A [2] | 54.58 | N/A |
| Reservoir pressure (psi) | 11,025 | 10,630 | 10,000 | 9300 | N/A |
| Reservoir temperature (°F) | 321 | 328 | 300 | 290 | 275 |
| Dew point (psi) | 4312 | 4165 | 4050 | 3892 | 3310 |
| $CO_2$ | 1.07 | 0.69 | 1.57 | 1.971 | 1.10 |
| $N_2$ | 0.15 | 0.07 | 0.09 | 0.092 | 0.10 |
| $C_1$ | 61.88 | 64.17 | 67.97 | 69.722 | 72.60 |
| $C_2$ | 11.64 | 11.22 | 11.61 | 11.791 | 12.95 |
| $C_3$ | 5.58 | 5.46 | 4.57 | 4.165 | 3.79 |
| I-$C_4$ | 1.32 | 1.53 | 1.38 | 1.316 | 1.25 |
| N-$C_4$ | 2.35 | 2.39 | 1.90 | 1.67 | 1.46 |
| I-$C_5$ | 1.20 | 1.35 | 1.24 | 1.188 | 1.14 |

[1] scf/STB: cubic feet of gas per barrel of oil or condensate. [2] N/A: not available.

In this study, the fluid type of drilling-associated hydrocarbons is classified by using the in situ hydrocarbon components based on the surface fluid classification. While $C_1$–$C_7$ advanced mud gas logging methods are employed during drilling, testing the gaseous hydrocarbon content of hybrid shale gas condensates is challenging. This difficulty arises because the liquid phase formed in the matrix typically remains below residual oil saturation. Moreover, the heavier hydrocarbon components ($C_{7+}$) will be permanently trapped in the matrix unless enhanced oil recovery techniques are applied. This is especially true under drilling conditions where only small quantities of $C_1$–$C_2$ hydrocarbons can flow to the gas content tester due to the density of the drilling fluid. Therefore, the hybrid condensates cannot be fully evaluated without considering the concomitant oils.

As a result, a new index, the Hydrocarbon Saturation Classification Index (*HSCI*), was proposed using the molar components of $C_1$–$C_7$ and $C_{7+}$ to classify the fluid type, as described in Equation (1) [50]. The ratio values are calculated according to the molar ratio of hydrocarbon components, as shown in Table 1. The *HSCI*, as defined in Equation (1), offers an advantage in representing the occurrence of equal and elevated maturity levels in both gaseous and liquid components of condensates, particularly for hybrid condensates. Since the $C_{7+}$ fraction is the only liquid-phase component available in gas condensate PVT analytical data, it is often spuriously increased by depressurization, making a reliable liquid index generally unavailable.

$$\begin{cases} \text{Liquid hydrocarbons}: \frac{n_{C_1-C_7}}{n_{C_{7+}}} < 10 \\ \text{Gas condensate}: 10 < \frac{n_{C_1-C_7}}{n_{C_{7+}}} < 60 \\ \text{Wet gas}: 60 < \frac{n_{C_1-C_7}}{n_{C_{7+}}} < 128 \\ \text{Dry gas}: \frac{n_{C_1-C_7}}{n_{C_{7+}}} > 128 \end{cases} \quad (1)$$

Here, some modifications were made to fit the light hydrocarbon component ratio as Equation (2), where *TGC* is used to indicate the volume content of desorbed gas, and $S_1$ is used to indicate the mass content of movable hydrocarbon in shales.

$$HSCI = \frac{n_{C_1-C_7}}{n_{C_{7+}}} = \frac{m_{C_1-C_7}M_{C_8-C_{33}}}{M_{C_1-C_7}m_{C_8-C_{33}}} = \frac{10^3 m_{rock}TGC}{M_{C_1-C_7}\rho_{C_1-C_7}} \cdot \frac{M_{C_8-C_{33}}}{m_{rock} \cdot S_1} = \frac{10^3 TGC \cdot M_{C_8-C_{33}}}{M_{C_1-C_7}\rho_{C_1-C_7}S_1} \quad (2)$$

where $n$ indicates the molar content, $m$ indicates the mass for special hydrocarbon components, $M$ indicates the molar mass, and $\rho$ indicates the hydrocarbon density. Additionally, the fitting parameters of $\rho_{C1-C7}$, $M_{C1-C7}$, and $M_{C8-C33}$ are settled as 0.739 g/cm$^3$, 16 g/mol, and 78 g/mol, respectively. Note that the *HSCI* is only proposed according to the fluid type classification during production, especially related to the hydrocarbon components. However, any kind of hydrocarbon components could not be used separately to illustrate the in situ fluid phase. Thus, a further hydrocarbon phase work using the *HSCI* and the geochemical pyrolysis logging data will be discussed in the following sections.

## 4. Results

### 4.1. Lacustrine Shale Components and Lithofacies

Quantitative mineral analysis utilizing RoqSCAN and rock pyrolysis reveals distinct compositional zonation in lacustrine shales (Figures 2 and 3). The ternary diagram in Figure 3a delineates bulk-rock mineral composition into three endmembers: carbonates, clay minerals, and non-clay mineral silicates. Detailed mineral speciation maps (Figure 3b,c) employ standardized symbology: solid circles denote high-*TOC* samples (>2 wt%), squares represent medium-*TOC* (0.5 to 2 wt%), and triangles mark low-*TOC* zones (<0.5 wt%). The associated clay mineral classification chart identifies seven species through diagnostic X-ray diffraction peaks: mixed clay (MC, 14 Å), illite/muscovite (I/M, 10 Å), kaolinite (K, 7 Å), mixed calcareous clays (MC(C), 12–14 Å composite), chlorite (Ch, 14 Å and 7 Å), glauconite (G, 10 Å and 4.5 Å), and biotite (Bio, 10 Å and 3.3 Å). Core analysis identifies three vertically stacked units marked by different mineral compositions: (1) argillaceous zone (3080–3115 m): clay mineral content > 75 vol% with homogenous fabric (lamination index < 0.15); (2) laminated and organic-rich unit (3115–3165 m): peak *TOC* (5.2 ± 0.8 wt%) coincident with pronounced bedding-parallel anisotropy (lamination index > 0.65); (3) calcite-fractured mudstone (3170–3200 m): 89 ± 4% calcite content exhibiting conjugate fracture sets (fracture density: 12 ± 3 fractures/m) and pressure solution crumples.

Three genetic classes are defined based on the following *TOC* thresholds:

- Type I (*TOC* > 2 wt%): $S_1$ ($C_7$–$C_{33}$) = 1.8 ± 0.3 mg/g, *TGC* ($C_1$–$C_2$) = 1.6 ± 0.2 m$^3$/t.
- Type II (0.5–2 wt%): $S_1$ = 0.9 ± 0.2 mg/g, *TGC* = 1.1 ± 0.3 m$^3$/t.
- Type III (<0.5 wt%): $S_1$ = 0.3 ± 0.1 mg/g, *TGC* = 0.7 ± 0.2 m$^3$/t.

Figure 3d demonstrates $S_1$ maxima (>1.5 mg/g) in shales of 50–75% clay mineral content, correlating with MC(C) abundance ($R^2$ = 0.82, $p$ < 0.01). Pore network modeling reveals MC(C)-rich zones develop 4.8 ± 1.2% intergranular porosity (Figure 4a, blue clusters) versus 1.2 ± 0.5% in I/M-dominated regions.

Furthermore, spatial correlation analysis (Figure 4b,c) shows *TGC* maxima (>1.5 m$^3$/t) overlap with high-$S_1$ intervals but lack *TOC* dependence (R = 0.18, $p$ = 0.32). SEM-EDS mapping confirms < 2% organic porosity, with 83 ± 7% gas stored in intercrystalline pores within the clay mineral matrix (Figure 4d, marked by yellow).

Integrated analysis identifies optimal reservoirs as (1) liquid phase hydrocarbon: MC(C) > 25%, I/M:K < 1.5:1, *TOC* > 2 wt% (Figure 4a, red zones); (2) gaseous phase hydrocarbon: clay mineral content = 50–75%, illite crystallinity < 0.35 Δ°2θ (Figure 4c).

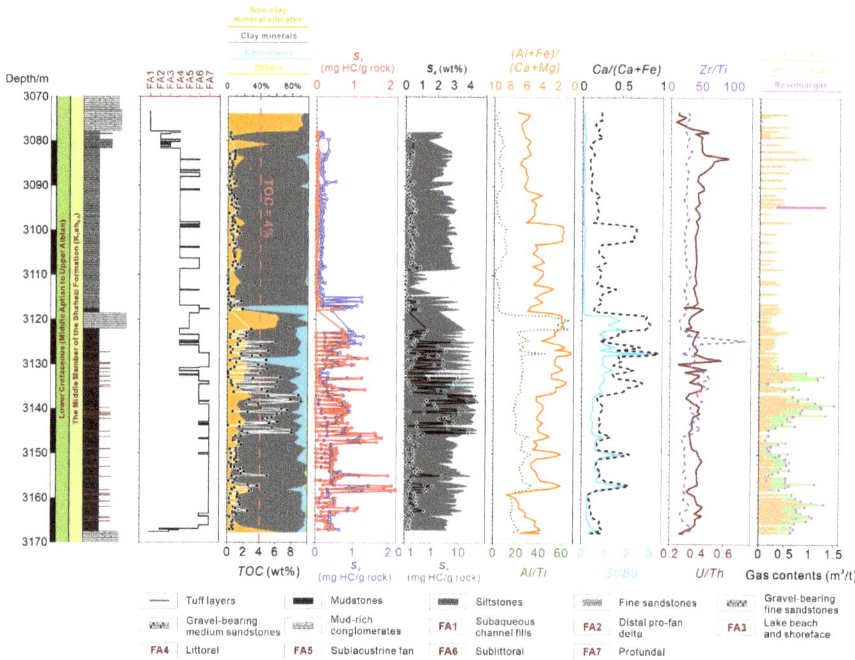


**Figure 2.** Comprehensive stratigraphic section of the lacustrine-dominated interval within the Shahezi Formation (JLYY1 borehole, Lishu Rift Depression), illustrating the vertical variations in lithology, facies associations (FA1–FA7, after Wang et al., 2025) [22], mineral composition, total organic carbon (*TOC*) content, pyrolysis logging proxies ($S_0$, $S_1$, $S_2$, and $S_4$), geochemical indicators, and total gas content (*TGC*). The *Al*/*Ti* ratio is used to infer paleoclimate conditions (humid vs. dry), while the (*Al* + *Fe*)/(*Ca* + *Mg*) proxy reflects paleowater depth. *Sr*/*Ba* and *Ca*/(*Ca* + *Fe*) ratios are employed to differentiate paleosalinity (freshwater vs. saline lake), the *Zr*/*Ti* ratio indicates the input intensity of volcanic materials, and the *U*/*Th* ratio serves as an indicator of bottom-water paleoredox conditions.

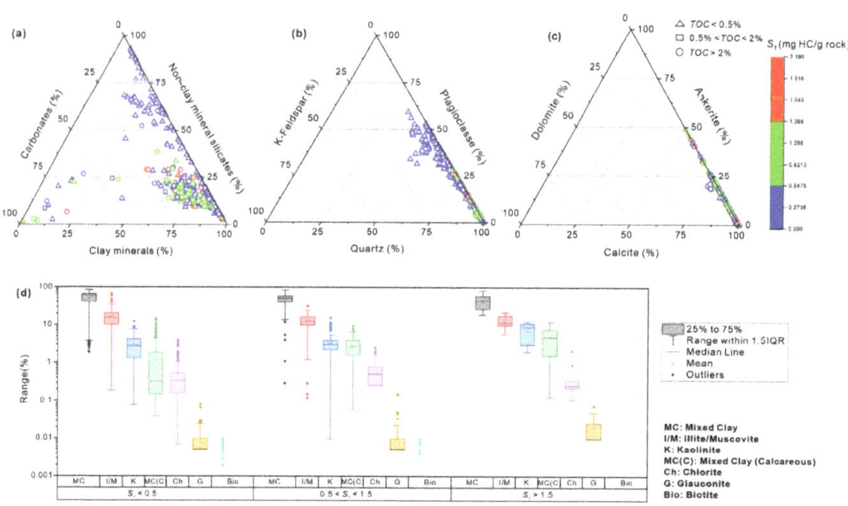


**Figure 3.** Ternary (**a**–**c**) and box (**d**) diagrams showing the correlation between RoqSCAN mineral composition, total organic carbon (*TOC*) concentrations, and pyrolysis $S_1$ values. $S_1$ values are used to indicate the contents of liquid hydrocarbons.

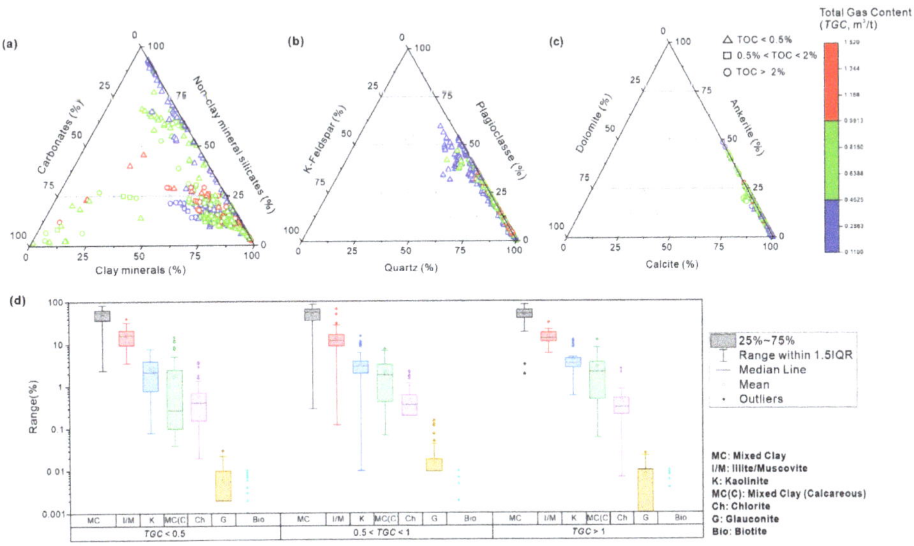

**Figure 4.** Ternary (**a–c**) and box (**d**) diagrams showing the correlation between RoqSCAN mineral composition, total organic carbon (*TOC*) concentrations, and total gas contents (*TGC*). *TGC* values are used to indicate the contents of liquid hydrocarbons.

### *4.2. Thermal Maturity and Hydrocarbon Phases*

$T_{max}$ values are used to indicate the thermal maturity between 450 °C to 510 °C in a mature and highly mature stage for the vertical shale section from 3080 m to 3165 m of JLYY1 well. The liquid friction, as mentioned in Section 3.4, is depicted by the *HSCI* in Equation (2), such that *HSCI* is valued between 0 to 140 for the vertical shale section. Generally, gas friction in gaseous and volatile phases controls the shales from 3080 m to 3115 m, where *HSCI* is valued between 60 to 140. Hydrocarbon content increases with the heavier components from 3115 m to 3165 m, which are controlled by the hybrid shale gas condensate phases.

To illustrate the influence of thermal maturity and hydrocarbon phases on the hydrocarbon content, parameters of $S_1$, *TGC*, and *HSCI* are used to make a statistical scatter chart where $T_{max}$ is settled as the x-axis, black points indicate the $S_1$ values, and blue points indicate the *TGC* values, as shown in Figure 5.

Here, $S_1$ and *TGC* are parameters tested just following the drilling process, indicating the hydrocarbon phases upon the ground, and $T_{max}$ and *HSCI* are the in situ indicators of the subsurface conditions for the JLYY1 well. In detail, $T_{max}$ is valued from 460 °C to 500 °C and associated with a gaseous and liquid hydrocarbon peak (highest *TGC* and $S_1$ values) around 490 °C. Here, *HSCI* was used to distinguish the shale hydrocarbon resources in the gas condensate, wet gas, and liquid states, respectively. Moreover, a further thermal maturity system was taken into consideration in this study. The hydrocarbon phases of the Shahezi Formation shale reservoir could be classified into six types: (1) the wet gas in mature shale ($T_{max}$ < 480 °C, *HSCI* > 60); (2) wet gas in highly mature shale ($T_{max}$ > 480 °C, *HSCI* > 60); (3) gas condensate in mature shale ($T_{max}$ < 480 °C, 10 < *HSCI* < 60); (4) gas condensate in highly mature shale ($T_{max}$ > 480 °C, 10 < *HSCI* < 60); (5) liquid in mature shale ($T_{max}$ < 480 °C, *HSCI* < 10); and (6) liquid in highly mature shale ($T_{max}$ > 480 °C, *HSCI* < 10). As indicated by the *HSCI* values in Figure 5, the red points indicate that shale gas condensate in mature and highly mature shale, acting as the major hydrocarbon phases, contributes mostly to the hydrocarbon content of the in situ hybrid shale gas condensate resources. Unlike marine shales (e.g., Haynesville, Marcellus), where biogenic silica and high thermal maturity dominate, the *HSCI* framework uniquely resolves fluid-phase ambiguities in lacustrine systems by accounting for volcanism–clay mineral interactions and lamina architectures. This approach reduces reliance on depressurization-distorted $C_{7+}$ data, offering a higher accuracy in sweet spot prediction compared to traditional *GOR*/*API* methods. Section 5 provides microscopic evidence for the influence of volcanism–clay mineral interactions and lamina architectures on hydrocarbon production in shale gas condensate reservoirs.

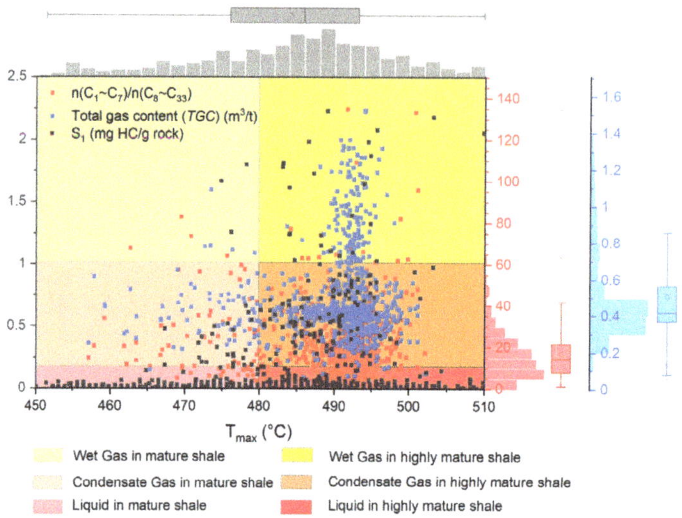


**Figure 5.** A scatter diagram of thermal maturation ($T_{max}$), Hydrocarbon Saturation Classification Index (*HSCI*), pyrolysis $S_1$, and total gas content (*TGC*), illustrating the changes in hydrocarbon fluid phase with evolving thermal maturation in the Shahezi Formation shales (JLYY1 borehole, Lishu Rift Depression, Songliao Basin). Frequency distribution histograms and box-and-whisker plots for $T_{max}$, *HSCI*, and *TGC* are also provided, showing data distribution, outliers, and confidence intervals for these variables.

### 4.3. Paleoenvironmental Conditions

Major elements (Na, K, Al, Si, Fe, Ca, Mg, Ti, Mn, S, P, Cl) and trace elements (Th, U, Ba, Sr, V, Ni, Cr, Zr) were measured to study the paleoenvironmental conditions during the deposition of the Shahezi Formation and their influences on hydrocarbon contents of shale gas condensate layers. The $Al/Ti$ ratios indicate paleoclimate conditions (humid vs. dry), while the $Sr/Ba$ and $Ca/(Ca + Fe)$ proxies reflect paleosalinity (freshwater vs. saline lake). The $(Al + Fe)/(Ca + Mg)$ ratio suggests paleowater depth, and the $Zr/Ti$ ratio indicates the input intensity of volcanic materials. The *U/Th* ratio helps assess bottom-water paleoredox conditions (oxic vs. anoxic) [51–56].

As shown in Figure 2, the paleoclimate shifts from humid to dry with decreasing *Al/Ti* values. Shales from 3080 m to 3120 m were deposited in a more arid climate, corresponding to lower hydrocarbon content. The paleo-lake environment transitions from freshwater to saline with increasing $Sr/Ba$ and $Ca/(Ca + Fe)$ values, generating shales of 3115–3135 m suitable for hydrocarbon generation. These sections have higher hydrocarbon contents due to the saline water geochemistry [57]. Similarly, increasing $(Al + Fe)/(Ca + Mg)$ values suggest high water depths, enhanced hydrocarbon generation in the section of (3120–3140 m).

Volcanic ash preserved in the paleo-lake sediments, indicated by fluctuating $Zr/Ti$ values of 3122–3142 m, also promotes hydrocarbon generation. As proposed by Li et al.

(2022), tuffaceous shale layers are key to the prediction of the hydrocarbon "sweet spot" created by water–rock interaction [50]. U is enriched in anoxic environments, while Th concentrations tend to be constant [58–60]. Thus, high *U/Th* values usually indicate anoxic bottom-water conditions. For the JLYY1 borehole, a high gas production is recorded in the shales from 3128 m to 3142 m, as well as 3155 m to 3167 m, where the highest hydrocarbon content corresponds to the most anoxic conditions.

Moreover, hybrid sedimentary factors contribute to the deposition of organic-rich shales from 3118 m to 3137 m. Under microscopes, organic matter particles are closely associated with terrigenous organic inputs (e.g., mineral particles covered by solid bitumen, Figure 6a,b), volcanogenic materials (e.g., organic-rich clay laminae derived from devitrified volcanic glass, Figure 6c–e), and terrestrial plant inputs (Figure 6f). Additionally, the migration of liquid hydrocarbon along volcanogenic silicate particles (Figure 6c,d) and the occurrence of solid bitumen within clay laminae (Figure 6e) suggest a hybrid sedimentary model contributing to hybrid condensate gas generation in lacustrine shale reservoirs.

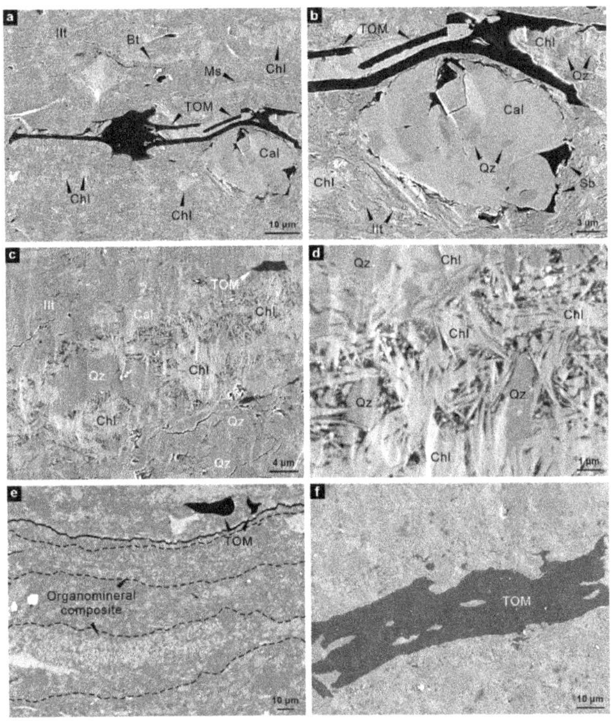


**Figure 6.** SEM photomicrographs of lacustrine shale samples from the Shahezi Formation (JLYY1 borehole, Lishu Rift Depression, Songliao Basin). (**a–d**) Secondary electron (SE) images, 3124.55 m depth. Panels (**b**) and (**d**) are magnified views of selected regions in (**a**) and (**b**), respectively. (**e**) Backscattered electron (BSE) image, 3128.67 m depth. (**f**) SE image, 3130.2 m depth. Images (**c–e**) show the microscopic structure and mineral composition of tuff laminae, consisting predominantly of authigenic chlorite blades, quartz, minor calcite, and pore-filling solid bitumen. These organomineral composites are interpreted as products of the devitrification of volcanic glass [22]. During this process, montmorillonite formed in early stages reacts with Fe oxide in shallow-water and alkaline conditions to produce chlorite, releasing $Ca^{2+}$ and $Si^{4+}$ ions that contribute to the generation of calcite and quartz. Abbreviations: Ilt = illite; Bt = biotite; Ms = muscovite; Chl = chlorite; Cal = calcite; Qz = quartz; Sb = solid bitumen; TOM = terrigenous organic matter.

Overall, paleoenvironmental conditions, such as volcanic materials input, a humid climate, saline lacustrine conditions, and anoxic bottom-water promoted the deposition of shale gas condensate reservoirs, enhancing hydrocarbon content in hybrid shale gas condensate reservoirs.

### 4.4. Shale Lamination and Lamina-Induced Fractures

Lamina-induced fracture systems constitute critical hydrocarbon storage spaces and migration pathways in lacustrine shale reservoirs, with four distinct lithological types identified through an integration of RoqSCAN and FEG-SEM approaches (Figures 6 and 7): Silicate laminae (Figure 7b), predominantly developed in 3080–3165 m clay mineral-rich intervals, characterized by submillimeter quartz and feldspar bands (<200 μm) and accounting for 78% of total hydrocarbon storage space ($S_1 = 1.8 \pm 0.4$ mg/g). These laminae enhance fracture connectivity through increasing fracture density and anisotropic permeability. Clay mineral laminae (Figure 7c) tend to create fractures with apertures of 50–150 μm in shales. Carbonate laminae (Figure 7a) within the same interval exhibit 20–80 μm calcite-filled fractures with 35 ± 7% spatial filling density, supported by crystalline infill textures.

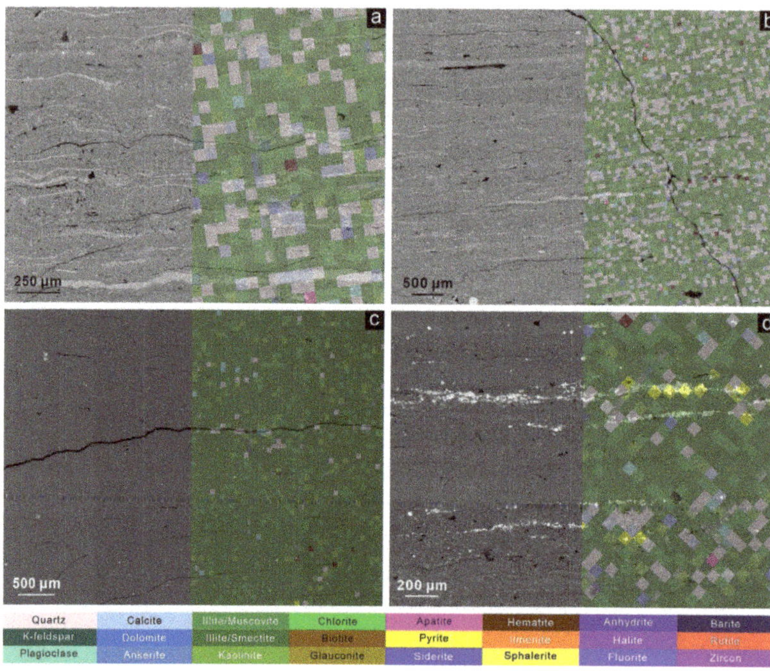


**Figure 7.** RoqSCAN SEM-EDS photomicrographs of lacustrine shale samples from the Shahezi Formation (JLYY1 borehole, Lishu Rift Depression, Songliao Basin), showing four types of laminae: (**a**) 3125.15 m depth, calcareous laminae; (**b**) 3125.15 m depth, non-clay mineral silicate laminae; (**c**) 3155.5 m depth, argillaceous laminae; and (**d**) 3138.4 m depth, pyrite laminae.

Pyrite laminae (Figure 7d) demonstrate a unique Fe-S diagenetic association with organic matter-hosted pore clusters (porosity $\theta = 8 \pm 3\%$), serving as robust indicators of the "sweet spot" through their strong correlation with *TOC* ($R^2 = 0.91$). Their development involves Fe migration along hydrocarbon flow paths and S enrichment in zones rich in solid bitumen (organic matter = 6 ± 2 vol%), particularly where silicate laminae fractures intersect pre-existing clay mineral-rich matrices. Spatial analysis confirms these lamina systems collectively create interconnected hydrocarbon storage networks, with silicate laminae governing bulk storage capacity while pyrite laminae mark high-*TOC* (>2 wt%) preservation zones. The observed lamina and fracture configurations—particularly the staggered propagation of pyrite and silicate laminae in organic-rich intervals—provide diagnostic criteria for identifying hybrid shale gas condensate sweet spots.

## 5. Discussion

An integrated model is proposed to illustrate the hydrocarbon generation and evolution of hybrid shale gas condensate resources, considering the four key factors, rock components and lithofacies, thermal maturity and hydrocarbon phases, paleoenvironmental conditions, and lamina-induced fractures, as shown in Figure 8.

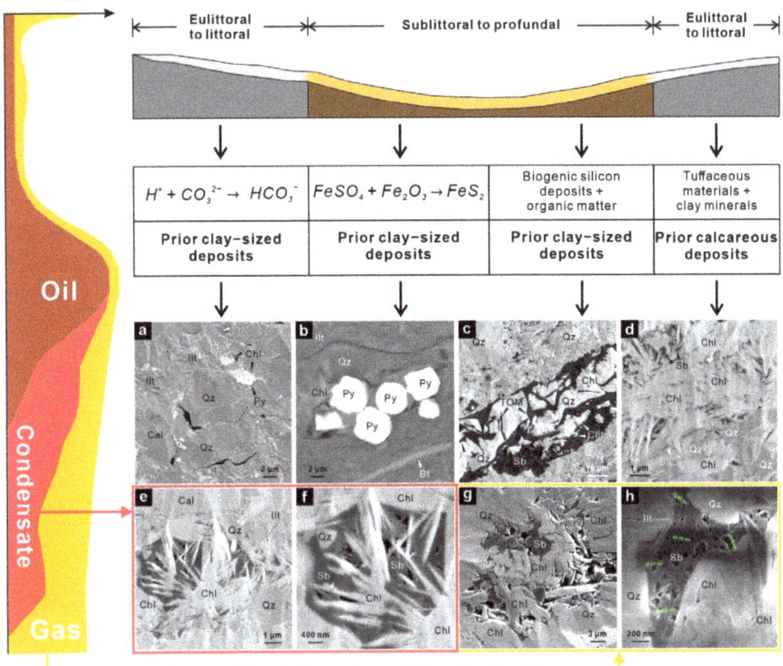


**Figure 8.** A schematic diagram showing the hydrocarbon generation and evolution model of the shale gas condensate reservoir in syn-rift lake basins. Four controlling factors are illustrated in this model: (1) rock components and lithofacies; (2) thermal maturity and hydrocarbon fluid phases; (3) paleoenvironmental conditions; and (4) shale lamination and lamina-induced fractures. (**a**) Second electron (SE) image, 3101.24 m depth. (**b**) Backscattered electron (BSE) image, 3166.52 m depth. (**c**) SE image, 3130.2 m depth. (**d**) SE image, 3124.55 m depth. (**e**,**f**) SE images, 3125.19 m depth. (**g**) SE image, 3122 m depth. (**h**) SE image, 3131.96 m depth. Panel (**f**) is an enlarged view of a region of (**e**).

### *5.1. Comparison of the Hybrid Sedimentation Model Between Hydrocarbon-Rich Shales and Hydrocarbon-Lean Shales*

Volcanism commonly occurs in the syn-rift Songliao Basin, where significant volcanic activity began during the Early Cretaceous due to the subduction of the Indian and Pacific plates beneath the Eurasian plate [34]. This led to the development of grabens, particularly in the central fault block of the Songliao Basin [61]. Based on basin geometric characteristics, volcanism, and early sedimentation, a hybrid sedimentation model is proposed in Li et al. (2022), illustrating the syn-rift lake basins, affected or unaffected by volcanism, respectively [50]. The hydrocarbon-rich shales occur between 3115 m and 3165 m, while the hydrocarbon-lean section is between 3080 m and 3115 m.

During the syn-rift period, volcanic activity raised temperatures, increased atmospheric $CO_2$, and caused mass mortality of organisms [22,62]. This created a unique environment for hydrocarbon-rich shales, including a quiescent water setting with volcanic inputs, a humid climate, saline lakes, and anoxic bottom-water conditions in the syn-rift lake. Both oxic and anoxic water layers contributed to hybrid organic-rich deposits [63], which contained terrigenous salt-tolerant plants, organic matter, and planktonic aerobic organisms [50]. These organic-rich deposits in the Shahezi Formation, primarily from oil-prone fractions of terrigenous organic matter [22,33], mixed with minerals and formed organic matter laminae, indicate a stable and anoxic water environment. Anaerobic bacteria helped degrade organic matter and form authigenic minerals. Dispersed terrigenous organic matter is deposited with detrital or authigenic quartz, generating the silicate- and organic-rich laminae possibly derived from the devitrification of volcanic glass in tuff [22].

The hydrocarbon-rich shales from 3115 m to 3165 m, which have higher hydrocarbon content than those from 3080 m to 3115 m (Figure 2), are key to defining "hybrid shale gas condensate". The gas and liquid hydrocarbon states are influenced not only by in situ conditions like temperature and pressure [64–66] but also by the input of terrestrial-sourced organic matter, which contributes to the production of gas and liquid hydrocarbons in the early stages of hydrocarbon evolution [67].

As the basin evolved, volcanic activity decreased, and secondary faults formed, leading to shallow water depth, a relatively cold and arid climate, and low salinity water conditions under weak evaporation conditions. The lake environment became more stable, with an oxic layer on top and a dysoxic layer below. In this stable period, aquatic organic matter preservation was difficult due to the weak reducing environment [68]. Instead, terrigenous organic matter sourced from land plants contributed to the enrichment of Type III kerogen, which performs relatively poorly in hydrocarbon generation. This explains the lower hydrocarbon content in the shales from 3080 m to 3115 m.

### 5.2. The Organic–Mineral Interactions During the Diagenesis of Hydrocarbon-Rich Shales

The hydrocarbon content of hybrid shale gas condensates is influenced by organic–mineral interactions during diagenesis (Figure 8). Here, organic–mineral interactions are examined to explain hydrocarbon storage and desorption, using SEM to describe the relationship between organic matter, mineral particles, and pores (Figure 8). The test samples are horizontally stratified and finely laminated mudstones.

Organic matter, primarily composed of solid bitumen, results from kerogen evolution and often intergrows with authigenic pyrite, calcite, and non-clay mineral silicate minerals (Figure 8). In calcareous laminae (Figure 8a), calcification occurs in the presence of prior clay-sized deposits, where $HCO_3^-$ combines with $Ca^{2+}$ and $Mg^{2+}$ ions to form calcareous diagenetic minerals in slightly alkaline, oxidizing conditions [69]. Intergranular pores of calcite particles contribute to solid bitumen storage (Figure 8a), where hydrocarbons are mainly stored in the intergranular pores of clay minerals and calcareous particles, as well as in organic matter-hosted pores.

Silicate laminae are often associated with pyrite laminae (Figures 7d and 8b) due to the symbiotic relationship between organic matter and pyrite. Pyrite can either separate from organic matter or become embedded in solid bitumen, showing flow traces (Figures 7d and 8b). Pyrite is typically surrounded by detrital quartz particles and authigenic quartz overgrowth, which are possibly generated during the devitrification of volcanic glass in tuff (Figure 7d). Pyrite forms in an anoxic bottom-water environment during early diagenesis [70].

An interesting phenomenon is seen in Figure 8e,f, where the clay mineral assemblages are associated with calcite particles within micro-scale organomineral composites. These formations resemble structures that likely result from the devitrification of volcanic glass into chlorite and illite. Although it remains uncertain whether this process directly influences organic matter evolution, it is clear that the combination of volcanic glass, organic matter, and calcite in a special diagenetic setting contributes to hydrocarbon-rich shales.

The most common mineral intergrowth with organic matter is authigenic quartz, followed by pyrite, illite, and carbonate minerals [71]. The complex microstructures formed by organic matter and authigenic minerals are similar to those found in intrusive igneous rock crystallization processes. This suggests that organic matter and authigenic minerals may have the same origin and evolutionary process according to their diagenetic pathways. The evolution of parent-rock material into solid bitumen and the crystallization of minerals hints at chemical differentiation between organic and inorganic components.

### 5.3. *The Liquid Hydrocarbon and Solid Bitumen of the Hybrid Shale Gas Condensate Reservoirs*

The organic matter in the cored shales of this study has evolved into condensate and wet gas stages, as discussed in Section 4.2. Hybrid shale gas condensate in mature and highly mature shale is the dominant fluid phase and contributes significantly to the in situ hydrocarbon content. Although the light hydrocarbons in gas or liquid states have been lost from the samples due to drilling and coring, the space left by hydrocarbon desorption, exudation, and diffusion remains detectable under SEM.

SEM images are used to indirectly illustrate the hydrocarbon state. The organic matter in the SEM photos is classified into three types: highly mature organic matter with a vesicular structure, mature organic matter with liquid mold holes and vesicular structures, and solid bitumen with no movable hydrocarbon traces (Figure 8e–h). Authigenic quartz and solid bitumen often coexist, with solid bitumen spherulites and pores present. In the SEM images, different pore types are visible, mainly formed by gas generation, accumulation, and escape of hydrocarbons from kerogen. These pores are typically round, with apertures ranging from tens to hundreds of nanometers, varying in size, depth, and shape. Solid bitumen development is particularly noticeable in intergranular pores of authigenic minerals, which protect the organic matter and allow hydrocarbons to generate and accumulate. The light hydrocarbons around these pores, along with the solid bitumen formed during hydrocarbon generation, as shown in Figure 8c–h, strongly support the accumulation of hybrid shale gas condensates.

## 6. Conclusions

This study establishes a multiscale framework for evaluating hybrid shale gas condensate reservoirs in rift lake basins affected by volcanism, integrating geochemical, mineralogical, and sedimentological analyses of the Shahezi Formation (Songliao Basin). By combining novel proxies (e.g., *HSCI* index) with paleoenvironmental diagnostics, a predictive model was proposed for lacustrine shale systems to distinguish them from marine analogs. Key findings include the following:

1. Argillaceous shales (50–75% clay minerals) containing mixed carbonate and clay minerals (MC(C) > 25%) and organic matter ($TOC > 2$ wt%) exhibit high potential of hydrocarbon storage, achieving $S_1 > 1.5 \pm 0.3$ mg/g ($C_7$–$C_{33}$) and $TGC > 1.5 \pm 0.2$ m$^3$/t ($C_1$–$C_2$). These intervals develop intergranular porosity of 4.8 ± 1.2%, which is higher than illite-dominated zones.
2. Six fluid phases of hydrocarbons were identified by the *HSCI* index, showing that the mature ($T_{max}$ = 460–480 °C) and highly mature ($T_{max}$ = 480–500 °C) shales are dominated by gas condensates. The production of gaseous and liquid hydrocarbon peaks at $T_{max}$ = 490 °C (highest $S_1$ and $TGC$ values), which corresponds to intervals with enhanced preservation of MC(C).

3. Volcanic material inputs ($Zr/Ti > 40$), humid paleoclimate ($Al/Ti > 10$), saline lake water ($Sr/Ba > 0.5$ and $Ca/(Ca + Fe) > 0.3$), and anoxic bottom-water conditions ($U/Th > 0.4$) promote hybrid silicate and pyrite laminations, elevating *TOC* to $5.2 \pm 0.8$ wt%, which is higher than intervals unaffected by volcanism.
4. Silicate laminae ($S_1 = 1.8 \pm 0.4$ mg/g) and pyrite laminae (porosity $\theta = 8\% \pm 3\%$) jointly form heterogenic seepage networks, promoting hydrocarbon generation in low-porosity shales (porosity $\theta = 1.2$–$4.8\%$). These systems differ from their marine analogs due to significant organic–mineral interactions.

Overall, these insights redefine "sweet spot" criteria for hybrid shale gas condensate reservoirs in rift lake basins, emphasizing volcanic–diagenetic pathways and lamina architecture as key discriminators from marine systems.

**Author Contributions:** Conceptualization, Y.L.; methodology, Y.L.; validation, Y.L., Y.X., Y.Y., L.T. and Y.T.; formal analysis, Y.L. and Q.W.; investigation, Y.L. and Q.W.; resources, Y.X., Y.Y., L.T., Y.T. and Q.W.; writing—original draft preparation, Y.L. and Q.W.; writing—review and editing, Y.L., C.F. and Q.W.; visualization, Y.L. and Q.W.; supervision, C.B. and C.F.; project administration, C.B. and C.F.; funding acquisition, Y.L., C.B. and C.F. All authors have read and agreed to the published version of the manuscript.

**Funding:** This research was financially supported by the National Natural Science Foundation of China (grant numbers U2244207 and 42202179), State Key Laboratory of Shale Oil and Gas Enrichment Mechanisms and Effective Development, State Key Laboratory of Petroleum Resources and Prospecting, China University of Petroleum (grant number PRP/open-2217), and China Geological Survey (grant numbers DD20190115 and DD20240201002).

**Data Availability Statement:** The data presented in this study are openly available in Mendeley Data: http://dx.doi.org/10.17632/gtgjcpyzcn.1 (accessed on 3 June 2025).

**Acknowledgments:** The authors would like to thank the Scanning Electron Microscopy Shared Research Facility (SEM-SRF) at the University of Liverpool for the ultra-high-resolution SEM-EDS equipment.

**Conflicts of Interest:** Author Chao Fu was employed by the Xinjiang Yaxin Coalbed Methane Investment and Development (Group) Co., Ltd. The remaining authors declare that the research was conducted in the absence of any commercial or financial relationships that could be construed as a potential conflict of interest.

## Abbreviations

The following abbreviations are used in this manuscript:

| | |
|---|---|
| TOC | Total organic carbon content (wt%) |
| TGC | Total gas content ($m^3/t$) |
| HSCI | Hybrid Shale Condensate Index |
| GOR | Gas-to-oil production ratio (Scf/STB) |
| API | American Petroleum Institute gravity (°API) |

## References

1. Speight, J. Chapter 7—Analysis of gas and condensate from tight formations. In *Shale Oil and Gas Production Processes*; Speight, J., Ed.; Gulf Professional Publishing: Houston, TX, USA, 2020; pp. 373–450.
2. Hassan, A.; Mahmoud, M.; Al-Majed, A.; Alawi, M.B.; Elkatatny, S.; BaTaweel, M.; Al-Nakhli, A. Gas condensate treatment: A critical review of materials, methods, field applications, and new solutions. *J. Pet. Sci. Eng.* **2019**, *177*, 602–613. [CrossRef]
3. Hassan, A.M.; Mahmoud, M.A.; Al-Majed, A.A.; Al-Shehri, D.; Al-Nakhli, A.R.; Bataweel, M.A. Gas production from gas condensate reservoirs using sustainable environmentally friendly chemicals. *Sustainability* **2019**, *11*, 2838. [CrossRef]
4. El Diasty, W.S.; Peters, K.E.; Moldowan, J.M.; Essa, G.I.; Hammad, M.M. Organic geochemistry of condensates and natural gases in the northwest Nile Delta offshore Egypt. *J. Pet. Sci. Eng.* **2020**, *187*, 106819. [CrossRef]

5. Clarkson, C.R.; Williams-Kovacs, J.D.; Qanbari, F.; Behmanesh, H.; Heidari Sureshjani, M. History-matching and forecasting tight/shale gas condensate wells using combined analytical, semi-analytical, and empirical methods. *J. Nat. Gas Sci. Eng.* **2015**, *26*, 1620–1647. [CrossRef]
6. Sheng, J.J.; Mody, F.; Griffith, P.J.; Barnes, W.N. Potential to increase condensate oil production by huff-n-puff gas injection in a shale condensate reservoir. *J. Nat. Gas Sci. Eng.* **2016**, *28*, 46–51. [CrossRef]
7. Wang, X.-W.; Xiao, P.; Yang, Z.-M.; Liu, X.-W.; Xia, Z.-Z.; Wang, L.-Q. Laboratory and field-scale parameter optimization of $CO_2$ huff–n–puff with the staged-fracturing horizontal well in tight oil reservoirs. *J. Pet. Sci. Eng.* **2020**, *186*, 106703.
8. Cardott, B.J. Thermal maturity of Woodford Shale gas and oil plays, Oklahoma, USA. *Int. J. Coal Geol.* **2012**, *103*, 109–119. [CrossRef]
9. Gherabati, S.A.; Browning, J.; Male, F.; Ikonnikova, S.A.; McDaid, G. The impact of pressure and fluid property variation on well performance of liquid-rich Eagle Ford shale. *J. Nat. Gas Sci. Eng.* **2016**, *33*, 1056–1068. [CrossRef]
10. Şen, Ş.; Kozlu, H. Impact of maturity on producible shale oil volumes in the Silurian (Llandovery) hot shales of the northern Arabian plate, southeastern Turkey. *AAPG Bull.* **2020**, *104*, 507–524. [CrossRef]
11. Strąpoć, D.; Jacquet, B.; Torres, O.; Khan, S.; Villegas, E.I.; Albrecht, H.; Okoh, B.; McKinney, D. Deep biogenic methane and drilling-associated gas artifacts: Influence on gas-based characterization of petroleum fluids. *AAPG Bull.* **2020**, *104*, 887–912. [CrossRef]
12. Chalmers, G.R.; Bustin, R.M.; Power, I.M. Characterization of gas shale pore systems by porosimetry, pycnometry, surface area, and field emission scanning electron microscopy/transmission electron microscopy image analyses: Examples from the Barnett, Woodford, Haynesville, Marcellus, and Doig units. *AAPG Bull.* **2012**, *96*, 1099–1119.
13. Merkel, A.; Fink, R.; Littke, R. The role of pre-adsorbed water on methane sorption capacity of Bossier and Haynesville shales. *Int. J. Coal Geol.* **2015**, *147–148*, 1–8. [CrossRef]
14. Gou, Q.; Xu, S. The controls of laminae on lacustrine shale oil content in China: A review from generation, retention, and storage. *Energies* **2023**, *16*, 1987. [CrossRef]
15. Chen, G.; Tang, Y.; Nan, Y.; Yang, F.; Wang, D. Paleo-sedimentary environments and controlling factors for enrichment of organic matter in alkaline lake sediments: A case study of the Lower Permian Fengcheng Formation in Well F7 at the western slope of Mahu Sag, Junggar Basin. *Processes* **2023**, *11*, 2483. [CrossRef]
16. Xiao, H.; Hu, T.; Pang, X.; Ding, C.; Xu, Y.; Zhang, S.; Hu, Y.; Li, C.; Xu, T.; Zheng, D.; et al. A novel method for identifying oil content and moveable thresholds in heterogeneous shales. *Fuel* **2025**, *397*, 135473. [CrossRef]
17. Clarkson, C.R.; Haghshenas, B.; Ghanizadeh, A.; Qanbari, F.; Williams-Kovacs, J.D.; Riazi, N.; Debuhr, C.; Deglint, H.J. Nanopores to megafractures: Current challenges and methods for shale gas reservoir and hydraulic fracture characterization. *J. Nat. Gas Sci. Eng.* **2016**, *31*, 612–657. [CrossRef]
18. Middleton, R.S.; Carey, J.W.; Currier, R.P.; Hyman, J.D.; Kang, Q.; Karra, S.; Jiménez-Martínez, J.; Porter, M.L.; Viswanathan, H.S. Shale gas and non-aqueous fracturing fluids: Opportunities and challenges for supercritical $CO_2$. *Appl. Energy* **2015**, *147*, 500–509. [CrossRef]
19. Zhang, C.P.; Cheng, P.; Ranjith, P.G.; Lu, Y.Y.; Zhou, J.P. A comparative study of fracture surface roughness and flow characteristics between $CO_2$ and water fracturing. *J. Nat. Gas Sci. Eng.* **2020**, *76*, 103188. [CrossRef]
20. Jiang, J.; Younis, R.M. Compositional modeling of enhanced hydrocarbons recovery for fractured shale gas-condensate reservoirs with the effects of capillary pressure and multicomponent mechanisms. *J. Nat. Gas Sci. Eng.* **2016**, *34*, 1262–1275. [CrossRef]
21. Sheng, J.J. Increase liquid oil production by huff-n-puff of produced gas in shale gas condensate reservoirs. *J. Unconv. Oil Gas Resour.* **2015**, *11*, 19–26. [CrossRef]
22. Wang, Q.; Li, Y.; Sanei, H.; Gardner, J.; Rudra, A.; Utley, J.E.P.; Worden, R.H. Lacustrine and fan-delta sediments in syn-rift lake basins. *Depos. Rec.* **2025**, *11*. [CrossRef]
23. Yang, W.; Wang, Q.; Wang, Y.; Jiang, Z.; Song, Y.; Li, Y.; Liu, D.; Zuo, R.; Gu, X.; Zhang, F. Pore characteristic responses to categories of depositional microfacies of delta-lacustrine tight reservoirs in the Upper Triassic Yanchang Formation, Ordos Basin, NW China. *Mar. Pet. Geol.* **2020**, *118*, 104423. [CrossRef]
24. Wang, Q.; Li, Y.; Yang, W.; Jiang, Z.; Song, Y.; Jiang, S.; Luo, Q.; Liu, D. Finite element simulation of multi-scale bedding fractures in tight sandstone oil reservoir. *Energies* **2020**, *13*, 131. [CrossRef]
25. Li, Y.; Song, Y.; Jiang, Z.; Yin, L.; Chen, M.; Liu, D. Major factors controlling lamina induced fractures in the Upper Triassic Yanchang formation tight oil reservoir, Ordos basin, China. *J. Asian Earth Sci.* **2018**, *166*, 107–119. [CrossRef]
26. Li, Y.; Wang, Q.; Xu, X.; Jiang, S.; Zhang, F. Permeability evolution of the lamina induced fractures (LFs) during the triaxial compression rupture phase. *J. Pet. Sci. Eng.* **2020**, *188*, 106870. [CrossRef]
27. Liu, D.; Li, Y.; Luo, Q.; Wang, Q.; Meng, X.; Ge, Y.; Luo, J.; Feng, H.; Li, Y.; Zhang, X. Dynamic mechanisms of lamina-induced fracture propagation in tight oil-sand formations. *Interpretation* **2019**, *7*, T179–T193. [CrossRef]

28. Li, Y.; Song, Y.; Jiang, S.; Jiang, Z.; Yang, W.; Wang, Q.; Liu, D. Tight reservoir oiliness numerical simulation based on a Markov chain Monte Carlo (MCMC) method: A case study of the upper Triassic Yanchang-6 formation (T3ch6 Fm.) outcrop of Ordos Basin. *J. Pet. Sci. Eng.* **2019**, *175*, 1150–1159. [CrossRef]
29. Li, Y.; Song, Y.; Jiang, S.; Wang, Q.; Jiang, Z.; Zhang, F. Influence of gas and oil state on oil mobility and sweet-spot distribution in tight oil reservoirs from the perspective of capillary force. *SPE Reserv. Eval. Eng.* **2019**, *23*, 824–842. [CrossRef]
30. Li, Y.; Song, Y.; Jiang, Z.; Yin, L.; Luo, Q.; Ge, Y.; Liu, D. Two episodes of structural fractures: Numerical simulation of Yanchang Oilfield in the Ordos basin, northern China. *Mar. Pet. Geol.* **2018**, *97*, 223–240. [CrossRef]
31. Wang, Q.; Yang, W.; Li, Y.; Jiang, Z.; Wen, M.; Zuo, R.; Wang, X.; Xue, Z.; Wang, Y. In-situ fluid phase variation along the thermal maturation gradient in shale petroleum systems and its impact on well production performance. *J. Earth Sci.* **2023**, *34*, 985–1001. [CrossRef]
32. Li, P.; Hao, F.; Zhang, B.; Zou, H.; Yu, X.; Wang, G. Heterogeneous distribution of pyrobitumen attributable to oil cracking and its effect on carbonate reservoirs: Feixianguan Formation in the Jiannan gas field, China. *AAPG Bull.* **2015**, *99*, 763–789. [CrossRef]
33. Wang, Q.; Li, Y.; Utley, J.E.P.; Gardner, J.; Liu, B.; Hu, J.; Shao, L.; Wang, X.; Gao, F.; Liu, D.; et al. Terrestrial dominance of organic carbon in an early cretaceous syn-rift lake and its correlation with depositional sequences and paleoclimate. *Sediment. Geol.* **2023**, *455*, 106472. [CrossRef]
34. Wang, P.-J.; Mattern, F.; Didenko, N.A.; Zhu, D.-F.; Singer, B.; Sun, X.-M. Tectonics and cycle system of the Cretaceous Songliao Basin: An inverted active continental margin basin. *Earth Sci. Rev.* **2016**, *159*, 82–102. [CrossRef]
35. Espitalié, J.; Marquis, F.; Barsony, I. 9—Geochemical Logging. In *Analytical Pyrolysis*; Voorhees, K.J., Ed.; Butterworth-Heinemann: Oxford, UK, 1984; pp. 276–304.
36. Wang, Q.; Li, Y.; Sanei, H.; Rudra, A.; Yuan, M.; Wang, Y.; Huang, Y.; Worden, R.H. Restoration of buried organic carbon for catagenesis-affected rocks using Rock-Eval thermal analysis: Assumptions, performance, and uncertainty analysis. *Earth Sci. Rev.* **2025**, *267*, 105155. [CrossRef]
37. Ashton, T.; Ly, C.V.; Spence, G.; Oliver, G. Portable technology puts real-time automated mineralogy on the well site. In Proceedings of the SPE Unconventional Resources Conference and Exhibition-Asia Pacific, Brisbane, Australia, 11–13 November 2013.
38. Oliver, G.M.; Ly, C.V.; Spence, G.; Rael, H. A new approach to measuring rock properties data from cores & cuttings for reservoir & completions characterization: An example from the Bakken Formation. In Proceedings of the SPE Unconventional Resources Conference and Exhibition-Asia Pacific, Brisbane, Australia, 11–13 November 2013.
39. Parian, M.; Lamberg, P.; Möckel, R.; Rosenkranz, J. Analysis of mineral grades for geometallurgy: Combined element-to-mineral conversion and quantitative X-ray diffraction. *Miner. Eng.* **2015**, *82*, 25–35. [CrossRef]
40. Shi, X.; Jiang, S.; Lu, S.; He, Z.; Li, D.; Wang, Z.; Xiao, D. Investigation of mechanical properties of bedded shale by nanoindentation tests: A case study on Lower Silurian Longmaxi Formation of Youyang area in southeast Chongqing, China. *Pet. Explor. Dev.* **2019**, *46*, 163–172. [CrossRef]
41. Zhai, G.; Li, J.; Jiao, Y.; Wang, Y.; Liu, G.; Xu, Q.; Wang, C.; Chen, R.; Guo, X. Applications of chemostratigraphy in a characterization of shale gas Sedimentary Microfacies and predictions of sweet spots—Taking the Cambrian black shales in Western Hubei as an example. *Mar. Pet. Geol.* **2019**, *109*, 547–560. [CrossRef]
42. Johnson, C.; Pownceby, M.I.; Wilson, N.C. The application of automated electron beam mapping techniques to the characterisation of low grade, fine-grained mineralisation; potential problems and recommendations. *Miner. Eng.* **2015**, *79*, 68–83. [CrossRef]
43. Li, G.; Li, G.; Wang, Y.; Qi, S.; Yang, J. A rock physics model for estimating elastic properties of upper Ordovician-lower Silurian mudrocks in the Sichuan Basin, China. *Eng. Geol.* **2020**, *266*, 105460. [CrossRef]
44. Arouri, K.R.; Jenden, P.D.; Al-Hajji, A.A. Petroleum inclusions atop Unayzah gas condensate reservoir: Signpost for an undocumented chapter of the Arabian Basin filling history? *Org. Geochem.* **2010**, *41*, 698–705. [CrossRef]
45. Cesar, J.; Eiler, J.; Dallas, B.; Chimiak, L.; Grice, K. Isotope heterogeneity in ethyltoluenes from Australian condensates, and their stable carbon site-specific isotope analysis. *Org. Geochem.* **2019**, *135*, 32–37. [CrossRef]
46. Huang, S.; Liu, D.; Wang, Z.; Feng, Z.; Huang, T. Genetic origin of gas condensate in Permian and Triassic strata in the southern Sichuan Basin, SW China. *Org. Geochem.* **2015**, *85*, 54–65. [CrossRef]
47. Huang, S.; Wang, Z.; Lv, Z.; Gong, D.; Yu, C.; Wu, W. Geochemical identification of marine and terrigenous condensates—A case study from the Sichuan Basin, SW China. *Org. Geochem.* **2014**, *74*, 44–58. [CrossRef]
48. Mei, M.; Bissada, K.K.; Malloy, T.B.; Darnell, L.M.; Liu, Z. Origin of condensates and natural gases in the Almond Formation reservoirs in southwestern Wyoming, USA. *Org. Geochem.* **2018**, *124*, 164–179. [CrossRef]
49. Milkov, A.V.; Faiz, M.; Etiope, G. Geochemistry of shale gases from around the world: Composition, origins, isotope reversals and rollovers, and implications for the exploration of shale plays. *Org. Geochem.* **2020**, *143*, 103997. [CrossRef]
50. Li, Y.; Xu, X.; Zhang, J.; Chen, S.; Bai, J.; Liu, W.; Wang, Q. Hybrid sedimentary conditions of organic-rich shales in faulted lacustrine basin during volcanic eruption episode: A case study of Shahezi Formation (K1sh Fm.), Lishu Faulted Depression, south Songliao Basin. *Earth Sci.* **2022**, *47*, 1728–1747. (In Chinese with English Abstract)

51. Hara, H.; Kurihara, T.; Kuroda, J.; Adachi, Y.; Kurita, H.; Wakita, K.; Hisada, K.-I.; Charusiri, P.; Charoentitirat, T.; Chaodumrong, P. Geological and geochemical aspects of a Devonian siliceous succession in northern Thailand: Implications for the opening of the Paleo-Tethys. *Palaeogeogr. Palaeoclimatol. Palaeoecol.* **2010**, *297*, 452–464. [CrossRef]
52. Hu, C.; Zhang, Y.; Tian, J.; Wang, W.; Han, C.; Wang, H.; Li, X.; Feng, S.; Han, C.; Algeo, T.J. Influence of paleo-Trade Winds on facies patterns of the Cambrian Shanganning Carbonate Platform, North China. *Palaeogeogr. Palaeoclimatol. Palaeoecol.* **2020**, *552*, 109556. [CrossRef]
53. Li, Y.; Yang, W.; Wang, Q.; Song, Y.; Jiang, Z.; Guo, L.; Zhang, Y.; Wang, J. Influence of the actively migrated diagenetic elements on the hydrocarbon generation potential in tuffaceous shale. *Fuel* **2019**, *256*, 115795. [CrossRef]
54. Schenk, B.; Gebhardt, H.; Wolfgring, E.; Zorn, I. Cyclic paleo-salinity changes inferred from benthic foraminiferal assemblages in the Upper Burdigalian (Lower Miocene) Korneuburg Basin, Austria. *Palaeogeogr. Palaeoclimatol. Palaeoecol.* **2018**, *490*, 473–487. [CrossRef]
55. Toyoda, K. Geochemical history of ancient Lake Biwa in Japan—Chemical indicators of sedimentary paleo-environments in a drilled core. *Palaeogeogr. Palaeoclimatol. Palaeoecol.* **1993**, *101*, 169–184. [CrossRef]
56. Wu, Y.; Tian, H.; Gong, D.; Li, T.; Zhou, Q. Paleo-environmental variation and its control on organic matter enrichment of black shales from shallow shelf to slope regions on the Upper Yangtze Platform during Cambrian Stage 3. *Palaeogeogr. Palaeoclimatol. Palaeoecol.* **2020**, *545*, 109653. [CrossRef]
57. Li, W.; Cao, J.; Shi, C.; Xu, T.; Zhang, H.; Zhang, Y. Shale oil in saline lacustrine systems: A perspective of complex lithologies of fine-grained rocks. *Mar. Pet. Geol.* **2020**, *116*, 104351. [CrossRef]
58. Algeo, T.J.; Liu, J. A re-assessment of elemental proxies for paleoredox analysis. *Chem. Geol.* **2020**, *540*, 119549. [CrossRef]
59. Breit, G.N.; Wanty, R.B. Vanadium accumulation in carbonaceous rocks: A review of geochemical controls during deposition and diagenesis. *Chem. Geol.* **1991**, *91*, 83–97. [CrossRef]
60. Tribovillard, N.; Algeo, T.J.; Lyons, T.; Riboulleau, A. Trace metals as paleoredox and paleoproductivity proxies: An update. *Chem. Geol.* **2006**, *232*, 12–32. [CrossRef]
61. Ji, Z.; Meng, Q.; Wan, C.; Ge, W.; Yang, H.; Zhang, Y.; Dong, Y.; Jin, X. Early Cretaceous adakitic lavas and A-type rhyolites in the Songliao Basin, NE China: Implications for the mechanism of lithospheric extension. *Gondwana Res.* **2019**, *71*, 28–48. [CrossRef]
62. Zhang, L.; Wang, C.; Wignall, P.B.; Kluge, T.; Wan, X.; Wang, Q.; Gao, Y. Deccan volcanism caused coupled $pCO_2$ and terrestrial temperature rises, and pre-impact extinctions in northern China. *Geology* **2018**, *46*, 271–274. [CrossRef]
63. Zou, C.; Zhu, R.; Chen, Z.-Q.; Ogg, J.G.; Wu, S.; Dong, D.; Qiu, Z.; Wang, Y.; Wang, L.; Lin, S.; et al. Organic-matter-rich shales of China. *Earth Sci. Rev.* **2019**, *189*, 51–78. [CrossRef]
64. Lerch, B.; Karlsen, D.A.; Abay, T.B.; Duggan, D.; Seland, R.; Backer-Owe, K. Regional petroleum alteration trends in Barents Sea oils and condensates as a clue to migration regimes and processes. *AAPG Bull.* **2016**, *100*, 165–190. [CrossRef]
65. Liu, Y.; Qiu, N.; Hu, W.; Li, H.; Shen, F.; Yao, Q. Temperature and pressure characteristics of Ordovician gas condensate reservoirs in the Tazhong area, Tarim Basin, northwestern China. *AAPG Bull.* **2019**, *103*, 1351–1381. [CrossRef]
66. Su, J.; Zhang, S.; Huang, H.; Wang, Y.; Wang, H.; He, K.; Wang, X.; Zhang, B.; Wang, H. New insights into the formation mechanism of high hydrogen sulfide–bearing gas condensates: Case study of Lower Ordovician dolomite reservoirs in the Tazhong uplift, Tarim Basin. *AAPG Bull.* **2016**, *100*, 893–916. [CrossRef]
67. Huang, R.; Li, J.; Xie, Z.; Li, J. Formation and distribution of condensate gas pools in China. *Oil Gas Geol.* **1996**, *17*, 237–242. (In Chinese with English Abstract)
68. Varela, S.; Lobo, J.M.; Hortal, J. Using species distribution models in paleobiogeography: A matter of data, predictors and concepts. *Palaeogeogr. Palaeoclimatol. Palaeoecol.* **2011**, *310*, 451–463. [CrossRef]
69. Cao, J.; Xia, L.; Wang, T.; Zhi, D.; Tang, Y.; Li, W. An alkaline lake in the Late Paleozoic Ice Age (LPIA): A review and new insights into paleoenvironment and petroleum geology. *Earth Sci. Rev.* **2020**, *202*, 103091. [CrossRef]
70. Kolker, A. Minor element distribution in iron disulfides in coal: A geochemical review. *Int. J. Coal Geol.* **2012**, *94*, 32–43. [CrossRef]
71. Kwiecińska, B.; Pusz, S.; Valentine, B.J. Application of electron microscopy TEM and SEM for analysis of coals, organic-rich shales and carbonaceous matter. *Int. J. Coal Geol.* **2019**, *211*, 103203. [CrossRef]

processes

MDPI

Article

# Three-Dimensional Geomechanical Modeling and Hydraulic Fracturing Parameter Optimization for Deep Coalbed Methane Reservoirs: A Case Study of the Daniudi Gas Field, Ordos Basin

**Xugang Liu [1,2], Xiang Wang [1,2], Fuhu Chen [1,2], Xinchun Zhu [1,2], Zheng Mao [1,2,*], Xinyu Liu [3] and He Ma [3,*]**

1 Research Institute of Petroleum Engineering and Technology, North China Oil & Gas Company, Sinopec, Zhengzhou 450006, China; liuxugang.hbsj@sinopec.com (X.L.); wangxiang.hbsj@sinopec.com (X.W.); chenfh.hbsj@sinopec.com (F.C.); zhuxchb.hbsj@sinopec.com (X.Z.)

2 Key Laboratory of Deep Coalbed Methane Exploration and Development, Inopec, Zhengzhou 450006, China

3 College of Petroleum Engineering, China University of Petroleum, Beijing 102249, China; 13659901447@163.com

* Correspondence: mzpds6@126.com (Z.M.); mahe010915@163.com (H.M.)

**Abstract:** Deep coalbed methane (CBM) resources represent a significant opportunity for future exploration and development. The combination of horizontal well drilling and hydraulic fracturing technology has emerged as the most efficient method for extracting deep CBM. By optimizing the fracturing parameters for horizontal wells, we can improve the effectiveness of reservoir stimulation even further. In this paper, taking the deep coalbed methane in the Daniudi gas field in the Ordos Basin as the research object, using Numerical simulation software such as Petrel, comprehensively considering the field logging, logging data and laboratory experimental data of rock mechanical parameters, the three-dimensional geomechanical and stress field model of deep coalbed methane is established, and on this basis, the numerical simulation research on the fracture network expansion and construction parameter optimization of single well and well group is carried out. Through the qualitative evaluation of fracture network morphology, the change of in situ stress field, the quantitative evaluation of post-pressure conductivity and fracture volume, the section spacing, construction fluid volume, and construction displacement under the conditions of single well and well group were optimized. The results show that under the condition of a certain well spacing, the fracture propagation of the well group is affected by stress shadowing and channeling, and the fracture pattern is more complex, and the construction scale is smaller than that of a single well. These findings provide critical insights for improving the efficiency of deep CBM recovery.

**Keywords:** deep coalbed methane (CBM); 3D geomechanics; geological modeling; hydraulic fracturing; parameter optimization

## 1. Introduction

With advancements in exploration technologies, deep subsurface energy resources have garnered increasing attention [1,2]. As a critical component of unconventional energy, China's deep CBM exploration has achieved significant breakthroughs [3]. Estimates suggest that CBM resources at depths exceeding 2000 m exceed $40.7 \times 10^{12}$ $m^3$ [4,5]. Deep CBM reservoirs are characterized by high in situ stress, formation pressure, and temperature ("triple-high" conditions) [6]. Challenges such as low permeability, structural complexity, and substantial heterogeneity hinder effective stimulation and produce low well productivity [7]. Horizontal wells with hydraulic fracturing aim to create complex fracture networks, yet optimizing fracture propagation in deep CBM reservoirs remains understudied.

CBM, a self-sourced and self-reserved unconventional natural gas adsorbed in coal seams, exhibits significant differences in genetic and development mechanisms compared to conventional oil and gas reservoirs [8]. This results in distinct variations in attribute modeling between CBM geological and conventional reservoir modeling [9]. Scholars have recently attempted to characterize coal reservoirs using geological modeling techniques quantitatively. Huai et al. achieved differentiation between coal seams and non-coal attributes through facies-controlled 3D modeling [10]. Ma et al. established permeability models by correlating coal permeability with stress and coal lithology [11]. Chen et al. constructed 3D coal structure models using single-well coal structure characterization [12]. Liu investigated pressure drop characteristics during CBM drainage using geological modeling [13]. Li et al. quantitatively analyzed the impact of coal structure on productivity based on coal structure models [14]. Lü developed multi-layer 3D models through characterization of coal seam roof and floor surfaces [15]. Zhou et al. established gas content models using relationships between logging parameters (depth, density, natural gamma, and resistivity) and gas content [16].

Numerous studies have been conducted on numerical simulation and parameter optimization of hydraulic fracture propagation in deep CBM reservoirs. Ren et al. utilized CT scanning and triaxial hydraulic fracturing experiments to investigate hydraulic fracture initiation and propagation in bedded deep coal seams, revealing that bedding planes with lower fracture energy significantly influence fracture behaviour [4]. Wang et al. conducted supercritical $CO_2$ fracturing experiments demonstrating its effectiveness in enhancing fracture complexity [17]. A Qi et al. proposed a volume fracturing technology for Shenfu deep CBM reservoirs in Ordos Basin, featuring "multi-cluster limited-entry perforation, composite fluid fracturing (high-viscosity fluid for rock breaking + low-viscosity fluid for complex fractures), and multi-size proppant combinations" [18]. Li et al. optimized horizontal well length and completion parameters through numerical simulation based on 79 production wells in Shenfu Block [19]. Yang et al. developed large-scale extreme-volume fracturing technology addressing challenges in deep coal seams, such as high plasticity and difficult proppant placement [20]. Liu Xiao conducted field tests comparing fracture networks under different treatment scales, establishing optimal stimulation practices [21]. Xiong et al. demonstrated enhanced stimulation effects through multi-stage diversion processes based on microseismic monitoring [22]. Tian et al. established a 2D fluid–solid coupling finite element model using ABAQUS to simulate fracture propagation [23]. Zhao et al. revealed weak structural surfaces' significant impacts on fracture patterns through actual triaxial experiments and acoustic emission monitoring [24].

To a certain extent, these studies have clarified the hydraulic fracture propagation law of coalbed methane, optimized the hydraulic fracturing parameters of deep coalbed methane, and proposed the corresponding hydraulic fracturing technology, but these studies have not carried out the geological modeling research of deep coalbed methane reservoirs, nor have they designed the optimization of hydraulic fracture parameters of well groups under the influence of stress shadow during deep coalbed methane fracturing. In this paper, the three-dimensional geological model of deep coalbed methane in the Ordos Basin-Yuyang favorable block is first established, and then the numerical simulation of construction parameters optimization under the conditions of hydraulic fracturing of a single well and a well group of deep coalbed methane is carried out. The results of this paper show that hydraulic fracturing of deep coalbed methane has certain guiding significance.

## 2. Geological Setting Area

The primary productive coal seams are located in deep coal reservoirs beneath 2000 m. Structurally, the field typically features a gentle monocline that slopes from the northeast (higher elevation) to the southwest (lower elevation), with an average gradient of 6–9 m per kilometer. The formation dip angles range from 0.3° to 0.6°. Additionally, there are locally developed, nose-shaped uplifts that exhibit orientations striking nearly east to west.

According to the comprehensive logging interpretation diagram of a pilot hole in the target block, the depth of the Tai 1 section is 2851.50–2874.00 m, the apparent thickness is 22.50 m, and the lithology is black coal, gray coal, gray-black mudstone, and limestone. The location of the target horizon 8# coal is 2851.50–2865.00 m, the apparent thickness is 14.50 m, and the mudstone is sandwiched with 2.5 m in the middle. The gas saturation result is 8.3~26.3%, and the average gas saturation is 22% (Table 1). The structure of 8# coal body is relatively complete, which is the primary-fractured type, and the macroscopic type of coal and rock is mainly the semi-bright type, followed by the bright type, a small amount of the semi-dark type and dull type, and is interbedded with multiple layers of carbonaceous mudstone and gray or black mudstone with a thickness of 2–3 m. The degree of thermal evolution of coal is moderate, and the Ro of coal rock is between 1.2~1.7%, which belongs to medium metamorphic degree, medium and low volatile bituminous coal; the cleavage and fractures formed by thermal metamorphism and tectonic action are relatively developed.

**Table 1.** Logging interpretation result table.

| Layer Number | Measuring Depth (m) | Visual Thickness (m) | Vertical Depth (m) | Porosity (%) | Permeability (mD) | Gas Saturation (%) | Interpret Conclusions |
|---|---|---|---|---|---|---|---|
| 1 | 2798.5–2801.7 | 3.2 | 2738.75–2741.85 | 5.6 | 0.29 | 26.4 | Gas-bearing layer |
| 2 | 2806.8–2811.9 | 5.1 | 2746.79–2751.74 | 7.7 | 0.57 | 33.8 | Differential gas layer |
| 3 | 2830.2–2834.2 | 4 | 2769.47–2773.34 | 4.8 | 0.19 | 28.3 | Gas-bearing layer |
| 4 | 2836.8–2838.5 | 1.7 | 2775.85–2777.50 | 7 | 0.42 | 28.9 | Gas-bearing layer |
| 5 | 2850.7–2855.1 | 4.4 | 2789.31–2793.57 | 7.7 | 0.54 | 46.3 | Differential gas layer |
| 6 | 2862.0–2865.4 | 3.4 | 2800.25–2803.54 | 5.2 | 0.02 | 12.4 | Carbonaceous mudstone |
| 7 | 2869.3–2870.7 | 1.4 | 2807.31–2808.67 | 6.7 | 0.35 | 22.3 | Coal |
| 8 | 2880.1–2882.0 | 1.9 | 2817.77–2819.61 | 7.2 | 0.32 | 27.8 | Coal |
| 9 | 2884.8–2897.1 | 12.3 | 2822.33–2834.25 | 4.6 | 0.19 | 22.1 | Dry layer |
| 10 | 2897.9–2904.6 | 6.7 | 2835.02–2841.52 | 5.9 | 0.42 | 31.3 | Coal |
| 11 | 2904.6–2909.4 | 4.8 | 2841.52–2846.17 | 2.8 | 0.71 | 17.4 | Dry layer |
| 12 | 2909.4–2911.5 | 2.1 | 2846.17–2848.21 | 4.5 | 0.04 | 11.2 | Carbonaceous mudstone |

To enhance production capacity, horizontal wells combined with large-scale volume fracturing are employed onsite to create complex fracture networks. However, optimizing key parameters for such large-scale volume fracturing operations requires further investigation. This study will conduct 3D geomechanical modeling of deep coalbed methane reservoirs and perform fracturing parameter optimization research based on field-derived parameters.

## 3. Three-Dimensional Geological and Mechanical Modeling

### 3.1. Geological Facies Modeling

Data were imported from 282 wells, including information on wellhead, deviation, stratification, logging, and wellbore trajectory, into Petrel. After defining the boundaries based on the logging interpretation results of the target block, the pay zones were identified as four layers: T2-3-T2-2-T2-1-T1-2-T1-1. The target area covers approximately

38 square kilometers, with a pay zone thickness of about 14 m. The grid step size was set to 25 m × 25 m in the horizontal plane and 1 m in the vertical direction. Figure 1 is a schematic diagram of the construction model.

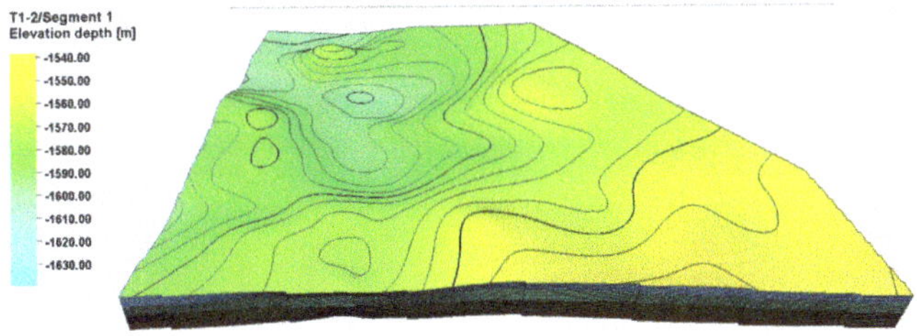


**Figure 1.** Construction model diagram.

The petrofacies division standard, shown in Table 2, was formulated by comparing the logging curves such as gamma, density, and deep lateral resistance with the logging data while drilling. The pay zones primarily consist of bright coal along with carbonaceous mudstone and shale, which are overlain by sandstone and limestone and underlain by mudstone.

**Table 2.** Standard table of lithofacies division.

| Lithofacies | Truncated Value |
|---|---|
| Bright coal | GR ≤ 60GAPI, DEN ≤ 1.6 g/cm$^3$, LLD > 3000 Ω.cm |
| Semi-bright coal | 60 < GR < 100GAPI, DEN ≤ 1.6 g/cm$^3$, LLD > 3000 Ω.cm |
| Semi-dark coal | 100 ≤ GR < 120GAPI, 1.6 < DEN ≤ 1.85 g/cm$^3$, LLD > 500 Ω.cm |
| Dull coal | 120 ≤ GR < 140GAPI, 1.6 < DEN ≤ 1.85 g/cm$^3$, LLD > 500 Ω.cm |
| Gangue | GR < 140GAPI, DEN ≤ 2 g/cm$^3$ |
| Limestone | GR < 60GAPI, DEN > 2 g/cm$^3$, CNL ≤ 7% |
| Semi-dark coal | 100 ≤ GR < 120GAPI, 1.6 < DEN ≤ 1.85 g/cm$^3$, LLD > 500 Ω.cm |
| Dull coal | 120 ≤ GR < 140GAPI, 1.6 < DEN ≤ 1.85 g/cm$^3$, LLD > 500 Ω.cm |

Figure 2 is the reservoir lithofacies profile. The roof is about 10 m thick and is a limestone–sandstone formation. The coal seam is about 14 m thick, with bright coal on top and 2–3 m of carbonaceous mudstone partings in the middle and lower part. The lower part is mainly semi-bright and semi-dark coal. The floor is about 8 m thick and mainly mudstone.

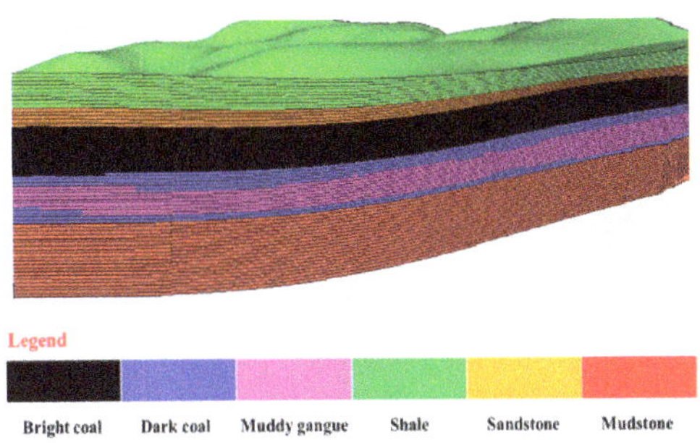


**Figure 2.** Reservoir lithofacies profile.

### *3.2. Geological Property Modeling*

Due to the strong correlation between coal seam density and porosity and permeability, the empirical formula for calculating coal seam porosity and permeability using density logging data can be obtained through regression, and the porosity and permeability logging interpretation curves in the target block are calculated by using the density logging curve. The porosity and permeability profiles of each single well are obtained after comparison and correction with the logging data.

Porosity:

$$POR = 38.725e^{-1.291*DEN} \tag{1}$$

Permeability:

$$PERM = 0.0002e^{-0.8492*POR} \tag{2}$$

where $POR$ is the porosity of the reservoir, %; $PERM$ is the reservoir permeability, mD; $DEN$ is the apparent density of the reservoir, g/cm$^3$.

New well logs were added for porosity, permeability, and gas saturation. The average algorithm was utilized to discretize these properties. Afterward, the harmonic mean, facies-controlled, and sequential Gaussian simulation algorithms were applied to create 3D geological property models for both porosity and permeability. The porosity model is shown in Figure 3 and the permeability model is shown in Figure 4.

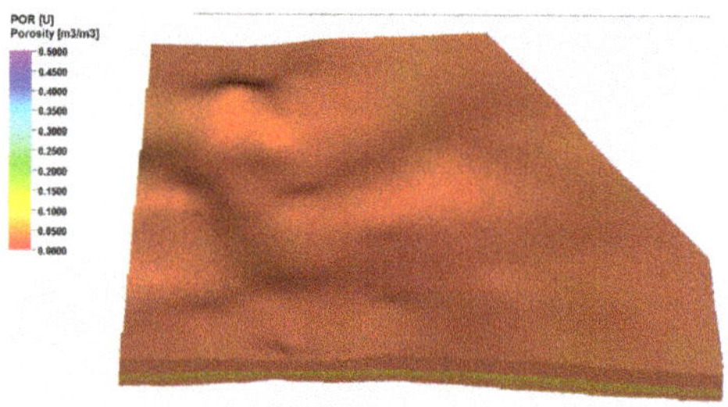


**Figure 3.** Geological model of porosity for the target block.

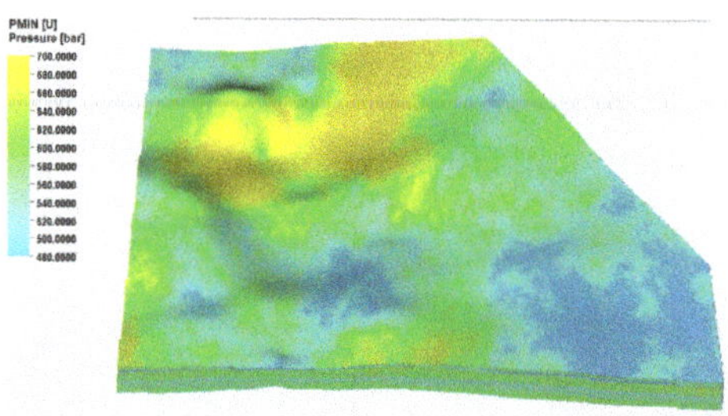


**Figure 4.** Geological model of permeability for the target block.

The porosity distribution of the uppermost roof ranges from 0.5% to 3.5%, with an average of 1.3%. The coal seam exhibits a porosity distribution of 1% to 10%, averaging 5%. The floor's porosity distribution spans from 0.5% to 8%, averaging 1.5%. The porosity plane distribution and histogram are shown in Figure 5.

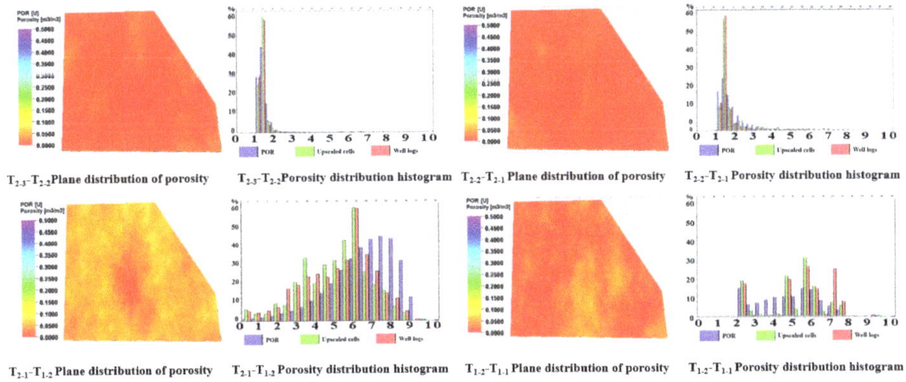


**Figure 5.** Histogram of porosity distribution for each small layer in the well group area.

The permeability range of the uppermost roof is 0.001 mD to 0.12 mD, with an average of 0.05 mD. For the coal seam, it ranges from 0.001 mD to 0.6 mD, averaging 0.085 mD. The floor has a permeability distribution of 0.001 mD to 0.12 mD, averaging 0.04 mD. This reservoir exhibits strong heterogeneity. The permeability plane distribution and histogram are shown in Figure 6.

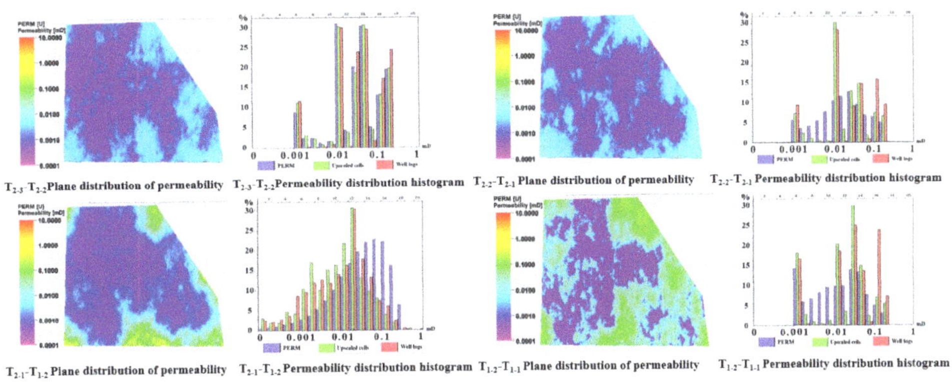


**Figure 6.** Histogram of permeability distribution for each small layer in the well group area.

### 3.3. Rock Mechanics and Stress Modeling

Acoustic travel time logs are utilized to determine the rock mechanics and stress parameters of each well. This is performed using regression formulas for P-to-S wave conversion, empirical rock mechanics equations, and the deep coal-seam-applicable Ge model. New well logs were created to include Young's modulus, Poisson's ratio, and the minimum and maximum horizontal and vertical principal stresses. An average algorithm was applied, and the properties were discretized. Additionally, harmonic mean, facies-controlled, and sequential Gaussian simulation algorithms were employed to generate 3D geological property models for these parameters. The geological model of the rock mechanics and stresses in the target area is shown in Figure 7.

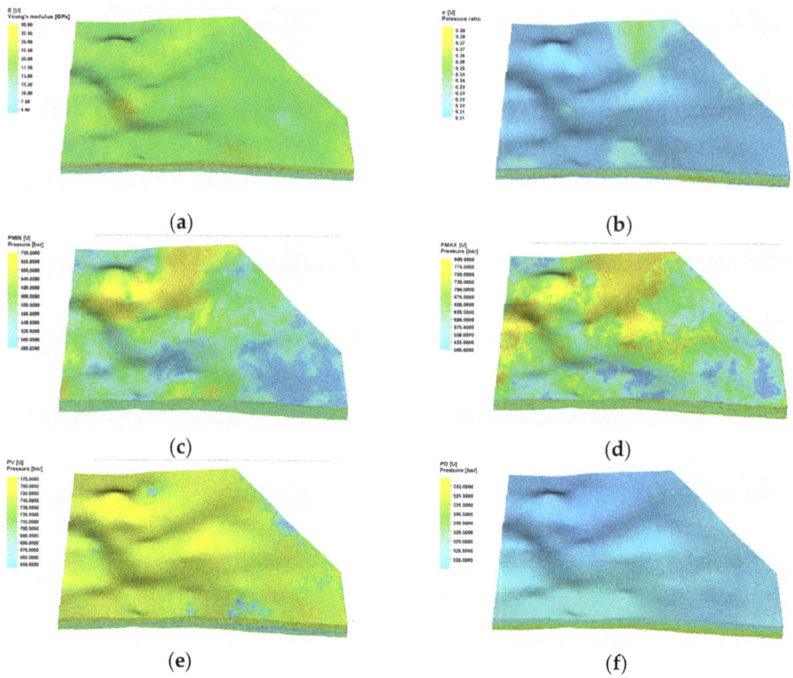


**Figure 7.** Geological model of rock mechanics and stress for the target area. (**a**) Young's modulus; (**b**) Poisson's ratio; (**c**) Minimum horizontal principal stress; (**d**) Maximum horizontal principal stress; (**e**) Vertical principal stress; (**f**) Pore pressure.

As shown in Figure 8, the Young's modulus of the roof ranges from 14 GPa to 32 GPa, with a mean value of 30 GPa. The Young's modulus of coal varies from 3 GPa to 12 GPa, averaging 7.5 GPa. The floor's Young's modulus spans from 10 GPa to 25 GPa, with a mean of 20 GPa.

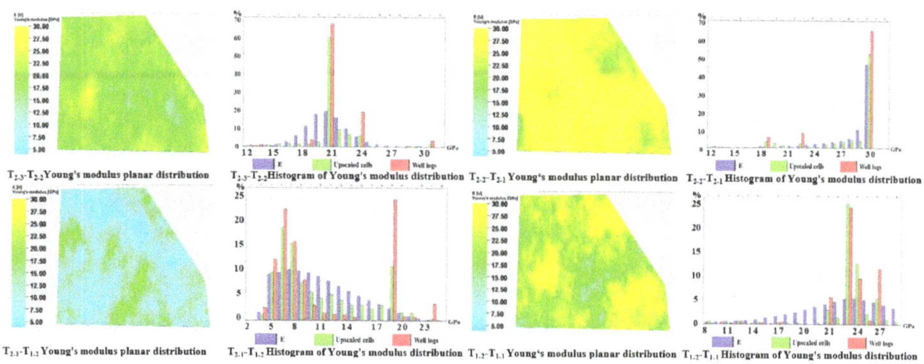


**Figure 8.** Distribution map of Young's modulus for each small layer.

As shown in Figure 9, the roof has a Poisson ratio of 0.21 to 0.28, with an average of 0.23. The coal has a Poisson ratio ranging from 0.26 to 0.29, averaging 0.28. The floor's Poisson ratio varies from 0.2 to 0.26, with a mean of 0.21.

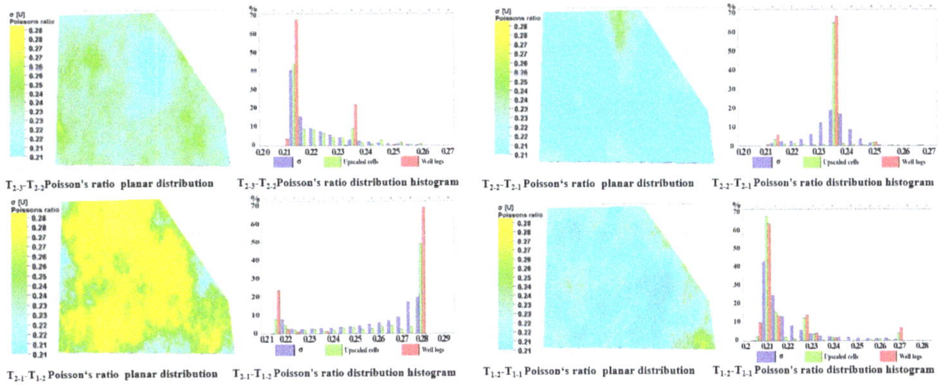


**Figure 9.** Distribution map of Poisson's ratio for each small layer.

As shown in Figure 10, the minimum horizontal principal stress for the roof ranges from 48 MPa to 70 MPa, with an average of 61 MPa. For the coal seam, the minimum horizontal principal stress is between 49 MPa and 64 MPa, averaging at 53 MPa. The floor exhibits a minimum horizontal principal stress ranging from 55 MPa to 70 MPa, with an average of 59 MPa.

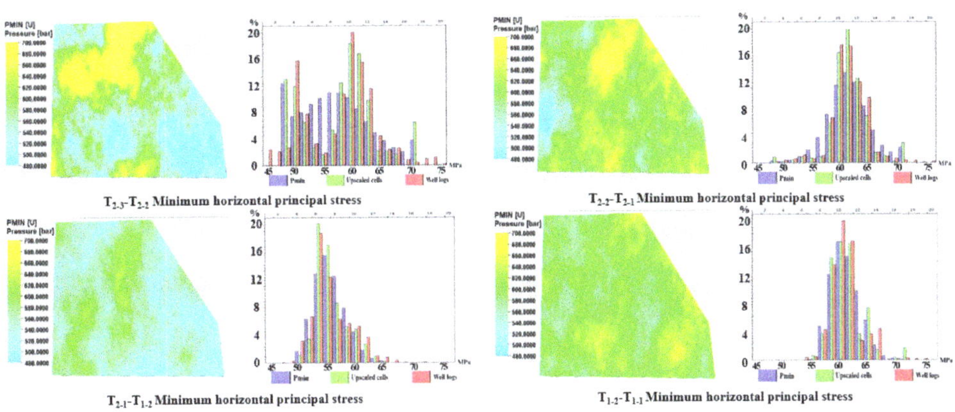


**Figure 10.** Distribution map of minimum horizontal principal stress for each small layer.

As shown in Figure 11, the maximum horizontal principal stress of the roof ranges from 50 MPa to 81 MPa, with an average value of 72 MPa. For the coal seam, the maximum horizontal principal stress varies from 56 MPa to 75 MPa, with a mean of 61 MPa. The floor exhibits a maximum horizontal principal stress that ranges from 61 MPa to 80 MPa, averaging 71 MPa.

As shown in Figure 12, the vertical principal stress of the roof and floor is approximately 75 MPa, while that of the coal seam is around 66 MPa. In the coal seam, the vertical stress (66 MPa) exceeds the maximum horizontal principal stress (61 MPa), which in turn is greater than the minimum horizontal principal stress (53 MPa). The difference between the horizontal stresses is about 8 MPa. Additionally, the stress difference among the coal seam, roof, and floor ranges from 8 to 12 MPa. The average pore pressure of the reservoir is 32.5 MPa. The pore pressure profile is shown in Figure 13.

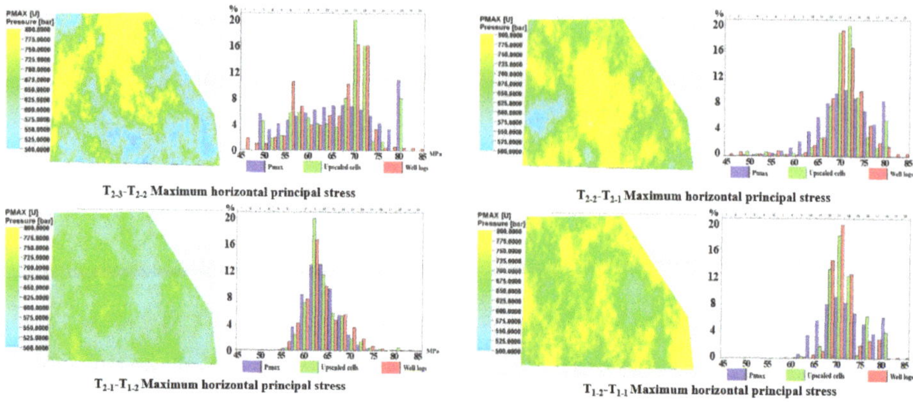


**Figure 11.** Distribution map of maximum horizontal principal stress for each small layer.

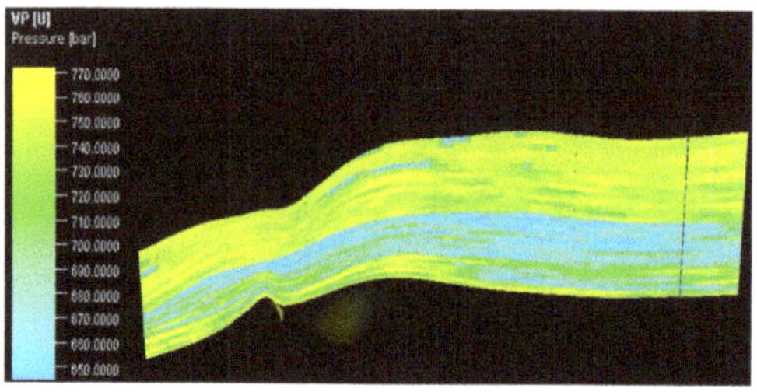


**Figure 12.** Vertical principal stress profile.

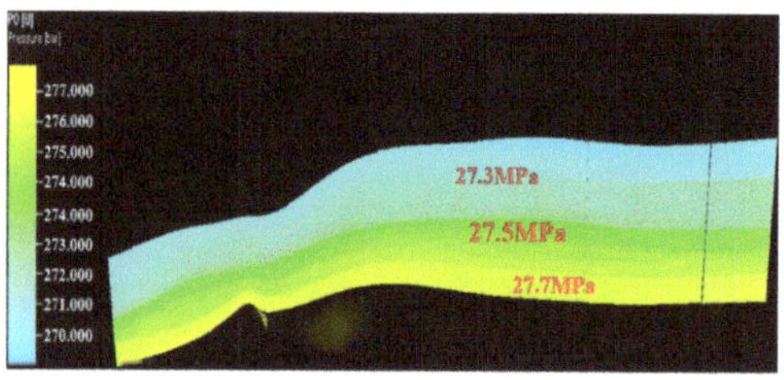


**Figure 13.** Pore pressure profile.

*3.4. Establishment and Calibration of Hydraulic Fracturing Model for Deep Coal Reservoirs*

Based on established 3D geological and stress field models, we developed a hydraulic fracturing model that takes into account natural fractures and coal seam cleats. The fracturing parameters and pumping program for Field X well were simulated for different azimuth fracture models. After adjusting for the loss coefficient of the fractures (including their length and height), we utilized the closest fracture network morphology as the natural fracture model for the healthy group area. Figure 14 compares the monitoring results of the fracturing fluid wavefront from well X with the results of numerical simulations. Figure 15 illustrates the construction pressure fitting of the established fracturing model using the construction parameters from well X, which aligns with the actual site conditions.

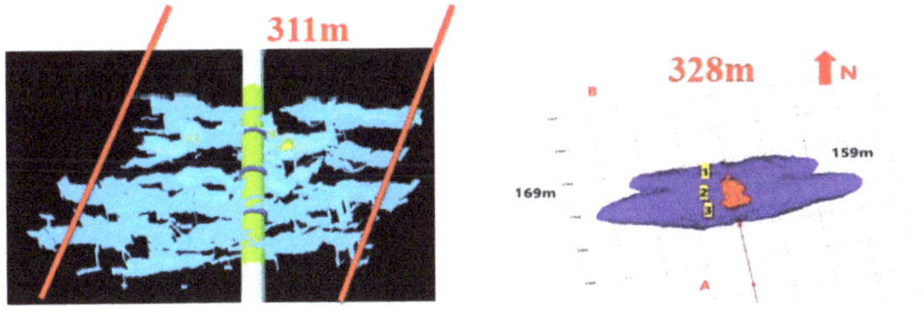


**Figure 14.** Comparison chart of X well's fracturing fluid wave monitoring results and numerical simulation results.

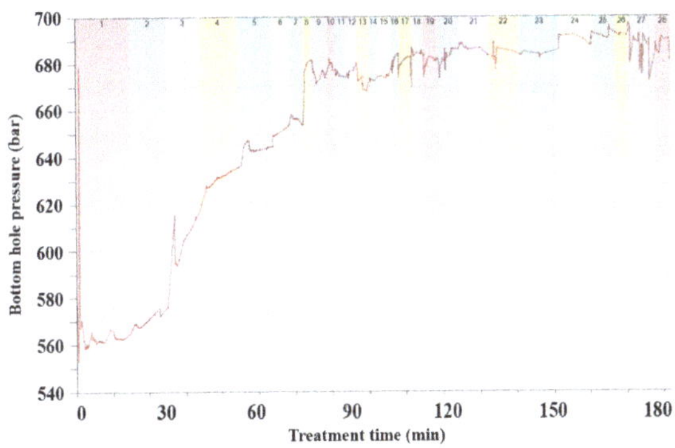


**Figure 15.** Numerical simulation of the construction pressure curve of a well.

The geological parameters of the coal seam in the fracturing model are as follows: average Poisson's ratio of 0.28, Young's modulus of 3–9 GPa, minimum horizontal principal stress of 53 MPa, maximum horizontal principal stress of 60 MPa, and vertical principal stress of 65.5 MPa. The stress difference between Taiyi coal seam and roof is 11–13 MPa, and the stress difference between Taiyi coal seam and bottom plate is 8–10 MPa; the main drilling encounter rock facies is bright coal. The numerical simulation uses the Y well group, which has a total of four horizontal wells, of which the horizontal section length of Y1, Y3 and Y4 wells is 1000 m, the horizontal section length of Y2 well is 800 m, the well spacing of Y1 and Y2 wells is 4000 m, and the adjacent well spacing of the rest of the wells is 350 m. Figure 16 is a schematic diagram of the borehole trajectory of the well group.

**Figure 16.** Schematic diagram of the borehole trajectory of the Y well group.

## 4. Fracturing Parameter Optimization

Considering the influences of intersection fracture spacing, construction fluid volume, and construction displacement on fracture propagation under single well conditions and well group conditions, the above construction parameters were optimized by comparing the fracture propagation morphology, in situ stress field change, conductivity, fracture making volume, and other factors. The specific process is as follows: Firstly, according to the construction data of the site, the control variable method is used to simulate the law of mesh propagation and crack parameters under different inter-section seam spacing, and the optimal inter-segment seam spacing is optimized by comprehensively considering the seam network ripple area, channeling situation, stress interference, conductivity and seam volume value. Then, the optimized inter-segment seam spacing was replaced by the inter-segment seam spacing used in the on-site construction scheme, and then the control variable method was used to simulate the seam network expansion law under different construction fluid quantities to optimize the construction fluid volume. Finally, the optimized inter-segment joint spacing and liquid volume are replaced by the inter-segment joint spacing and liquid volume used in the on-site construction scheme, and the construction displacement is optimized according to the above process.

The construction parameters of the foundation are a section of 3 clusters: cluster spacing 20 m; each cluster perforation is 1 m, and the hole density is 16 rounds/m; phase angle 60°; the comprehensive sand ratio is 14%; the viscosity of fracturing fluid is 31 mPa.s; the single well simulation uses a perforated well section of 600 m (3720~4320 m) of Well Y1, and the gap spacing between the sections is 30–90 m. Construction fluid volume 2500–4500 m$^3$; construction row 10–30 m$^3$/min. The perforated sections of the well group were a 600 m horizontal section (3720~4320 m) of Well Y3 and a 600 m horizontal section (3670~4270 m) of Well Y4, and the gap spacing between sections was 30–90 m; construction fluid volume: 2500–4500 m$^3$; the construction displacement is 10–30 m$^3$/min. For the specific pumping procedure, please refer to the on-site construction data of Well X1.

### *4.1. Single Well Fracturing Parameter Optimization*

#### 4.1.1. Inter-Stage Fracture Spacing

The inter-stage fracture spacing is the distance between the previous stage's last perforation hole and the next stage's first perforation hole. Figure 17 shows the fracture network morphology under different inter-stage fracture spacings. Results show that at 30 m spacing, inter-cluster and inter-stage interference are severe. At 50 m, there is almost no interference. No interference occurs when spacing exceeds 50 m, but the unfractured area between stages increases, leading to insufficient reservoir stimulation.

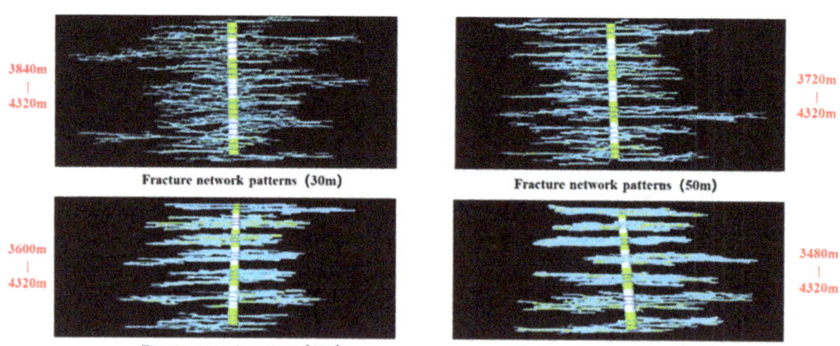


**Figure 17.** Fracture network patterns under different inter-segment fracture spacings.

Figure 18 illustrates the variation of horizontal stress with different inter-stage fracture spacings. When the spacing is 30 m, there is a significant interference in horizontal stress. At 50 m spacing, the area of stress change is the largest, and the interference between stages is minimal. When the spacing exceeds 50 m, the interference in horizontal stress becomes weak. As the inter-stage fracture spacing increases, the rate of increase in horizontal stress decreases. After fracturing, the minimum horizontal principal stress increases by approximately 3.9 MPa, and the maximum horizontal principal stress increases by 2.5 MPa. Geological structure, rock properties, and fracturing parameters influence the variation in horizontal stress with inter-stage fracture spacing.

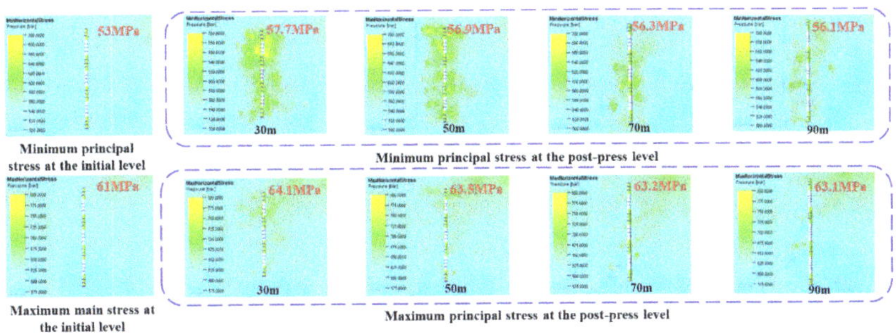


**Figure 18.** Ground stress variations under different inter-segment fracture spacings.

Figure 19 illustrates the total fracture volume and average conductivity curves for various inter-stage fracture spacings in a 1000 m horizontal coalbed methane well. When the spacing is less than 50 m, the total fracture volume experiences a slight decrease as the spacing increases, while the average conductivity per cluster improves. However, when the spacing exceeds 50 m, the total fracture volume drops significantly, and the average conductivity per cluster increases at a slower rate. These findings align with research on fracture conductivity in tight reservoirs, where factors such as stress closure and proppant properties have a significant impact on conductivity.

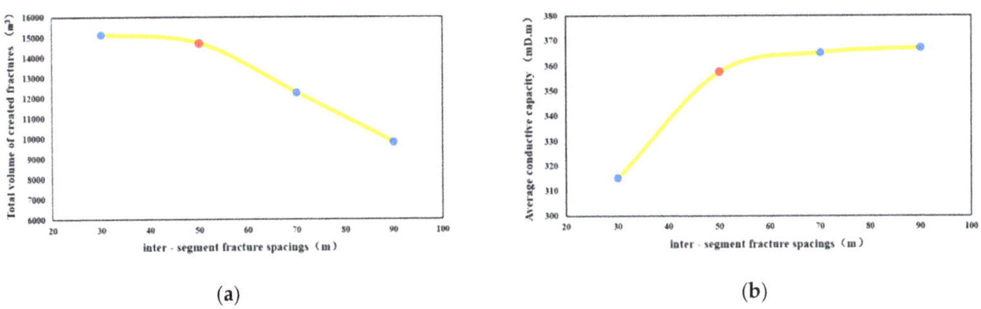


**Figure 19.** Total fracture volume and average conductive capacity under different inter-segment fracture spacings. (**a**) Total volume of created fractures under different inter-segment fracture spacings; (**b**) Average conductive capacity under different inter-segment fracture spacings.

A comprehensive evaluation of fracture network morphology, total fracture volume, and conductivity under different inter-stage fracture spacings shows that around 50 m spacing yields the best results for the overall fracture network morphology, total fracture volume, and conductivity.

#### 4.1.2. Fracturing Liquid Volume

Figure 20 illustrates the fracture network morphology under different pumping liquid volumes. The volume of liquid used significantly impacts fracture network coverage. During the early fracturing stages, the stimulated volume increases rapidly with increasing liquid volume, then slows down after reaching a specific volume. For a coalbed methane horizontal well with a 50 m inter-stage fracture spacing and 18 $m^3$/min discharge rate, the fracture coverage is low at 2500 $m^3$ and 3000 $m^3$ liquid volumes. The fracture coverage is better at 3500 $m^3$ and 4000 $m^3$, with minor inter-stage interference. At 4500 $m^3$, the fracture coverage is high, but there is significant inter-stage interference. These findings are consistent with research indicating that excessive liquid volume can lead to rapid pressure drops and sustainability challenges in fracture networks.

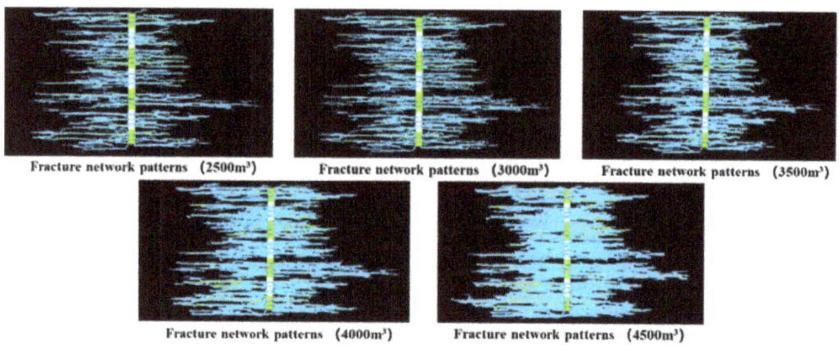


**Figure 20.** Fracture network patterns under different pumping liquid volumes.

Figure 21 shows the stress changes under different pumping liquid volumes. At 2500 $m^3$ and 3000 $m^3$ liquid volumes, the stress change area is small, with weak inter-stage and cluster interference, but there are still many unfractured areas. Stress increases with liquid volume. The stress change area is large at liquid volumes above 3500 $m^3$, but there is substantial inter-well and inter-stage interference. Stress increases slowly with liquid volume. After fracturing, the minimum horizontal principal stress increased by about 4.5 MPa, and the maximum horizontal principal stress increased by about 2.3 MPa.

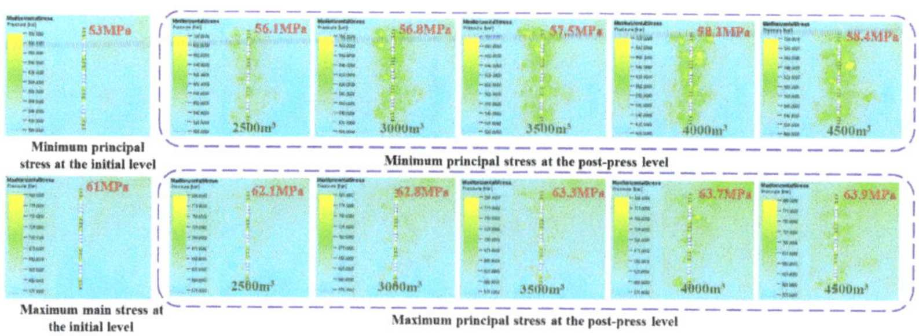


**Figure 21.** Formation stress variations under different treatment liquid volumes.

Figure 22 illustrates the relationship between total fracture volume and average conductivity across different pumping liquid volumes. When the liquid volume per stage is below 4000 $m^3$, both the total fracture volume and average conductivity increase as the liquid volume rises. However, once the liquid volume per stage exceeds 4000 $m^3$, while both the total fracture volume and average conductivity continue to increase, the rate of increase slows significantly.

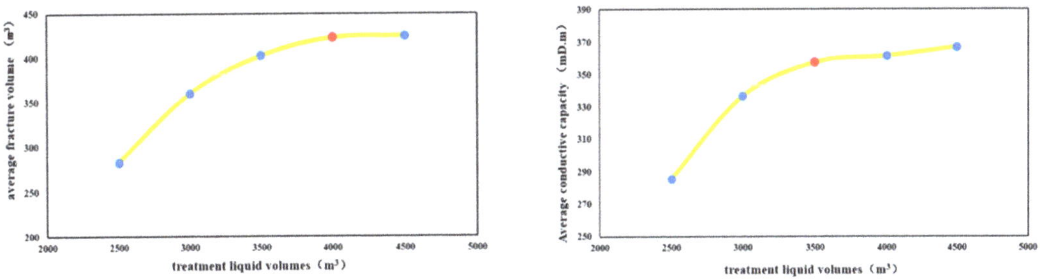


(a) (b)

**Figure 22.** Fracture volume and average conductive capacity under different treatment liquid volumes. (**a**) Total fracture volume under different treatment liquid volumes; (**b**) Average conductive capacity under different treatment liquid volumes.

A comprehensive evaluation of fracture network morphology, stress changes, total fracture volume, and conductivity under varying volumes of fracturing liquid indicates that approximately 3500–4000 $m^3$ per stage yields optimal results for fracture network morphology and conductivity.

### 4.1.3. Pumping Rate

Figure 23 shows fracture network morphologies under different pumping rates. With an inter-stage fracture spacing of 50 m and a liquid volume of 4000 $m^3$, fracture coverage is low at rates of 10–15 $m^3$/min. At 20 $m^3$/min, there is minor inter-stage interference and better fracture coverage. Fracture coverage is high at 25–30 $m^3$/min, but there is significant inter-stage and cluster interference.

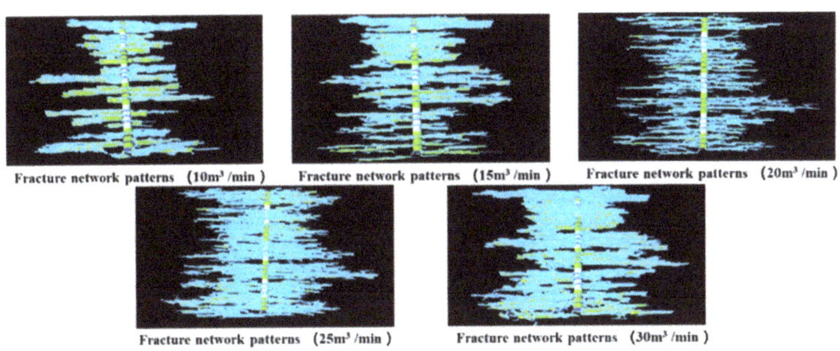


**Figure 23.** Fracture network patterns under different pumping rates.

Figure 24 shows the stress changes under different pumping rates. When the pumping rate is below 20 $m^3$/min, the area of stress change is small, with some unfractured regions remaining between stages. As the pumping rate increases, the stress change also increases. However, when the rate exceeds 20 $m^3$/min, the area of stress change becomes significantly larger, leading to noticeable interference between stages, and the rate of stress change increases more slowly. After the fracturing process, the minimum horizontal principal stress increases by approximately 3.6 MPa, while the maximum horizontal principal stress rises by about 2.2 MPa.

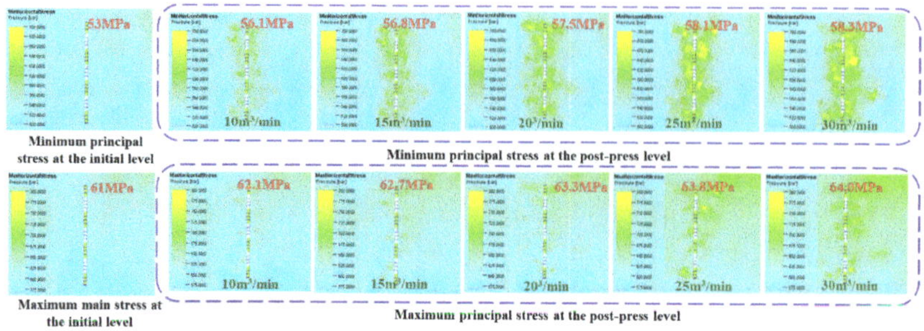


**Figure 24.** Formation stress variations under different pumping rates.

Figure 25 illustrates the total fracturing volume and average conductivity at varying pumping rates. When the rate is below 22 $m^3/min$, both volume and conductivity increase with the rate. Above 22 $m^3/min$, the rate of increase in conductivity slows down.

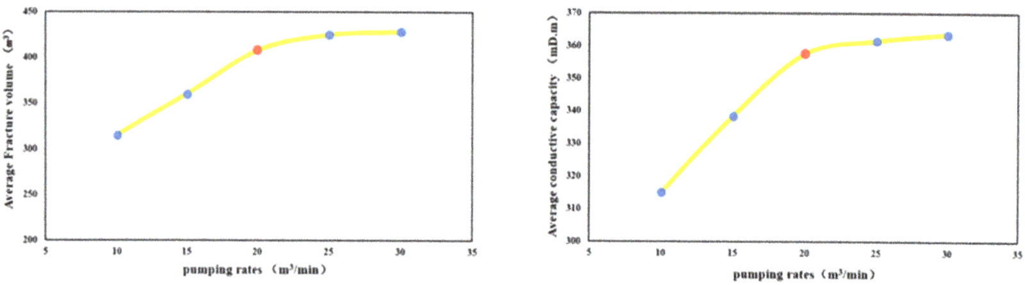


(a) (b)

**Figure 25.** Total fracture volume and average conductive capacity under different treatment liquid volumes. (**a**) Fracture volume under different pumping rates; (**b**) Average conductive capacity under different pumping rates.

A comprehensive evaluation of fracture network morphology, stress variation, and fracture volume and conductivity shows that the overall fracture network morphology and fracture parameter results are optimal at a pumping rate of around 20–22 $m^3/min$.

### *4.2. Well Group Fracturing Parameter Optimization*

#### 4.2.1. Inter-Stage Fracture Spacing

Figure 26 illustrates the morphology of the fracture network under various inter-stage fracture spacings, using a liquid volume of 3500 $m^3$ and a discharge rate of 22 $m^3/min$. At a spacing of 30 m, there is significant inter-cluster and inter-well interference. When the spacing is adjusted to 50 m, only minor inter-stage and inter-well interference is observed. However, when the spacing exceeds 70 m, inter-well interference is minimal, but this results in low fracture coverage and insufficient stimulation of the reservoir.

Figure 27 illustrates how stress varies with inter-stage fracture spacing. As the spacing increases, inter-stage interference decreases, leading to a reduced post-fracturing increase in stress. At 30 m of spacing, both inter-stage and inter-well stress interference are significant. At 50 m of spacing, the area of stress change is large, but inter-well interference becomes minimal. Beyond 50 m, the unfractured areas between stages and wells increase. After fracturing, the minimum horizontal principal stress rises by approximately 5 MPa, while the maximum horizontal principal stress increases by about 2.7 MPa.

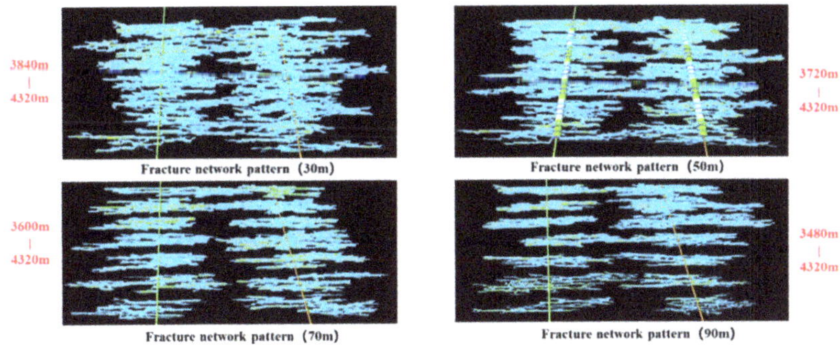


**Figure 26.** Fracture network patterns under different inter-segment fracture spacings.

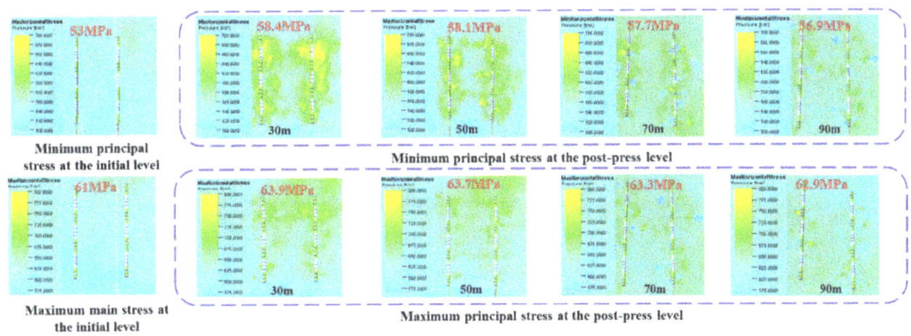


**Figure 27.** Formation stress variations under different inter-segment fracture spacings.

Figure 28 illustrates the total fracture volume and average conductivity curves for varying inter-stage fracture spacings. When the spacing is less than 60 m, the total fracture volume decreases gradually as the spacing increases, while the average conductivity per cluster improves. However, when the spacing exceeds 60 m, the total fracture volume drops significantly, and the average conductivity per cluster increases at a slower rate.

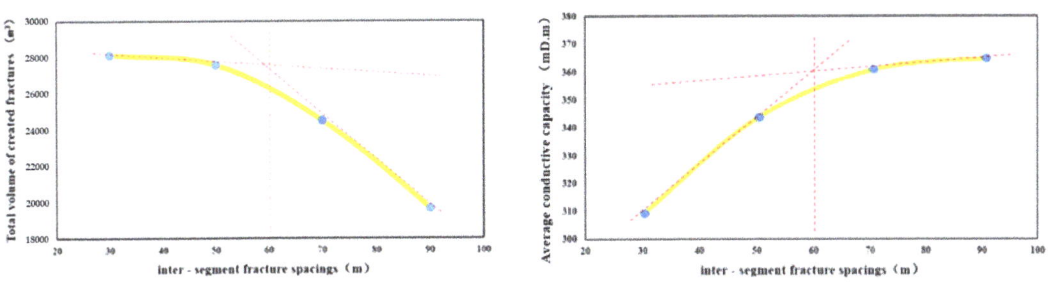


(**a**) (**b**)

**Figure 28.** Curve chart of total fracture volume and average conductive capacity under different inter-segment fracture spacings. (**a**) Total fracture volume under different inter-segment fracture spacings; (**b**) Average conductive capacity under different inter-segment fracture spacings.

A comprehensive evaluation of fracture network morphology, stress changes, total fracture volume, and conductivity shows that around 60 m inter-stage fracture spacing yields the best overall results for fracture network morphology, total fracture volume, and conductivity.

Based on the qualitative evaluation of the mesh morphology, the change of in situ stress, the total fracture volume and the quantitative evaluation of the conductive capacity under different intersection seam spacing, when the inter-section seam spacing is about 60 m, the results of the overall seam network morphology, total seam volume and conductive capacity are optimal.

#### 4.2.2. Fracturing Liquid Volume

Figure 29 presents the fracture network morphology under different fracturing liquid volumes with an inter-stage fracture spacing of 60 m and a 22 m$^3$/min discharge rate. When the liquid volume is 2500–3000 m$^3$, the fracture coverage between stages and wells is low. At 3500–4000 m$^3$, coverage improves. At 4500 m$^3$, coverage is high but with significant inter-stage and inter-well interference.

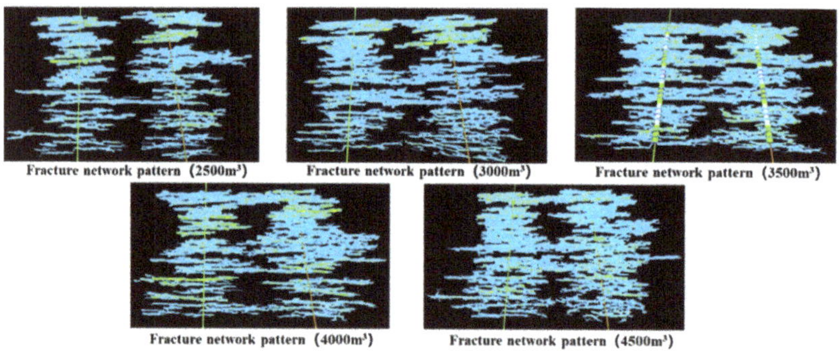


**Figure 29.** Fracture network patterns under different treatment liquid volumes.

Figure 30 shows the stress variation under different fracturing liquid volumes. At 2500–3000 m$^3$, the stress change area is small, with weak inter-well and inter-stage interference, but many unfractured areas exist. Stress increases with liquid volume. Above 3500 m$^3$, the stress change area is large but with strong inter-well and inter-stage interference, and stress increases slowly with liquid volume. Post-fracturing, the minimum horizontal principal stress rises by about 5 MPa, and the maximum horizontal principal stress increases by 2.7 MPa.

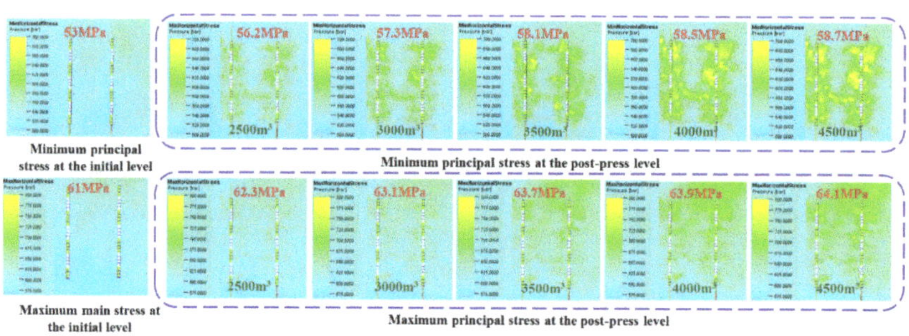


**Figure 30.** Formation stress variations under different treatment liquid volumes.

Figure 31 illustrates the fracture volume and average conductivity at varying volumes of fracturing liquid. When the liquid volume per stage is less than 3500 m$^3$, both fracture volume and conductivity increase with the volume. However, beyond 3500 m$^3$, the rate of increase slows down.

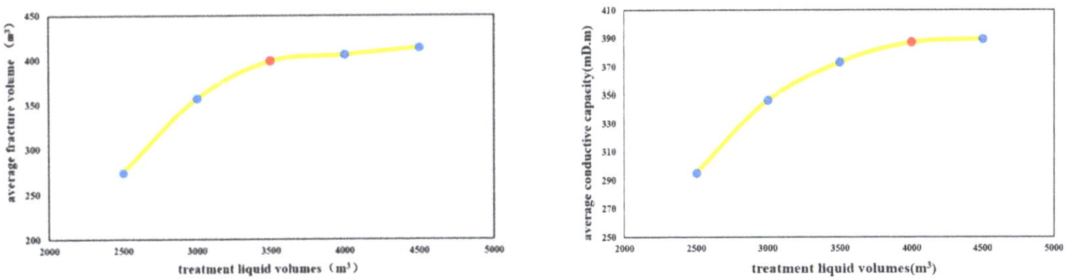


(**a**) (**b**)

**Figure 31.** Total fracture volume and average conductive capacity under different treatment liquid volumes. (**a**) Average fracture volume per cluster under different liquid volumes; (**b**) Average conductive capacity under different liquid volumes.

A comprehensive evaluation of fracture network morphology, stress variation, fracture length, volume, and conductivity shows that around 3500 $m^3$ per stage yields the best overall results.

Based on the qualitative evaluation of fracture network morphology, in situ stress change, and fracture length, fracture volume, and conductivity under different fracturing fluid volumes, when the single-stage fracturing fluid volume is about 3000–3500 $m^3$, the overall fracture network morphology and conductivity are optimal.

#### 4.2.3. Pumping Rate

Figure 32 shows the fracture network morphology under different pumping rates. With an inter-stage fracture spacing of 60 m, a liquid volume of 3500 $m^3$ per stage, and pumping rates of 10–15 $m^3$/min, fracture coverage between stages and wells is low. At 20 $m^3$/min, there is minor inter-stage and inter-well interference and better fracture coverage. At 25–30 $m^3$/min, fracture coverage is high but with significant inter-stage and inter-well interference.

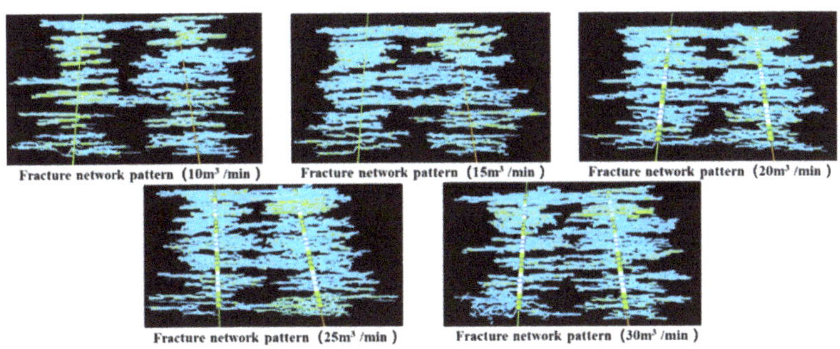


**Figure 32.** Fracture network patterns under different pumping rates.

Figure 33 illustrates the variations in stress under different pumping rates. Below a rate of 20 $m^3$/min, the area experiencing stress changes is small, and there are some unfractured regions between the wells. As the pumping rate increases, the stress changes also increase. However, when the rate exceeds 20 $m^3$/min, the area of stress change becomes significantly larger, leading to noticeable interference between the wells. In this case, the increase in stress changes occurs at a slower rate. After the fracturing process, the minimum horizontal principal stress increases by approximately 4 MPa, while the maximum horizontal principal stress rises by about 2.5 MPa.

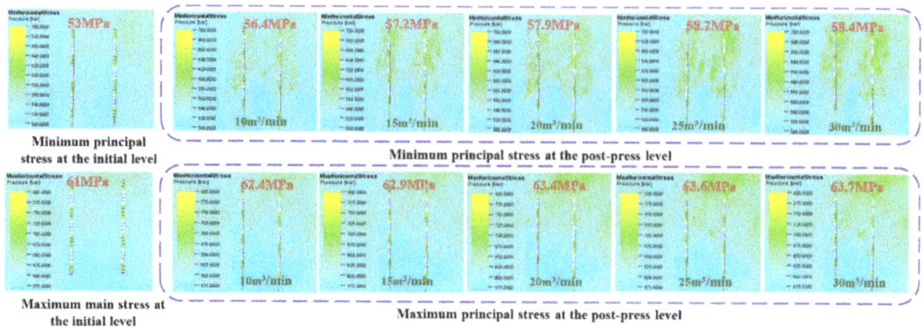


**Figure 33.** Formation stress variations under different pumping rates.

Figure 34 presents the fracturing volume and average conductivity under different pumping rates. When the rate is below 20 $m^3/min$, the volume and conductivity rise. Above 20 $m^3/min$, the increase in conductivity slows down.

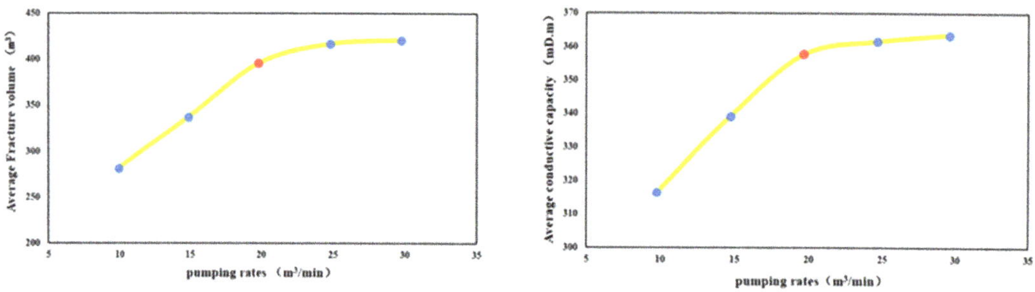


(**a**) (**b**)

**Figure 34.** Total fracture volume and average conductive capacity under different pumping rates. (**a**) Average fracture volume under different pumping rates; (**b**) Average conductive capacity under different pumping rates.

A comprehensive evaluation of fracture network morphology, stress variation, and fracture volume and conductivity shows that the overall fracture network morphology, fracture parameters, fracture volume, and conductivity are optimal at a pumping rate of around 18–20 $m^3/min$.

### 4.3. Sensitivity Analysis

The optimization results of the construction parameters of the single well show that the construction parameters of Well Y1 are 50 m spacing, the construction fluid volume of the single section is 3500–4000 $m^3$, the construction displacement is 20–22 $m^3/min$, and the m0.ain seam length is about 315 m. The results of the optimization of the construction parameters of the well group show that the optimal construction parameters of wells Y3 and Y4 are 60 m spacing, 3000–3500 $m^3$ of single-stage construction fluid, 18–20 $m^3/min$, and about 303 m of main seam length.

Combined with the analysis of geological factors and engineering factors, it is concluded that the size of the fracture spacing between the sections is closely related to the degree of hydraulic fracture penetration, too large the fracture spacing between the sections will cause the hydraulic fractures to not be able to affect the middle area of the two adjacent sections, too small the spacing of the sections will cause the fracture network to be unable to extend outward due to the interference between the sections, and when the fracture spacing between the fracturing sections of the target coal seam reaches 50–60 m, the fracture network will be fully affected and the channeling phenomenon will be slight. When the amount of construction fluid is small, it is not conducive to the extension of hydraulic cracks and the formation of effective support joints, the larger the amount of construction fluid, the better the fracturing effect, but when the scale of construction fluid reaches a critical value, the increase of filtration loss and the phenomenon of channeling will lead to the continuous increase of scale and the effect of improving fracturing effect is not obvious, so when the construction fluid volume in the target area is 3000–4000 $m^3$, a more ideal fracture length is formed. It also avoids the adverse effects of channeling and filtration, and saves costs. Although increasing the displacement is conducive to the expansion of the fracture network, the excessively high displacement may penetrate the roof and bottom plate of the reservoir, resulting in the obstruction of the expansion of the fracture network, so the fracturing effect is optimal when the current coal seam displacement reaches 18–22 $m^3$/min.

## 5. Conclusions

(1) Based on the logging data of the well area, the three-dimensional natural fracture geology and in situ stress model of the well group area was calculated and established by modifying the calculation formula of deep coalbed methane mechanics and in situ stress, and the hydraulic fracturing simulation results of Well X were compared with the field fracture monitoring results. The results show that the mean value of Young's modulus of the coal seam is 7.5 GPa, the average Poisson's ratio is 0.28, the vertical principal stress is 66 MPa, the minimum horizontal principal stress is 53 MPa, the maximum horizontal principal stress is 61 MPa, and the stress difference between the top and bottom plate is 8–12 MPa.

(2) Based on the qualitative evaluation of the seam network morphology, the change of in situ stress, the quantitative evaluation of the fracture length, the volume of the fracture and the conductivity under different construction parameters, the results show that the optimal construction parameters of a single well are 50 m between sections, 3500–4000 $m^3$ of construction fluid volume for a single section, 20–22 $m^3$/min for construction displacement, and 315 m for the average main seam length.

(3) Based on the qualitative evaluation of the seam network morphology, the change of in situ stress, the quantitative evaluation of the fracture length, the fracture volume and the conductivity under different construction parameters, the results show that the optimal construction parameters are 60 m between sections when the well spacing is 350 m, the construction fluid volume of a single section is 3000–3500 $m^3$, the construction displacement is 18–20 $m^3$/min, and the average main seam length is 305 m. It can be seen that under the influence of stress shadow and fracture disturbance, the construction scale is smaller than that of a single well, and the fracture length is shorter.

(4) In the study area with an initial minimum horizontal principal stress of 53 MPa and a maximum of 57 MPa, the optimal parameters increase the single well's minimum horizontal principal stress by 4.5 MPa and maximum by 2.3 MPa. The post-fracturing stress increase for good groups is more significant due to inter-well stress interference.

**Author Contributions:** Conceptualization, H.M.; Methodology, X.L. (Xugang Liu), X.W. and X.L. (Xinyu Liu); Software, X.Z.; Validation, X.L. (Xugang Liu) and F.C.; Investigation, X.L. (Xugang Liu), X.W. and F.C.; Resources, X.Z.; Data curation, Z.M.; Writing—original draft, Z.M. and X.L. (Xinyu Liu); Writing—review & editing, H.M. All authors have read and agreed to the published version of the manuscript.

**Funding:** This research received no external funding.

**Data Availability Statement:** The original contributions presented in this study are included in the article. Further inquiries can be directed to the corresponding author.

**Conflicts of Interest:** Authors Xugang Liu, Xiang Wang, Fuhu Chen, Xinchun Zhu and Zheng Mao was employed by the company Sinopec. The remaining authors declare that the research was conducted in the absence of any commercial or financial relationships that could be construed as a potential conflict of interest.

## References

1. Pang, X.Q.; Jia, C.Z.; Wang, W.Y. Petroleum geology features and research developments of hydrocarbon accumulation in deep petroliferous basins. *Pet. Sci.* **2015**, *12*, 1–53. [CrossRef]
2. Xu, C.G.; Zhu, G.H.; Ji, H.Q. Exploration progress and reserve increase strategy of onshore natural gas of CNOOC. *China Pet. Explor.* **2024**, *29*, 32–46.
3. Zhang, Q.; Jiang, W.P.; Jiang, Z.B. Present situation and technical research progress of coalbed methane surface development in coal mining areas of China. *Coal Geol. Explor.* **2023**, *51*, 139–158.
4. Qing, S.R.; Cun, Z.; Guang, W. Hydraulic fracture initiation and propagation in deep coalbed methane reservoirs considering weak plane: CT scan testing. *Gas Sci. Eng.* **2024**, *125*, 205286.
5. Zhou, D.H.; Chen, G.; Chen, Z.L. Exploration and development progress, key evaluation parameters and prospect of deep CBM in China. *Nat. Gas Ind.* **2022**, *42*, 43–51.
6. Xie, H.P.; Gao, F.; Ju, Y. Research and exploration of deep rock mass mechanics. *Chin. J. Rock Mech. Eng.* **2015**, *34*, 2161–2178.
7. Xiong, D.; He, J.Y.; Ma, X.F. Fracturing simulation with different perforation positions atdeep coal seam and roof/floor rock—case study at the No.8 deep coal seam of a gas-field in the Ordos basin. *J. China Coal Soc.* **2024**, *49*, 4897–4914. [CrossRef]
8. Lei, N.Z. Predication and Stochastic Modeling of Coalbed Methane Reservoir Properties in Luling Well Field. Ph.D. Thesis, China University of Mining and Technology, Xuzhou, China, 2012.
9. Qu, L.C. Geological Modeling Technology and Application of Coalbed Methane. Ph.D. Thesis, China University of Geosciences (Beijing), Beijing, China, 2013.
10. Huai, Y.C.; Zhang, M.; Yang, L.W. 3D geological modeling of coalbed methane reservoirs based on facies control. *J. Earth Sci. Environ.* **2017**, *39*, 275–282.
11. Ma, P.H.; Shao, X.J.; Huo, M.Y. Geological modeling ideas and methods for coal reservoirs. *Pet. Geol. Oilfield Dev. Daqing* **2018**, *39*, 601–610.
12. Chen, B.; Tang, D.Z.; Zhang, Y.P. Logging inversion and 3D geological modeling of coal body structure. *Coal Sci. Technol.* **2019**, *47*, 88–94.
13. Liu, W.; Zhang, S.A.; Peng, C. Research on pressure drawdown characteristics during coalbed methane drainage. *Coal Technol.* **2019**, *38*, 18–21.
14. Li, Y.; Jiang, Z.B.; Liu, Y.H. Analysis of production characteristics of coalbed methane wells based on 3D geological modeling. *Coal Mine Saf.* **2020**, *51*, 190–194.
15. Lü, J.T.; Zhang, M.; Huai, Y.C. Wang Dianao Coal Reservoir Logging Analysis and Fine Geological Modeling Technology: A Case Study of Coalbed Methane Area in Surat Basin, Australia. *J. China Coal Soc.* **2020**, *45*, 1824–1834.
16. Zhou, Y.; Zhang, S.H.; Tang, S.H. 3D modeling of gas content in coalbed methane reservoirs. *Coal Geol. Explor.* **2020**, *48*, 96–103.
17. Xiang, L.W.; Jie, N.P.; Yi, J. Morphology and propagation of supercritical carbon dioxide-induced fractures in coal based on a non-destructive surface extraction method. *Fuel* **2025**, *384*, 133948.
18. An, Q.; Yang, F.; Yang, R.Y. Volume fracturing practice and understanding of deep coalbed methane in the Shenfu area, Ordos Basin. *J. Coal Sci. Eng.* **2024**, *49*, 2376–2393.
19. Li, B.; Yang, F.; Zhang, H.J. Research on efficient development technology of deep coalbed methane in the Shenfu area. *Coal Geol. Explor.* **2024**, *52*, 57–68.
20. Yang, F.; Li, B.; Wang, Q.J. Large-scale limit volume fracturing technology for deep coalbed methane horizontal wells. *Pet. Explor. Dev.* **2024**, *51*, 389–398.
21. Liu, X. Comparison of fracture network morphology in coal reservoirs under different fracturing scales. *Reserv. Eval. Dev.* **2024**, *14*, 510–518.
22. Xiong, X.Y.; Zhen, H.B.; Li, S.G. Multi-round turn fracturing technology and application for deep coalbed methane in the Da Ning-Ji County area. *Coal Geol. Explor.* **2024**, *52*, 147–160.

23. Zhong, Z.T.; Zhuang, X.; Yu, H.W. Numerical simulation of fracture propagation in deep coalseam reservoirs. *Energy Sci. Eng.* **2023**, *11*, 3559–3574.
24. Zhi, H.Z.; Tian, Y.W.; Jian, C.G. Study on the Spatial and Temporal Evolution Law of Large-Size Fracture Propagation in Deep Coalbed Based on Acoustic Emission Technology. *Rock Mech. Rock Eng.* **2025**, *58*, 3799–3814. [CrossRef]

processes

MDPI

Article

# Development of Volumetric Adsorption Isotherms for Volcanic Fly Ash from Egypt for Carbon Dioxide Capture Under Elevated Pressure and Temperature

**Sherif Fakher *, Abdelaziz Khlaifat, Ann Maria Salib and Ali Elsayed**

Department of Petroleum and Energy Engineering, The American University in Cairo, Cairo 11835, Egypt; abdelaziz.khlaifat@aucegypt.edu (A.K.); annmaria@aucegypt.edu (A.M.S.); aelsa526@aucegypt.edu (A.E.)
* Correspondence: sherif.fakher@aucegypt.edu

**Abstract:** One of the most promising methods for direct carbon dioxide ($CO_2$) capture from the atmosphere is using material-based adsorption. Fly ash, a solid waste material, has been found to have good adsorption potential for $CO_2$. Since different fly ashes have different properties, their $CO_2$ adsorption behaviors differ; therefore, it is important to develop separate isotherms for each fly ash to quantify its $CO_2$ adsorption capacity. This research investigates the adsorption capacity of an extremely abundant volcanic fly ash in Egypt that is currently being researched for use in $CO_2$ capture applications in Egypt. Adsorption was measured using the volumetric adsorption principle. Four adsorption isotherms for the volcanic fly ash were developed at different temperatures, including 23, 40, 60, and 80 °C. The adsorption capacity was found to be impacted by the temperature of the experiment, the pressure of the $CO_2$, and the interactions occurring between the $CO_2$ and the fly ash. As the temperature increased, the adsorption capacity increased significantly. This was primarily due to the expansion of fly ash particles at elevated temperatures, which resulted in a higher contact surface area between the fly ash and the $CO_2$. This created more adsorption sites for the $CO_2$, therefore increasing the $CO_2$ adsorption potential significantly. This research can help facilitate the use of volcanic fly ash for $CO_2$ capture applications in Egypt in the near future, hence reducing the overall $CO_2$ emissions to the atmosphere.

**Keywords:** volumetric adsorption; volcanic fly ash; carbon dioxide capture

## 1. Introduction

$CO_2$ capture is a vital research topic to reduce greenhouse gas emissions into the atmosphere [1–9]. There are many methods for carbon capture; however, the majority of these methods are extremely costly, and therefore, their upscaling to real field plants is limited. $CO_2$ adsorption to the surface of different materials is one of the most effective methods for direct carbon capture from the atmosphere [10–16]. The key to $CO_2$ adsorption is the utilization of a material that has a high affinity for $CO_2$ and also has a considerably large surface area for $CO_2$ adsorption to occur in order to maximize $CO_2$ capture [16–23].

Different materials have been studied for $CO_2$ adsorption. Karimi [1] conducted a comprehensive review of $CO_2$ separation and capture using different materials via adsorption. Karimi [1] conducted a comprehensive review of $CO_2$ separation and capture using different materials via adsorption. Reddy [2] highlighted the significance of different physical materials on $CO_2$ direct adsorption. The material researched had a high porosity and also a large contact surface area to allow for a high adsorption capacity. Boer [3] studies the



https://doi.org/10.3390/pr13051570

use of zeolites as $CO_2$ adsorbent material for carbon capture and storage. The zeolites were modified in a way that they can selectively adsorb the $CO_2$, thus maximizing the adsorption capacity. Mehrmohammadi [4] utilized artificial intelligence through neural networks to model the $CO_2$ adsorption capacity on different porous structures through the utilization of preliminary experimental results obtained using different materials. Giraldo [5] investigated the $CO_2$ adsorption on carbon particles with modified surfaces leached through the utilization of nitric acid and ammonium aqueous solution. The modified surfaces had a larger overall adsorption surface area, thus improving $CO_2$ adsorption capacity. Zeng [6] studied the ability of different porous materials conventionally used for $CO_2$ capture to be utilized with flue gas for selective $CO_2$ adsorption. The flue gas was composed of different gases, primarily carbon monoxide and $CO_2$. Yu [7] performed a comprehensive review of $CO_2$ capture using both adsorption methods and absorption methods. The research compared different methods and highlighted the advantages and limitations of each method.

Fly ash is considered a waste material that occurs as a byproduct of the combustion of other materials. Fly ash has been utilized extensively as a filler in many different industries. The main applications of fly ash are as cementing materials, either as a percentage in concrete or as a full replacement for conventional cement. Fly ash has also been researched for use in pavements instead of cement due to its high durability and significantly lower cost compared to conventional cement. Fly ash has also been researched as a $CO_2$ capture material via adsorption [5–12]. The main limitation of fly ash in this area is the different properties of different fly ash types due to their widespread origin. Fly ash is a byproduct of the combustion of other materials. The type of material and its properties generate a fly ash with specific properties as well. Since the fly ash has different properties, their adsorption behavior differs significantly; therefore, there is a large variation in the adsorption capacity and behavior of fly ash worldwide [3].

Fly ash has a variety of properties depending on its origin. The type of biomass/coal used to generate the fly ash therefore has a strong impact on the physical and chemical properties of the ash [20–25]. Fly ash usually exists as a fine powder with colors ranging from brown to black, depending on the volume of unburnt carbon [26–28]. The particles are mostly spherical in shape with an amorphous nature and are, in most cases, solid, with some fly ash particles being hollow due to rapid formation, such as volcanic ash [29–34]. For coal-based fly ash, if the source of the coal is bituminous or subbituminous coal, the silicon content is usually high, reaching up to 60%, whereas lignite-based fly ash is lower, reaching a maximum of 40%. The iron content in the bituminous coal is much higher than in the other types, reaching 40%. The sulfur content in all three types is low, with the highest resulting from lignite. Volcanic fly ash has a considerable concentration of sulfur, reaching up to 10% in some cases [35–39]. Fly ash is classified into two broad categories including Class F and Class C. Class F has a low sodium, potassium, and calcium concentration, whereas Class C has a calcium concentration of up to 40% and a high concentration of sodium and potassium, making it more favorable in terms of usage as a binding material for cementing and other applications [7,39–46].

This research investigates the adsorption capacity of an extremely abundant volcanic fly that has been purified from heavy metal ash in Egypt, which is currently being researched for use in $CO_2$ capture applications in Egypt. Volcanic ash has multiple advantageous properties compared to coal-based ash due to its large surface area and its hollow structure that can allow for large adsorption capacities. Four adsorption isotherms were developed at different temperatures and pressures in order to assess the performance of this fly ash under different conditions. The selected temperatures were chosen based on the different expected applications and different governorates where the volcanic fly ash will be used in Egypt.

## 2. Experimental Description

The experimental material, setup, and procedure followed to conduct all the experiments are mentioned and explained.

### 2.1. Experimental Material

The material used to conduct the experiments is as follows:

- **Fly Ash**: The fly ash used to conduct all the experiments in this research was provided as a greyish powder with a high aluminosilicate concentration. The volcanic fly ash provided was purified by the supplier to remove any traces of heavy metals and sulfur. The fly ash had 30% aluminum oxide, 10% iron oxide, 10% calcium oxide, 10% silicon oxide, and 25% sodium oxide, with the remaining concentration composed of sulfur oxide, potassium, and magnesium oxide, and loss on ignition. The surface area of the fly ash ranged between 3 and 7 $m^2/g$;
- **Water Bath**: A water bath with distilled water was used to provide a constant temperature to the experimental setup across the experiments. The water bath had a slotted lid on it to ensure minimal water evaporation, with no pressure buildup;
- **Precision Balance**: A precision scale was used to weigh the fly ash packed into the adsorption vessel to ensure that all the experiments used the same weight of fly ash;
- **$CO_2$ Cylinder**: The $CO_2$ cylinder was used to provide a source of $CO_2$ for the experiment. It was provided as a $CO_2$ cylinder with a purity of 99.99%;
- **Helium Cylinder**: Helium was used to measure the void space volume between the fly ash particles prior to conducting the adsorption experiment;
- **Pressure Transducers**: Pressure transducers were connected to the sample cell and the reference cell to measure the pressure differential and the pressure equilibrium values for adsorption calculation.

### 2.2. Experimental Setup

An illustration of the experimental setup used to conduct all the experiments is shown in Figure 1. The setup is composed of a volumetric adsorption setup composed of a sample cell that houses the fly ash sample, a reference cell that houses the $CO_2$ initially before expansion, pressure transducers to record the pressure, and a water bath to maintain constant temperature for the duration of the experiment. The transducers were calibrated based on the recommendation of the manufacturer by zeroing the transducer before conducting the experiment. This was conducted through the software. Once zeroed, the temperature reading was ensured to be similar to that of the room temperature to ensure that the transducers were functioning correctly. The $CO_2$ is provided through a high-pressure $CO_2$ cylinder with a pressure regulator to control the flow of $CO_2$. The pressure transducers are connected directly to a computer to log and record the pressure values for the duration of the experiment. The logging system has an accuracy of four decimal places and can log four pressure readings every second.

### 2.3. Experimental Procedure

The procedure followed to conduct all the experiments in this research is as follows:

- The sample cell was fully packed with the volcanic fly ash sample. The weight of the fly ash sample was determined to be the same for all experiments. The sample cell was then sealed and vacuumed. The fly ash was used in the same manner in which it was provided, without the utilization of any additional alkaline activators. This was due to the high initial concentration of alkaline components in the fly ash;

- The experimental setup is placed in the water bath and is left for 12 h until the temperature homogenizes;
- The sample cell was filled with helium at the design pressure of the experiment. Following this, the void space volume was measured using helium expansion. This was used in the adsorption calculations to remove errors in the calculation due to the void space present between the fly ash particles. When the shale particles are placed in the reference cell, some small spaces will exist between the particles. This volume is referred to as the void volume and must be accounted for during each experiment, since this is considered excess volume that will be occupied by the $CO_2$ and will not contribute to the adsorption. The void space therefore signifies the pores present between the grains and is thus similar to the total porosity of the system. The void volume is also an essential factor in the adsorption calculations, as was shown previously. The void volume is usually measured using a gas with extremely low adsorption; the gas used in this study to measure void volume is helium, which is the most widely used gas to measure adsorption. Helium has many advantages that make it extremely suitable to use when measuring the void space;
- Once the void space is measured, the $CO_2$ adsorption experiment commences. The pressure transducers begin recording the pressure in both cells. Initially, the sample cell reads a negative gauge pressure due to vacuum, while the reference cell reads the reference pressure of the experiment;
- The valve connecting the sample cell and the reference cell is opened, and the $CO_2$ is allowed to expand in the vacuumed sample cell. The pressure is then left to equilibrate between both cells. This can take between 8 h to 3 days, depending on the pressure and temperature conditions;
- Once equilibrium pressure is reached, the experiment is concluded, and the pressure is used to calculate the adsorption capacity for the design pressure. The experiment is then repeated at the same temperature for different pressure values. By measuring the $CO_2$ adsorption at the same temperature at different pressure values, the adsorption isotherm for the fly ash can be determined for this specific temperature;
- Once the isotherm is developed at a specific temperature, the temperature is then changed, and the process is repeated to construct the isotherm for the elevated temperature.

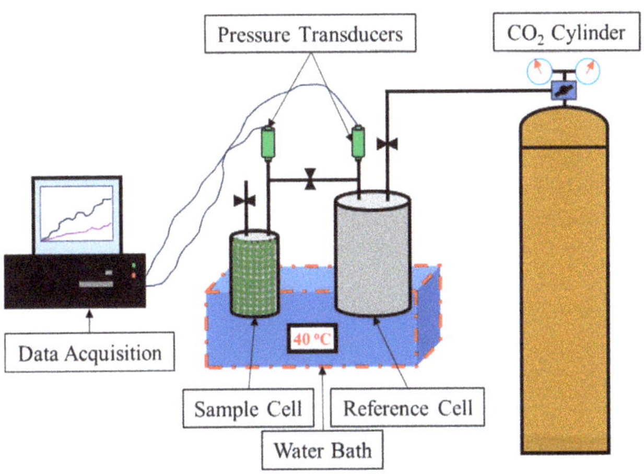


**Figure 1.** Volumetric adsorption experimental setup.

## 3. Volumetric Adsorption

Adsorption is classified as physical adsorption (physisorption) and chemical adsorption (chemisorption). Physical adsorption is a low energy adsorption that occurs in multiple layers. It does not require high energy for the adsorbate to adhere to the adsorbent and does not require high pressure to desorb the adsorbate. It is a relatively high capacity adsorption due to its ability to form layers of adsorbate on the surface of the adsorbent. $CO_2$ adsorption to volcanic fly ash exhibited physical adsorption based on the analysis of the adsorption behavior during experimentation [7,42–48].

Chemical adsorption is a high energy single layer adsorption. The bond between the adsorbate and the adsorbent form under high energy is considered a much stronger bond compared to physical adsorption. This bond requires high energy to dissociate and is therefore also more stable than physical adsorption bonds. Chemical adsorption does not exhibit the stacking effect found in physical adsorption and therefore cannot accommodate a large adsorption capacity compared to physical adsorption [28,49–52]. The conditions in which the experiments were conducted in this research were not suitable for chemical adsorption to occur; therefore, all the adsorption isotherms developed in this study followed physical adsorption.

There are many methods by which adsorption can be measured experimentally. The two most common methods for adsorption measurement are gravimetric adsorption and volumetric adsorption. Gravimetric adsorption relies on mass change to determine the volume of adsorption. It requires the utilization of an extremely high precision scale to be able to detect the very low weight adsorption value of the gas, especially in small experimental setups. As the volume of adsorption increases, the ability of gravimetric adsorption measurement to produce more reliable readings increases. Volumetric adsorption measurement relies on measurement of volume change due to adsorption. This method requires prior knowledge of the void space volume in order to be able to differentiate between the fluid present in the void spaces and the fluid that is actually adsorbed to the surface. Volumetric adsorption relies on knowledge of the properties of the fluid used, especially if the fluid is compressible, such as gases. Not accounting for the compressibility of gases would produce highly erroneous adsorption results. The compressibility of gases is usually accounted for using a gas super compressibility factor, determined experimentally or using correlation. The Dranchuk–Abou–Kassem correlation was used in this research to determine the value of super compressibility factors for the $CO_2$. Based on this, the volume of adsorbed $CO_2$ was calculated using the following equation:

$$n_{ads} = \left(\frac{P\Delta V}{zRT}\right)_{pump} - \frac{P_{af} - V_{void}}{z_{af}RT}$$

where $n_{ads}$ is the number of moles of adsorbed $CO_2$, $P_{af}$ is the pressure of the experiment after $CO_2$ expansion from the reference cell to the sample cell, $V_{void}$ is the void space volume measured using helium, and $z$ is the $CO_2$ super compressibility factor determined using Dranchuk–Abou–Kassem correlation.

This research utilizes the concept of volumetric adsorption to measure the adsorption capacity of the $CO_2$ onto the surface of the volcanic fly ash sample [28,48–52]. During $CO_2$ adsorption, multiple adsorption/desorption hysteresis cycles are conducted on the same material. Based on the initial composition of the fly ash, this can result in a change in the properties of the fly ash, especially with constant repetition of the cycles. This research conducted only single adsorption desorption cycles, which is a limitation for the actual utilization of the fly ash in real tests. Future experiments will be conducted to assess the stabilized adsorption potential through cyclic stability testing [28,48–68].

## 4. Results and Analysis

$CO_2$ adsorption isotherms for the volcanic fly ash were developed at different temperatures including 23, 40, 60, and 80 C. All adsorption isotherms can be used for $CO_2$ adsorption capacity prediction through extension of the experimental results for the isotherms. All isotherms were developed using pure $CO_2$ with no additional gases. The research did not take into consideration gas selectivity, which is an important parameter for real life adsorption applications.

### 4.1. Adsorption Isotherm

An adsorption isotherm is a plot that can be used to determine the adsorption behavior for a specific material at different pressures. It can also be used to determine the maximum adsorption capacity for a material at a specific temperature regardless of the increase in pressure. At different temperatures, different adsorption isotherms are developed.

The isotherm presents a relationship between the number of moles of gas adsorbed per unit mass of the adsorbent on the y-axis, the volcanic fly ash in this research, and the pressure used to measure the adsorption value on the x-axis. The pressure on the x-axis is usually represented as a pressure ratio, as is used in this research. This is a ratio between the pressure used in the experiment, denoted P, and the saturation pressure of the $CO_2$ at the specified temperature condition, denoted $P_o$. The saturation pressure is a standard value that is determined experimentally and varies with temperature. Once experimental pressure reaches the saturation pressure of the $CO_2$, the maximum adsorption capacity is reached. Exceeding the saturation pressure will result in not significant increase in the adsorption capacity, as long as the fly ash properties remain unchanged and the $CO_2$ phase remains unchanged.

### 4.2. Adsorption Isotherm: 23 °C

The initial $CO_2$ adsorption isotherm was developed at a temperature close to the average room temperature in Egypt, 23 °C. The saturation pressure of the $CO_2$ at 23 °C was determined mathematically to be 903.6 psi. Based on this, adsorption was measured at different pressures and then divided by the saturation pressure to develop the isotherm. The $CO_2$ adsorption isotherm using the volcanic fly ash sample at 23 °C is presented in Figure 2. Initially, the adsorption behavior is linear until a pressure ratio value of 0.44. Beyond this, the adsorption capacity seems to normalize at a value of 0.875 mmol/g. Based on these results, initially, as the pressure increased, more $CO_2$ could adsorb to the fly ash sample. Once the maximum adsorption capacity at 23 °C was reached, increasing the pressure had no impact on $CO_2$ adsorption; therefore, the adsorption isotherm plot normalized at the maximum adsorption value for the volcanic fly ash at this temperature.

### 4.3. Adsorption Isotherm: 40 °C

The second adsorption isotherm developed in this research was at 40 °C, shown in Figure 3. This temperature resembles the average high temperature in different governorates in Egypt. The saturation pressure of $CO_2$ at this temperature was determined to be 1391.3 psi. Increasing the temperature resulted in an increase in the $CO_2$ adsorption capacity of the fly ash. This was due to the expansion of the fly ash particles in the sample cell, which resulted in an increase in the overall surface area available for adsorption. This was verified when the void space volume, the volume not used for adsorption, was measured to be lower than that of the 23 °C experiment. The adsorption trend for pressure at 40 °C exhibited a more linear behavior compared to that of 23 °C, mainly due to the higher adsorption capacity. The maximum value for $CO_2$ adsorption reached was more than 3 mmol per g of fly ash. The line begins to deviate from its linear behavior at elevated temperatures, indicating that the maximum adsorption capacity at this temperature is close to being reached. Since the same injection pressures were used across all experiments and due to the variation in $CO_2$ saturation pressure with temperature, the pressure ratio on the x-axis reached a lower maximum value for every increase in temperature.

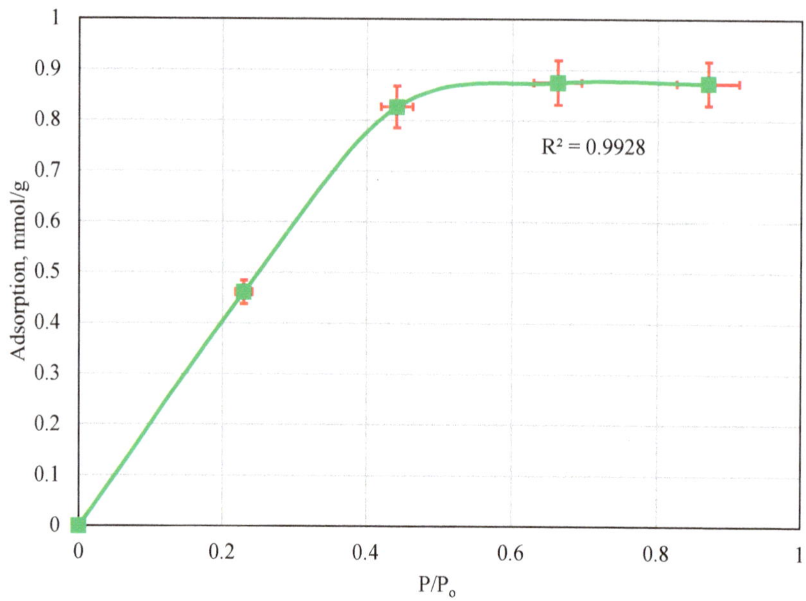


**Figure 2.** $CO_2$/volcanic fly ash adsorption isotherm at 23 °C.

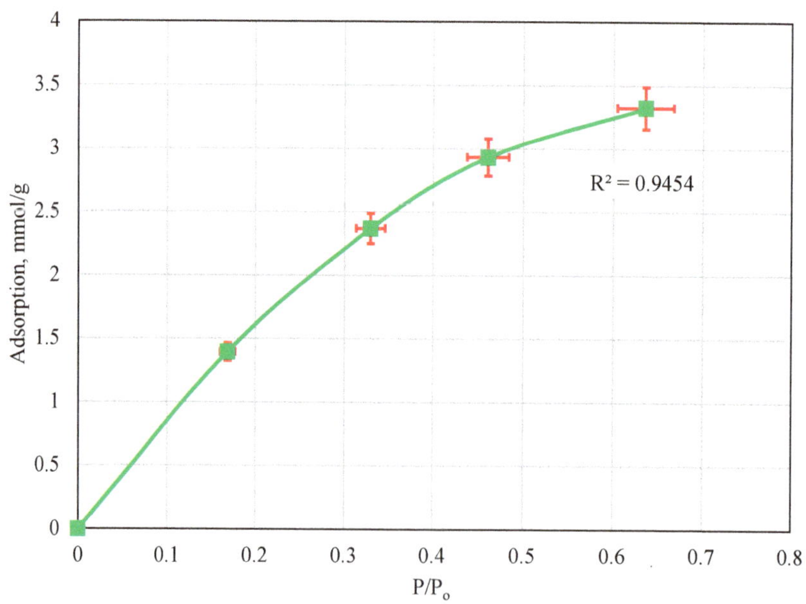


**Figure 3.** $CO_2$/volcanic fly ash adsorption isotherm at 40 °C.

### 4.4. Adsorption Isotherm: 60 °C

When the temperature of the experiment increased to 60 °C, the $CO_2$ saturation pressure was determined to be 2245.2 psi. Based on this, the adsorption isotherm for the 60 °C experiment was developed as shown in Figure 4. The adsorption capacity increased further with the increase in temperature, which supports the findings in the 40 °C adsorption isotherm. The void space was also found to decrease with the increase in temperature. The maximum adsorption capacity exceeded 5 mmol/g for the highest-pressure value used in the experiment. It is important to note that the behavior is linear with no obvious deviation in the line due to the much higher adsorption capacity still possible. This can be determined by generating a trendline for the adsorption isotherm following the same trend behavior as the 23 °C experiment. Using this, the maximum adsorption capacity at 60 °C can be estimated.

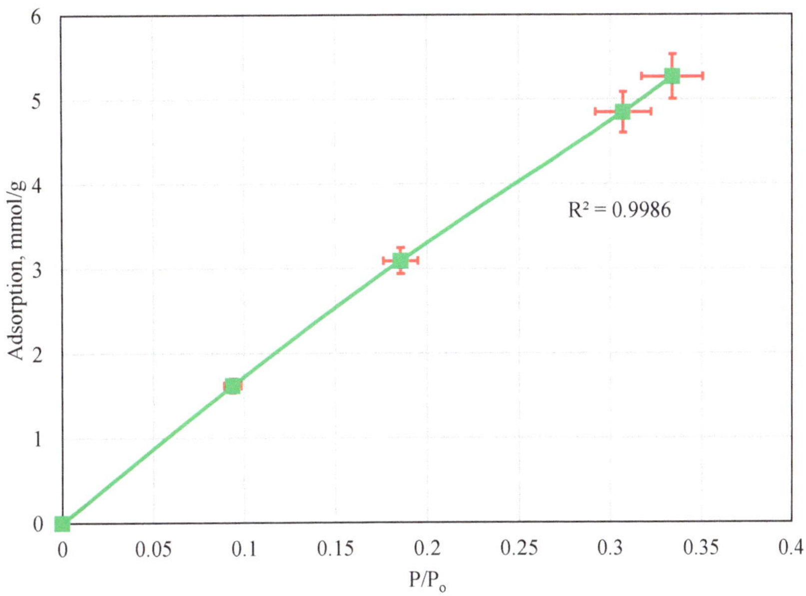


**Figure 4.** $CO_2$/volcanic fly ash adsorption isotherm at 60 °C.

### 4.5. Adsorption Isotherm: 80 °C

The highest temperature isotherm generated in this research was at 80 °C. This temperature was selected since many exhaust flue gases are generated from factories and different facilities at temperatures near this value. The $CO_2$ adsorption isotherm for the fly ash is presented in Figure 5. At 80 °C, the highest adsorption capacity for the fly ash was reached: more than 9 mmol/g. The $CO_2$ saturation temperature was calculated to be 3523.5 psi. The linear behavior of the line indicates a much higher predicable adsorption capacity at higher pressure values. Since the 80 °C experiment was used to model flue gas being produced from exhausts, lower pressure values are more realistic, since the average pressures of flue gas from exhausts is 150 psi. Overall, the volcanic fly ash exhibits a good potential to be used as a $CO_2$ capture material due to its high adsorption potential, especially at elevated temperatures.

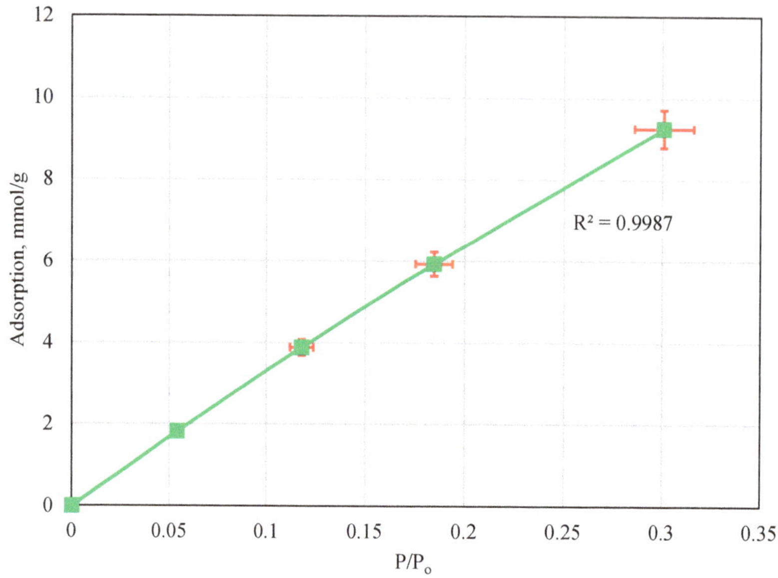


**Figure 5.** $CO_2$/volcanic fly ash adsorption isotherm at 80 °C.

## 5. Discussion

In order to show the difference between mass-based adsorption behavior (gravimetric adsorption) and volume-based adsorption behavior (volumetric), the adsorption isotherms were converted from moles to cubic centimeters (mL) and grams of $CO_2$ per gram of fly ash. For all the results, the mass behavior follows the same behavior as the molar mass, while the volume behavior follows an opposite trend.

Figure 6 shows the mass and volume of $CO_2$ adsorbed per unit mass of fly ash for the 23 °C and 40 °C experiments. As the temperature increases, the total mass of $CO_2$ absorbed also increases. This was also observed in the adsorption isotherms, in mmol instead of grams. Increasing the pressure resulted in a decrease in the volume of $CO_2$ per gram of fly ash due to the compression of the $CO_2$ molecules at elevated pressures. Although the volume decreased, the overall mass increased with the increase in pressure, thus indicating an overall increase in the adsorption capacity at elevated pressure conditions. The conversion between the mass and volume was performed by using the density of $CO_2$ at each pressure and temperature condition.

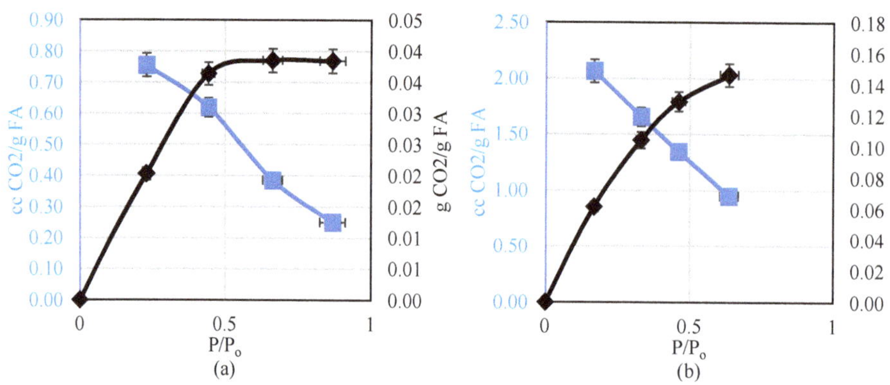


**Figure 6.** $CO_2$ adsorption volume and mass per gram of fly ash at (**a**) 23 °C and (**b**) 40 °C.

Figure 7 shows the mass and volume of $CO_2$ adsorbed per unit mass of fly ash for the 60 °C and 80 °C experiments. Increasing the temperature increased the volume of $CO_2$ adsorbed per unit mass of fly ash, as is evident from the comparison between the adsorption values in Figures 6 and 7. With the increase in pressure, the volume of $CO_2$ decreased due to the compression of the $CO_2$ molecules. The rate of decrease in volume of the $CO_2$ in the elevated temperature experiments is higher than the lower temperature experiments. This is mainly due to the larger volume of $CO_2$ available, and the higher $CO_2$ saturation pressure at the higher temperature experiments compared to the low temperature experiments. Although the volume decreased with pressure, the overall mass increased, which is an indication that the total amount of $CO_2$ adsorbed increased.

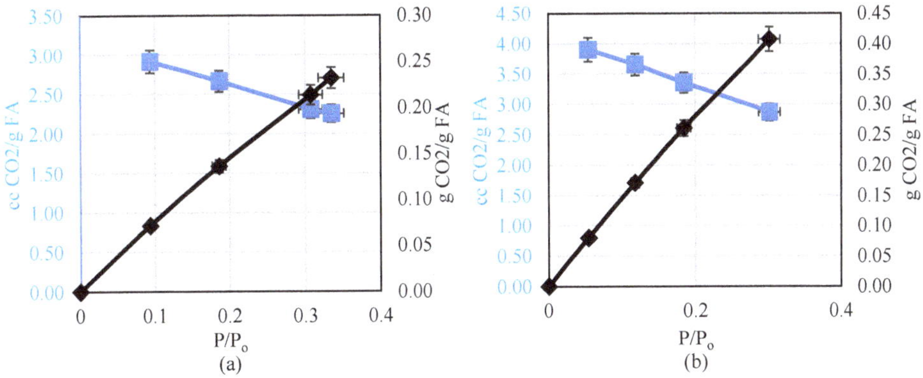


**Figure 7.** $CO_2$ adsorption volume and mass per gram of fly ash at (**a**) 60 °C and (**b**) 80 °C.

Compared to other types of fly ash, the volcanic fly ash used in this research had a relatively higher adsorption capacity. This is primarily due to the conditions in which the volcanic fly ash was formed, which resulted in higher adsorption sites and hence a larger adsorption capacity. The adsorption values for coal-based fly ash, biomass (plant)-based fly ash, and the volcanic fly ash used in this research are presented in Table 1.

**Table 1.** Adsorption volume of different types of fly ash.

| Fly Ash Type | Maximum Adsorption Volume | Reference |
|---|---|---|
| Coal-Fired Power Plant | 0.6 g/g FA | [37] |
| Pulversized Coal | 5 mmol/g FA | [34] |
| Burnt Biomass | 8 $cm^3$/g FA | [49] |
| Biomass | 0.28 kg/kg FA | [28] |
| Coal | 8 mmol/g FA | [52] |
| Volcanic | 9 mmol/g FA | Current Study |

## 6. Conclusions

This research develops $CO_2$ adsorption isotherms for a volcanic fly ash that is currently being researched for use as a $CO_2$ capture agent in Egypt. Isotherms were developed at 23, 40, 60, and 80 °C. Based on these isotherms, the $CO_2$ adsorption capacity under different conditions can be predicted. The main findings obtained from this research are as follows:

- The volcanic fly ash used in this research has a good potential for $CO_2$ storage applications based on its ability to adsorb a large volume of $CO_2$ on its surface;
- Increasing the temperature of the fly ash resulted in expansion of the particles, which in turn increased the available surface area for adsorption, thus increasing $CO_2$ capture;
- Increasing the pressure of the injected $CO_2$ resulted in an increase in the adsorption capacity. This trend was observed until the maximum available adsorption sites were occupied by $CO_2$;
- Increasing pressure allows for the compression of a larger volume of $CO_2$ in the available adsorption sites. This was observed by comparing the change in volume and mass of $CO_2$ with pressure at each temperature.

**Author Contributions:** Conceptualization: S.F. and A.K.; Methodology: S.F., A.K. and A.E.; Software: S.F. and A.M.S.; Validation: S.F., A.K., A.E. and A.M.S.; Formal Analysis: S.F.; Writing—Original Draft: S.F. and A.K.; Supervision: S.F. and A.K.; Project Administration: S.F.; Funding Acquisition: S.F. and A.K. All authors have read and agreed to the published version of the manuscript.

**Funding:** Funding for this research was provided by The American University in Cairo through the Faculty Support Grant.

**Data Availability Statement:** The raw data supporting the conclusions of this article will be made available by the authors on request.

**Acknowledgments:** The authors would like to thank The American University in Cairo for funding this research project through their Faculty Support Grant.

**Conflicts of Interest:** The authors declare no conflicts of interest.

## References

1. Karimi, M.; Shirzad, M.; Silva, J.A.C.; Rodrigues, A.E. Carbon dioxide separation and capture by adsorption: A review. *Environ. Chem. Lett.* **2023**, *21*, 2041–2084. [CrossRef] [PubMed]
2. Reddy, M.S.B.; Ponnamma, D.; Sadasivuni, K.K.; Kumar, B.; Abdullah, A.M. Carbon dioxide adsorption based on porous materials. *RSC Adv.* **2021**, *11*, 12658–12681. [CrossRef] [PubMed]
3. Boer, D.G.; Langerak, J.; Pescarmona, P.P. Zeolites as Selective Adsorbents for $CO_2$ Separation. *ACS Appl. Energy Mater.* **2023**, *6*, 2634–2656. [CrossRef]
4. Mehrmohammadi, P.; Ghaemi, A. Investigating the effect of textural properties on $CO_2$ adsorption in porous carbons via deep neural networks using various training algorithms. *Sci. Rep.* **2023**, *13*, 21264. [CrossRef]
5. Giraldo, L.; Vargas, D.P.; Moreno-Piraján, J.C. Study of $CO_2$ Adsorption on Chemically Modified Activated Carbon with Nitric Acid and Ammonium Aqueous. *Front. Chem.* **2020**, *8*, 543452. [CrossRef]
6. Zeng, H.; Qu, X.; Xu, D.; Luo, Y. Porous Adsorption Materials for Carbon Dioxide Capture in Industrial Flue Gas. *Front. Chem.* **2022**, *10*, 939701. [CrossRef]
7. Yu, C.-H.; Huang, C.-H.; Tan, C.-S. A Review of $CO_2$ Capture by Absorption and Adsorption. *Aerosol Air Qual. Res.* **2012**, *12*, 745–769. [CrossRef]
8. Gunawardene, O.H.P.; Gunathilake, C.A.; Vikrant, K.; Amaraweera, S.M. Carbon Dioxide Capture through Physical and Chemical Adsorption Using Porous Carbon Materials: A Review. *Atmosphere* **2022**, *13*, 397. [CrossRef]
9. Maia, R.A.; Louis, B.; Gao, W.; Wang, Q. $CO_2$ adsorption mechanisms on MOFs: A case study of open metal sites, ultra-microporosity and flexible framework. *React. Chem. Eng.* **2021**, *6*, 1118–1133. [CrossRef]
10. Ren, F.; Liu, W. Review of $CO_2$ Adsorption Materials and Utilization Technology. *Catalysts* **2023**, *13*, 1176. [CrossRef]
11. Elsayed, A.; Fakher, S. Investigating Asphaltene Damage Reduction in Wellbores, and Pipelines Using Alkaline and Surfactant Chemical Agents. In Proceedings of the Mediterranean Offshore Conference, Alexandria, Egypt, 20–22 October 2024. [CrossRef]
12. Fakher, S.; Tamer, S.; Eltohamy, M.; Ali, M.; Alsakkaf, A.; Gehad, M.; Haydar, R.; Mourad, K. Development and Optimization of Epoxy-Resin Based Cement Reinforced with Low-Cost Fly Ash for High Durability Environmentally Friendly Cement. In Proceedings of the 57th U.S. Rock Mechanics/Geomechanics Symposium, Atlanta, GA, USA, 25–28 June 2023. [CrossRef]
13. Haydar, R.; Fakher, S. Harsh Environmental Effects on Low Density Fly Ash Proppants. In Proceedings of the Mediterranean Offshore Conference, Alexandria, Egypt, 20–22 October 2024. [CrossRef]

14. Youssef, H.; Fakher, S. Evaluating the Performance of Class F Fly Ash Compared to Class G Cement for Hydrocarbon Wells Cementing: An Experimental Investigation. *Materials* **2024**, *17*, 2710. [CrossRef]
15. Ebied, A.; Fakher, S.; Kayed, H. Experimental Investigation of Inclusion of Various Nanocarbon Black Concentrations on Mechanical Characteristics of Oil-Well Cement Slurries in High-Pressure High-Temperature Conditions. *Int. J. Concr. Struct. Mater.* **2025**, *19*, 10. [CrossRef]
16. Raz, H.; Fakher, S. Investigating and Evaluating Novel Fly Ash-Based Proppant Compressive Strength Under Various Environmental Conditions. *Materials* **2025**, *18*, 399. [CrossRef] [PubMed]
17. Sherif, F.; Khlaifat, A.; Mokhtar, K.; Abdelsamei, M. Assessment of Two Crosslinked Polymer Systems Including Hydrolyzed Polyacrylamide and Acrylic Acid–Hydrolyzed Polyacrylamide Co-Polymer for Carbon Dioxide and Formation Water Diversion Through Relative Permeability Reduction in Unconsolidated Sandstone Formation. *Polymers* **2024**, *16*, 3503. [CrossRef]
18. Sherif, F.; Khlaifat, A.; Hassanien, A. Carbon Dioxide Capture under Low-Pressure Low-Temperature Conditions Using Shaped Recycled Fly Ash Particles. *Gases* **2024**, *4*, 117–132. [CrossRef]
19. Sherif, F.; El-Sayed, A.; Sameh, L.; Abdeltawab, B. Evaluating a Novel Fly Ash Resin-Reinforced Cement's Interactions under Acidic, Basic, High-Salinity, and High-Temperature Conditions. *Polymers* **2023**, *15*, 3404. [CrossRef]
20. Sherif, F.; Khlaifat, A.L. Experimental Investigation of Polymer Injection in High Permeability Conduits for Material Sustainability and Behavior in Oil Reservoirs. *Polymers* **2023**, *15*, 2950. [CrossRef] [PubMed]
21. Sherif, F.; Yousef, A.; Al-Sakkaf, A.; Eldakar, S. Asphaltene onset pressure measurement and calculation techniques: A review. *Petroleum* **2024**, *10*, 191–201. [CrossRef]
22. Wang, Y.; Hu, X.; Guo, T.; Hao, J.; Si, C.; Guo, Q. Efficient $CO_2$ adsorption and mechanism on nitrogen-doped porous carbons. *Front. Chem. Sci. Eng.* **2021**, *15*, 493–504. [CrossRef]
23. Sukkathanyawat, H.; Bampenrat, A.; Jarunglumlert, T.; Prommuak, C. Modification of waste sugarcane bagasse fly ash for $CO_2$ capture application. *Mater. Renew. Sustain. Energy* **2022**, *11*, 267–276. [CrossRef]
24. Majchrzak-Kucęba, I.; Nowak, W. Development of Fly Ash-Based Sorbent to Capture $CO_2$ from Flue Gas. In Proceedings of the 20th International Conference on Fluidized Bed Combustion, Xi'an, China, 18–21 May 2009; Yue, G., Zhang, H., Zhao, C., Luo, Z., Eds.; Springer: Berlin/Heidelberg, Germany, 2009. [CrossRef]
25. Islam, R.; Biswas, B.; Naidu, R. $CO_2$ Capture Using Zeolite Synthesized from Coal Fly Ash and Its Subsequent Utilization for Fire Retardation and Dye Removal. *ACS Sustain. Resour. Manag.* **2024**, *1*, 799–809. [CrossRef]
26. Shen, X.; Yan, F.; Li, C.; Zhang, Z.; Zhang, Z. A green synthesis of PEI-nano-$SiO_2$ adsorbent from coal fly ash: Selective and efficient $CO_2$ adsorption from biogas. *Sustain. Energy Fuels* **2021**, *5*, 1014–1025. [CrossRef]
27. Guo, F.; He, J.; Johnson, P.A.; Aryana, S.A. Stabilization of $CO_2$ foam using by-product fly ash and recyclable iron oxide nanoparticles to improve carbon utilization in EOR processes. *Sustain. Energy Fuels* **2017**, *1*, 814–822. [CrossRef]
28. Boycheva, S.; Marinov, I.; Zgureva-Filipova, D. Studies on the $CO_2$ Capture by Coal Fly Ash Zeolites: Process Design and Simulation. *Energies* **2021**, *14*, 8279. [CrossRef]
29. Zhao, Z.; Zhang, K.; Luo, J.; Wu, M.; Wang, X.; Wang, K.; Liu, S. A Study on the Mechanisms of Coal Fly Ash to Improve the $CO_2$ Capture Efficiency of Calcium-Based Adsorbents. *Sustainability* **2024**, *16*, 8139. [CrossRef]
30. Wu, W.; Zhang, M.; Wang, C.; Tao, L.; Bu, J.; Zhu, Q. Harnessing Ash for Sustainable $CO_2$ Absorption: Current Strategies and Future Prospects. *Chem.—Asian J.* **2024**, *19*, e202400180. [CrossRef]
31. Czuma, N.; Casanova, I.; Baran, P.; Szczurowski, J.; Zarębska, K. $CO_2$ sorption and regeneration properties of fly ash zeolites synthesized with the use of differentiated methods. *Sci. Rep.* **2020**, *10*, 1825. [CrossRef]
32. Chen, L.; Lu, S.; Zhang, L.; Zhang, C.; Liu, L.; Kang, G.; Tang, H.; Chen, S. Solid Waste of Fly Ash toward Energy-Efficient $CO_2$ Capture. *ACS Sustain. Chem. Eng.* **2023**, *11*, 8281–8293. [CrossRef]
33. Liu, R.; Shi, X.; Wang, C.; Gao, Y.; Xu, S.; Hao, G.; Chen, S.; Lu, A. Advances in Post-Combustion $CO_2$ Capture by Physical Adsorption: From Materials Innovation to Separation Practice. *ChemSusChem* **2021**, *14*, 1428–1471. [CrossRef]
34. Lai, J.Y.; Ngu, L.H.; Hashim, S.S. A review of $CO_2$ adsorbents performance for different carbon capture technology processes conditions. *Greenh. Gases: Sci. Technol.* **2021**, *11*, 1076–1117. [CrossRef]
35. Madden, D.; Curtin, T. Carbon dioxide capture with amino-functionalised zeolite-β: A temperature programmed desorption study under dry and humid conditions. *Microporous Mesoporous Mater.* **2016**, *228*, 310–317. [CrossRef]
36. Wang, Q.; Luo, J.; Zhong, Z.; Borgna, A. $CO_2$ capture by solid adsorbents and their applications: Current status and new trends. *Energy Environ. Sci.* **2011**, *4*, 42–55. [CrossRef]
37. Mouhtaris, T.; Charistos, D.; Kantiranis, N.; Filippidis, A.; Kassoli-Fournaraki, A.; Tsirambidis, A. GIS-type zeolite synthesis from Greek lignite sulphocalcic fly ashes promoted by NaOH solutions. *Microporous Mesoporous Mater.* **2003**, *61*, 57–67. [CrossRef]
38. Medek, J. Possibility of micropore analysis of coal and coke from the carbon dioxide isotherm. *Fuel* **1977**, *56*, 131–133. [CrossRef]
39. Derkowski, A.; Franus, W.; Waniak-Nowicka, H.; Czímerová, A. Textural properties vs. CEC and EGME retention of Na-X zeolite prepared from fly ash at room temperature. *Int. J. Miner. Process.* **2007**, *82*, 57–68. [CrossRef]

40. Chen, X.B.; Tang, Y.T.; Ke, C.C.; Zhang, C.Y.; Ding, S.C.; Ma, X.Q. $CO_2$ capture by double metal modified CaO-based sorbents from pyrolysis gases. *Chin. J. Chem. Eng.* **2022**, *43*, 40–49. [CrossRef]
41. Modekwe, H.U.; Moothi, K.; Daramola, M.O.; Mamo, M.A. Corn cob char as catalyst support for developing carbon nanotubes from waste polypropylene plastics: Comparison of activation techniques. *Polymers* **2022**, *14*, 2898. [CrossRef]
42. Lafmejani, M.K.A.; Parsa, A.; Mirmohammadi, M.; Ahmadi, T.; Mirmohammadi, H. A novel and facile synthesis of calcium silicate nanoparticles as a base for root canal cement/sealer under constant potential: Compared to chemical synthesis. *Mater. Chem. Phys.* **2024**, *315*, 128924. [CrossRef]
43. Brunauer, S.; Emmett, P.; Teller, E. Adsorption of gases in multimolecular layers. *J. Am. Chem. Soc.* **1938**, *60*, 309–319. [CrossRef]
44. Castañeda, J. Estudio de las Propiedades de Trasporte en Materiales Porosos Mediante Espectroscopía Infrarrojo con Trasformada de Fourier (FTIR). Master's Thesis, Universidad Nacional de Colombia, Medellin, Colombia, 2012.
45. Da Gama, C.; Navarro-Torres, V.; Falcao-Neves, A. Techological innovations on underground coal gasification and $CO_2$ sequestration. *Dyna* **2010**, *77*, 101–108.
46. Lira-Zúñiga, S.; Sáez-Navarrete, C.; Rodríguez-Córdova, L.; Herrera-Zeppelin, L.; Herrera-Urbina, R. $CO_2$ adsorption on agricultural biomass combustion ashes. *Maderas-Cienc. Tecnol.* **2016**, *18*, 607–616. [CrossRef]
47. Jiang, L.; Cheng, L.; Zhang, Y.; Liu, G.; Sun, J. A Review on $CO_2$ Sequestration via Mineralization of Coal Fly Ash. *Energies* **2023**, *16*, 6241. [CrossRef]
48. Verrecchia, G.; Cafiero, L.; de Caprariis, B.; Dell'Era, A.; Pettiti, I.; Tuffi, R.; Scarsella, M. Study of the parameters of zeolites synthesis from coal fly ash in order to optimize their $CO_2$ adsorption. *Fuel* **2020**, *276*, 118041. [CrossRef]
49. Ragipani, R.; Sreenivasan, K.; Anex, R.P.; Zhai, H.; Wang, B. Direct Air Capture and Sequestration of $CO_2$ by Accelerated Indirect Aqueous Mineral Carbonation under Ambient Conditions. *ACS Sustain. Chem. Eng.* **2022**, *10*, 7852–7861. [CrossRef]
50. Zgureva, D.; Boycheva, S. Experimental and model investigations of $CO_2$ adsorption onto fly ash zeolite surface in dynamic conditions. *Sustain. Chem. Pharm.* **2020**, *15*, 100222. [CrossRef]
51. International Zeolite Association (IZA). Available online: https://www.iza-online.org (accessed on 1 April 2025).
52. Uppili, S.; Thomas, K.J.; Crompton, E.M.; Ramamurthy, V. Probing zeolites with organic molecules: Supercages of X and Y zeolites are superpolar. *Langmuir* **2000**, *16*, 265–274. [CrossRef]
53. Osatiashtiani, A.; Puertolas, B.; Oliveira, C.C.S.; Manayil, J.C.; Barbero, B.; Isaacs, M.; Michailof, C.; Heracleous, E.; Pérez-Ramírez, J.; Lee, A.F.; et al. On the influence of Si:Al ratio and hierarchical porosity of FAU zeolites in solid acid catalysed esterification pretreatment of bio-oil. *Biomass Conv. Bioref.* **2017**, *7*, 331–342. [CrossRef]
54. Nikolakis, V.; Xomeritakis, G.; Abibi, A.; Dickson, M.; Tsapatsis, M.; Vlachos, D.G. Growth of a faujasite-type zeolite membrane and its application in the separation of saturated/unsaturated hydrocarbon mixtures. *J. Membr. Sci.* **2001**, *184*, 209–219. [CrossRef]
55. Julbe, A.; Drobek, M. Zeolite X: Type. In *Encyclopedia of Membranes*; Drioli, E., Giorno, L., Eds.; Springer: Berlin/Heidelberg, Germany, 2014.
56. Su, F.; Lu, C. $CO_2$ capture from gas stream by zeolite 13X using a dual-column temperature/vacuum swing adsorption. *Energy Environ. Sci.* **2012**, *5*, 9021–9027. [CrossRef]
57. Thommes, M.; Kaneko, K.; Neimark, A.V.; Olivier, J.P.; Rodriguez-Reinoso, F.; Rouquerol, J.; Sing, K.S.W. Physisorption of gases, with special reference to the evaluation of surface area and pore sizedistribution (IUPAC Technical Report). *Pure Appl. Chem.* **2015**, *87*, 1051–1069. [CrossRef]
58. Bardestani, R.; Patience, G.S.; Kaliaguine, S. Experimental methods in chemical engineering: Specific surface area and pore size distribution measurements—BET, BJH, and DFT. *Can. J. Chem. Eng.* **2019**, *97*, 2781–2791. [CrossRef]
59. Bolis, V. Fundamentals in adsorption at the solid-gas interface. Concepts and thermodynamic. In *Calorimetry and Thermal Methods in Catalysis*; Springer Series in Materials Science; Auroux, A., Ed.; Springer: Berlin/Heidelberg, Germany, 2013; Volume 154, pp. 3–50.
60. Ruthven, D.M. Adsorption (Chemical Engineering). In *Encyclopedia of Physical Science and Technology*, 3rd ed.; Meyers, R.A., Ed.; Academic Press: Cambridge, MA, USA, 2003; pp. 251–271.
61. Chou, C.-T.; Wu, B.-C.; Wu, T.-L.; Yang, H.-S.; Shen, C.-H. Concentrating high purity $CO_2$ from syngas after oxy-fuel combustion by pressure swing adsorption process. *Comp. Aided Chem. Eng.* **2018**, *43*, 1425–1431. [CrossRef]
62. Perry, R.H.; Green, D.W.; Maloney, J.O. *Perry's Chemical Engineers Handbook*, 7th ed.; McGraw-Hill: New York, NY, USA, 2007.
63. Herraiz, L.; Palfi, E.; Sánchez, E.F.; Mathieu, L. Rotary Adsorption: Selective recycling of $CO_2$ in combined cycle gas turbine power plants. *Front. Energy. Res.* **2020**, *8*, 308. [CrossRef]
64. Kalvachev, Y.; Zgureva, D.; Boycheva, S.; Barbov, B.; Petrova, N. Synthesis of carbon dioxide adsorbents by zeolitization of fly ash. *J. Therm. Anal. Calorim.* **2016**, *124*, 101–106. [CrossRef]
65. Boycheva, S.; Zgureva, D. Studies on the $CO_2$ adsorption onto coal fly ash zeolites at elevated Pressures. In Proceedings of the 17th International Conference on Environmental Science and Technology (CEST 2021), Athens, Greece, 1–4 September 2021.
66. Kareem, F.A.A.; Shariff, A.M.; Ullah, S.; Dreisbach, F.; Keong, L.K.; Mellon, N.; Garg, S. Experimental measurements and modeling of supercritical $CO_2$ adsorption on 13X and 5A zeolites. *J. Nat. Gas Sci. Eng.* **2018**, *50*, 115–127. [CrossRef]

67. Siemons, N.; Busch, A. Measurement and interpretation of supercritical $CO_2$ sorption on various coals. *Int. J. Coal Geol.* **2007**, *69*, 229–242. [CrossRef]
68. Sircar, S. Measurement of Gibbsian surface excess. *AIChE J.* **2001**, *47*, 1169–1176. [CrossRef]

processes MDPI

*Article*

# Carbon Isotope and Sterane Records of Biological Diversity in the Fortunian Stage of the Early Cambrian Tarim Basin, Northwest China

**Wenhao Li [1,2,*], Yifan Chen [1,2] and Longwei Wang [1,2]**

[1] National Key Laboratory of Deep Oil and Gas, China University of Petroleum (East China), Qingdao 266580, China; z22010078@s.upc.edu.cn (Y.C.); s22010133@s.upc.edu.cn (L.W.)
[2] School of Geosciences, China University of Petroleum (East China), Qingdao 266580, China
* Correspondence: liwh@upc.edu.cn; Tel.: +86-15253220962

**Abstract:** Carbon isotope of the kerogen ($\delta^{13}C_{org}$), steranes/hopanes (S/H), and $C_{28}/C_{29}$ sterane ratios in the source rocks from the SARK section at the Early Cambrian Yurtus Formation in the Fortunian Stage in the Tarim Basin of Northwest China reveal a positive excursion that is associated with biological diversity. The enrichment of vanadium/(vanadium + nickel) (V/(V + Ni)) ratios (0.64~0.99, averaging 0.87) for the Yurtus Formation of the Fortunian Stage provide evidence for predominant anoxic bottom water conditions. A sharply decreased V/(V + Ni) ratio in the middle Yurtus Formation suggests enhanced oxygen content of the water column in this interval. However, the total organic carbon (TOC) values in the sedimentary rocks show a marked increase in the middle Yurtus Formation, which is due to the enhanced productivity suggested by a positive $\delta^{13}C_{org}$ increase of ~2.0‰ and enhanced S/H and $C_{28}/C_{29}$ sterane ratios. We suggest that the enhanced oxygen content may have contributed to the biological diversity during the Fortunian Stage in the Tarim Basin. The $\delta^{13}C_{org}$ excursion first reported here associated with biological diversity can be correlated with that in South China and possibly elsewhere in this interval.

**Keywords:** $\delta^{13}C_{org}$; sterane; biological diversity; redox conditions; Tarim Basin

## 1. Introduction

The $\delta^{13}C$ excursions and biomarkers are commonly used to discuss the environment and its control on the well-known "Cambrian explosion" [1–3], which is characterized by the appearance of skeletal metazoan [4–6]. It is reported that $\delta^{13}C$ excursions may have been associated with the biological diversity events at the Nemakit-Daldynian Stage in Siberia and South China [7,8]. A positive organic carbon isotope shift reported by Chen et al. [9] has been found in the upper Liuchapo Formation, which is correlated to the positive isotopic excursion in the upper Xiaowaitoushan Member (peak S2) of the Laolin section and the Nemakit-Daldynian Stage in Siberia [10,11]. In this interval, carbon isotope and fossil data in feature D in the Mongolian section also record this Cambrian diversification [12].

The biological diversity event at the Nemakit-Daldynian Stage in South China has been paid extensive attention. However, the biological diversity in this interval when Yurtus black shales were developed in the Tarim Basin of northwest China has been poorly understood mainly due to the lack of research, although many scholars have investigated the black shales in the Early Cambrian Yurtus Formation [13–16]. The Yurtus black shales are mainly developed in an anoxic environment [17–19], and the high primary productivity, anoxic environment, upwelling, and hydrothermal fluids may have contributed to the formation of the Yurtus black shales [13,19–22]. Li et al. [14] reported that some crude oils and Cambrian source rocks in the basin have a high content of $C_{28}$ regular steranes with $C_{28}/C_{29}$ sterane ratios greater than 0.50. However, Grantham and Wakefield found that the ratio of the $C_{28}/C_{29}$ sterane is less than 0.5 for the Lower Paleozoic oils by analyzing 400 oil samples corresponding to 600 Ma of evolution covering the entire Phanerozoic and believed that the increase in $C_{28}$ steranes might be related to the increased diversification of phytoplanktonic assemblages [23].

In this study, we report the co-evolution of carbon isotope of the kerogen ($\delta^{13}C_{org}$), steranes/hopanes (S/H), and $C_{28}/C_{29}$ sterane ratios ($C_{28}/C_{29}$) from the SARK section to record the change to the evolution and diversification of algal assemblages in the Early Cambrian. Our results may hold important implications for the Cambrian explosion during the Fortunian Stage. This biodiversification event is not only the first found in northwest China but also has enhanced its global correlation.

## 2. Geological Settings and Samples

### 2.1. Geological Settings

The Tarim Block assembled in Australia formed part of the Rodinia supercontinent during the Neoproterozoic [19,24,25]. With the collapse of the Rodinia supercontinent in the Late Neoproterozoic, it was disrupted from the supercontinent [19,24–26]. In the early Cambrian, the Tarim Block rotated and rifted apart from northwestern Australia, then drifted from the high-latitude region to near the equator [27].

The Tarim Block is composed of a Precambrian basement and a Late Neoproterozoic to Cambrian cover series [25]. Cambrian outcrops are mainly in the northwestern Tarim Basin, especially in the Wushi–Keping area [25]. The Lower Cambrian Yurtus Formation unconformably overlies the Ediacaran Qegebulake Formation in this region [19,28]. The samples numbered N1 to N13 in the Lower Cambrian Yurtus Formation were collected from the SARK section in the Tarim Basin, NW China (Figures 1 and 2). The Yurtus Formation in the Tarim Basin is at the Fortunian Stage to Stage 2 (Figure 2). Based on the small shelly fossils and acritarch assemblage data, the Siliceous shales intercalated with thin dolomites, black shales intercalated with some straticulate siliceous shales were developed at the Fortunian Stage [29], while dolomites were mainly developed at the Stage 2 (Figure 2). The total organic carbon (TOC) values of the black shales in the Yurtus Formation in the Tarim Basin are high, the average of which can reach up to >6% [13].

### 2.2. Sample Pre-Treatment and Analytical Methods

In this study, the source rock samples were measured using a Leco CS230 analyzer (LECO Corporation, USA) for TOC analysis. The samples were milled in an agate mortar and pestle for 200-mesh particles, and some details are in Li et al. [30]. The experimental conditions and procedure of the trace elements are listed in Li et al. [30].

The carbon isotopic compositions of the kerogens were measured using a Finnigan $Delta^{Plus}$ XL IRMS instrument connected to a CE $flash^{1112}$ EA via a ConfloIII interface (Finnigan MAT, Bremen, Germany). The samples were converted to $CO_2$ through high-temperature combustion using helium as the carrier gas, and then separated by chromatographic column and entered into a mass spectrometer for analysis. The $\delta^{13}C$ value was reported relative to Vienna PeeDee Belemnite (VPDB). Each sample was analyzed at least twice, and the deviations for the results were no more than 0.3‰. The average for the two runs was accepted as the final carbon isotopic result for the sample. The analytical precision of these analyses is better than 0.2‰.

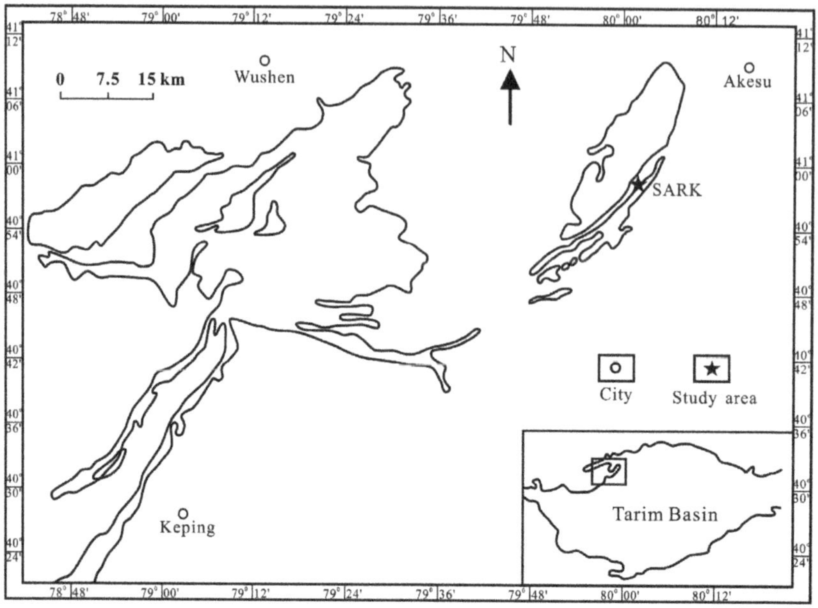


**Figure 1.** Location of the SARK section in the Tarim Basin, NW China.

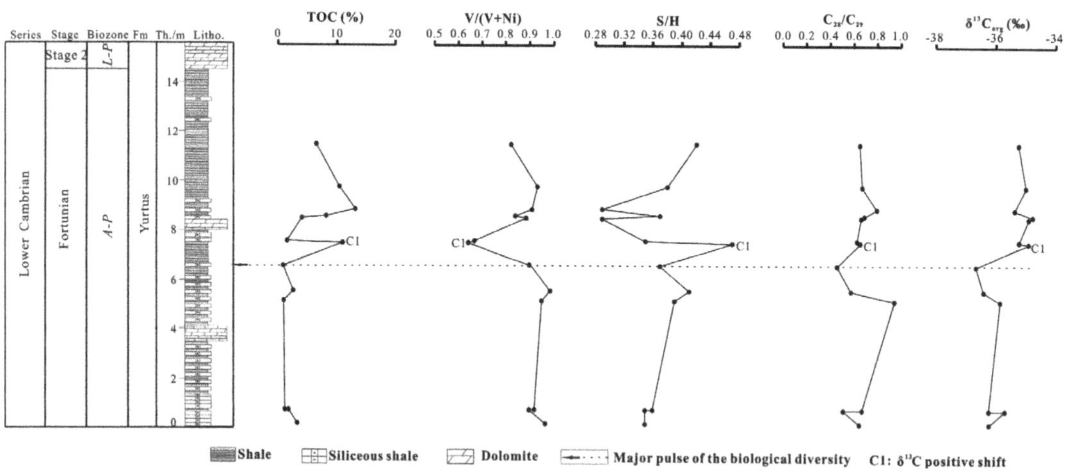


**Figure 2.** Total organic carbon (TOC), V/(V + Ni), S/H, $C_{28}/C_{29}$, and $\delta^{13}C_{org}$ values for Early Cambrian sediments in the Yurtus Formation at the SARK section, Tarim Basin. Note: *A-P* = *Anabarites–Protohertzina*; *L-P* = *Lapworthella–Paragloborilus*; S/H = steranes/hopanes ($C_{27}$–$C_{29}$ steranes/$C_{27}$–$C_{35}$ hopanes); $C_{28}/C_{29}$ = $C_{28}/C_{29}$ steranes.

Before the analysis for gas chromatography–mass spectrometry (GC-MS), the samples were crushed into 80 meshes and then extracted in a Soxhlet apparatus with dichloromethane ($CH_2Cl_2$) for 24 h. The extracts were evaporated, handled with n-hexane, and then separated through column chromatography into the saturated, aromatic, and polar fractions, respectively. The saturated and aromatic fractions were analyzed by gas chromatography–mass spectrometry (GC-MS). The experimental conditions were listed in Zhan et al. [31]. The trace elements data were measured using an inductively coupled plasma mass spectrometer (ICP-MS). The experimental conditions of these analyses were reported by Li et al. [30].

## 3. Results

### 3.1. Total Organic Carbon and Redox Conditions

The TOC values are stable from 0.93~3.37% (samples from N1 to N6), with an average of 1.85%, in the lower to middle Yurtus Formation, which increase to 1.61~13.17% (samples from N7 to N13), with an average of 7.88%, from the middle to upper Yurtus Formation (Table 1, Figure 2). A significant increase in the TOC value in the sample N7 has occurred (Table 1, C1 in Figure 2).

**Table 1.** Distribution of several organic geochemical parameters of Early Cambrian sediments in the Yurtus Formation at the SARK section, Tarim Basin. Note: S/H = steranes/hopanes ($C_{27}$~$C_{29}$ steranes/$C_{27}$~$C_{35}$ hopanes); $C_{28}/C_{29}$ = $C_{28}/C_{29}$ steranes.

| Sample | TOC (%) | V (ppm) | Ni (ppm) | V/(V + Ni) | S/H | $C_{28}/C_{29}$ | $\delta^{13}C_{org}$ (‰) |
|---|---|---|---|---|---|---|---|
| N1 | 3.37 | 315 | 11.1 | 0.97 | 0.35 | 0.64 | −36.20 |
| N2 | 1.87 | 1026 | 116 | 0.90 | 0.35 | 0.58 | −35.67 |
| N3 | 1.25 | 441 | 40.0 | 0.92 | 0.36 | 0.67 | −36.21 |
| N4 | 1.03 | 391 | 21.4 | 0.95 | 0.39 | 0.95 | −35.84 |
| N5 | 2.65 | 2617 | 25.0 | 0.99 | 0.41 | 0.59 | −36.40 |
| N6 | 0.93 | 249 | 27.9 | 0.90 | 0.37 | 0.46 | −36.67 |
| N7 | 11.00 | 1614 | 913 | 0.64 | 0.47 | 0.65 | −34.92 |
| N8 | 1.61 | 140 | 67.7 | 0.67 | 0.35 | 0.62 | −35.24 |
| N9 | 4.15 | 164 | 19.4 | 0.89 | 0.29 | 0.66 | −34.91 |
| N10 | 8.27 | 903 | 169 | 0.84 | 0.37 | 0.69 | −34.77 |
| N11 | 13.17 | 593 | 57.1 | 0.91 | 0.29 | 0.80 | −35.37 |
| N12 | 10.43 | 421 | 32.1 | 0.93 | 0.38 | 0.67 | −35.01 |
| N13 | 6.55 | 393 | 85.9 | 0.82 | 0.42 | 0.64 | −35.26 |

The vanadium/(vanadium + nickel) (V/(V + Ni)) ratios can be used to infer redox conditions. When the V/(V + Ni) ratio > 0.60, it may suggest anoxic bottom water conditions [32–34]. The V/(V + Ni) ratios are in the range from 0.64 to 0.99, with an average of 0.87, indicating predominant anoxic bottom water conditions. The V/(V + Ni) ratios are stable from 0.90 to 0.99 in the samples N1~N6, and the ratio sharply decreased to 0.64 in the sample N7 (Table 1, C1 in Figure 2), suggesting the increase in oxygen content in the water column in this interval. Iron speciation and Mo isotope data have also supported the similar conclusion that the paleo-environmental conditions of the source rocks of the Lower Cambrian Yurtus Formation is a predominant anoxic bottom water condition, but transient oxic conditions may have existed [19]. The V/(V + Ni) ratio shows a decreasing trend from the bottom to the top, which is not coordinated with the change in the TOC value. Thus, the enhanced TOC values in the middle to upper Yurtus Formation (especially in sample N7) have no correlation with the redox conditions.

### 3.2. Biomarkers Related to Biological Diversity

The ratio of steranes to hopanes can record the relative contribution of eukaryotic and bacterial lipids to marine sediments [35]. Brocks et al. [36] discussed the relative contribution of eukaryotic and bacterial lipids to sedimentary organic matter from 850 Myr ago to the present estimated by the sterane/hopane (S/H) ratio and believed that S/H ratios in 820 to 720 Myr old sediments (0.003 to 0.300, 0.06 ± 0.10 (mean ± s.d.)) are three orders of magnitude lower than in typical Phanerozoic marine sediments (~0.5 to >2, 0.75 ± 1.1). The S/H ratios in the marine sediments range from 0.29 to 0.47, with an average of 0.37 (Table 1), suggesting that the bacterial still had a major role. A positive S/H ratio increase of ~0.1 has been observed in the middle Yurtus Formation (C1 in Figure 2), which is associated with biological diversity. The increase in dissolved oxygen content in the water column suggested by V/(V + Ni) promoted the algal blooms (C1 in Figure 2).

The $C_{28}/C_{29}$ ratios range from 0.46 to 0.95, with an average of 0.66 (Table 1). This result does not match the results reported by Grantham and Wakefield [23] that oils generated from the Lower Paleozoic and older marine source rocks have relatively lower $C_{28}/C_{29}$ regular sterane values (less than 0.5). However, there is a great increase in the $C_{28}/C_{29}$ ratio from <0.55 to >0.70 at the Devonian/Carboniferous boundary relating to the enhanced prasinophytes [37]. Considering that $C_{28}$ steranes are associated with the increased diversification of phytoplanktonic assemblages [23], the high $C_{28}/C_{29}$ ratio in our study may imply the presence of biological diversity in the early Cambrian in the Tarim Basin. A positive $C_{28}/C_{29}$ ratio increase of ~0.2 in the middle Yurtus Formation (C1 in Figure 2) suggests the algal blooms due to enhanced photosynthesis.

### 3.3. $\delta^{13}C_{org}$ Chemostratigraphy and Its Global Correlation Related to Biological Diversity

The $\delta^{13}C_{org}$ values from the SARK section range from −36.67‰ to −34.77‰, with a mean value of −35.57‰ (Table 1). The $\delta^{13}C_{org}$ shows a stable low value from the bottom to the middle Yurtus Formation (Samples from N1 to N6 in Figure 2). A positive excursion of the $\delta^{13}C_{org}$ is exhibited toward the Anabarites–Protohertzina biozone in the middle Yurtus Formation during the Fortunian Stage (C1 in Figure 2), which is associated with biological diversity in the Tarim Basin.

The similarity positive excursions of the $\delta^{13}C$ at the Anabarites–Protohertzina biozone also are reported in Siberia and South China (Figure 3). Brasier et al. [7] believed that positive $\delta^{13}C_{carb}$ excursions up to 3‰ in the Lower Cambrian of Siberia at the Nemakit-Daldynian Stage indicate the Cambrian explosion. Microfossils including macroscopic algae, conotubular life forms, and presumed metazoans have also been reported in the interval to record this biological diversity event [8]. Further studies on detailed fossil information during the Fortunian Stage will enhance the understanding of the globally biological diversity event in this interval.

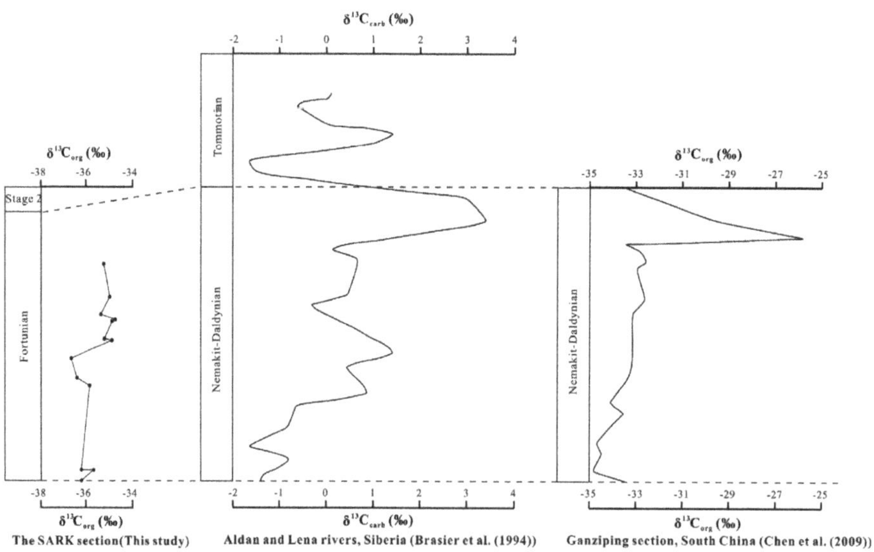


**Figure 3.** A correlation of the $\delta^{13}C$ profiles among representative sections in the world in the Early Cambrian [7,9].

## 4. Discussion

A positive TOC increase of ~10% in the middle Yurtus Formation is exhibited (C1 in Figure 2), which is caused by phytoplankton blooms. A positive $\delta^{13}C_{org}$ increase of ~2‰ occurred in the middle Yurtus Formation shortly after the biological diversity (C1 in Figure 2). The increase of ~0.2 and ~0.1, respectively, in $C_{28}/C_{29}$ and S/H of the marine sediments in the middle Yurtus Formation also (C1 in Figure 2) supported the event of biological diversity. We believe the oxygen increase in the water column (supported by the V/(V + Ni) ratio) is the environmental trigger for Cambrian biological diversity in this study. Although the redox conditions are mainly anoxic in the Fortunian Stage, the oxygen content is close to suboxic suggested by the V/(V + Ni) ratio shortly after the biological diversity (C1 in Figure 2). A similar point has also been reported that short-lived oxic conditions may have existed in the Lower Cambrian Yurtus Formation in the northwest Tarim Basin [19]. The V/(V + Ni) ratios in the marine sediments from the middle to upper Yurtus Formation fluctuate frequently (Figure 2), suggesting changeable redox conditions, which may trigger the small shelly animals to flourish at the Tommotian Stage.

We noticed that the positive $\delta^{13}C_{org}$ excursion at the Fortunian Stage is with global correlation. This positive $\delta^{13}C_{org}$ excursion event commonly occurred in South China, and Laolin and Songtao sections from the Yangtze Platform [11,38], Longbizui section from Western Hunan [39], and Ganziping section from Dayong of Hunan Province [9] record this excursion. The $\delta^{13}C_{org}$ excursion ranges in the above sections (except the Ganziping section with a range over 7.0‰) are close to the section in this study. $\delta^{13}C_{carb}$ data in the sections from Mongolian and Siberia also indicated the Cambrian diversification in this interval [7,12].

## 5. Conclusions

The Yurtus source rocks in the Fortunian Stage in the Tarim Basin of Northwest China were mainly deposited in anoxic conditions. Although the oxygen content is increasing, a positive TOC increase of ~10% in the middle Yurtus Formation has occurred, which is attributed to phytoplankton blooms. A positive C isotope excursion and increase in the S/H and $C_{28}/C_{29}$ ratios in this interval record the biological diversity. The enhanced oxygen content may have contributed to the biological diversity. The $\delta^{13}C_{org}$ excursion first reported here associated with biological diversity event in the Fortunian Stage can be correlated with that in South China and possibly elsewhere in this interval. Further studies on detailed fossil information during the Fortunian Stage in the Tarim Basin will improve our understanding of biological evolution.

**Author Contributions:** W.L. contributed to the design of the study, wrote the manuscript, and was the principal author of the manuscript. Y.C. contributed to the methodology. L.W. contributed to the discussion of the results and manuscript refinement. All authors have read and agreed to the published version of the manuscript.

**Funding:** This study was financially supported by the National Natural Science Foundation of China (42472193) and the Natural Science Foundation of Shandong Province (ZR2022YQ30).

**Data Availability Statement:** All the data analyzed during this study are included in Table 1 of this article, which has been uploaded.

**Conflicts of Interest:** The authors declare that they have no known competing financial interest or personal relationships that could have appeared to influence the work reported in this paper.

## References


1. Ishikawa, T.; Ueno, Y.; Komiya, T.; Sawaki, Y.; Han, J.; Shu, D.G.; Li, Y.; Maruyama, S.; Yoshida, N. Carbon isotope chemostratigraphy of a Precambrian/Cambrian boundary section in the Three Gorge area, South China: Prominent global-scale isotope excursions just before Cambrian Explosion. *Gondwana Res.* **2008**, *14*, 193–208. [CrossRef]
2. Ishikawa, T.; Ueno, Y.; Shu, D.G.; Li, Y.; Han, J.; Guo, J.F.; Yoshida, N.; Maruyama, S.; Komiya, T. The $\delta^{13}$C excursions spanning the Cambrian explosion to the Canglangpuian mass extinction in the Three Gerges area, South China. *Gondwana Res.* **2014**, *25*, 1045–1056. [CrossRef]
3. Wang, C.J.; Zhao, Y.L.; Peng, J.; Yang, X.L.; Bai, J.; Liu, Y.; Chen, T. Biomarker evidence for biotic and environmental change across the Cambrian Series 2-Series 3 boundary at the Wuliu-Zengjiayan section, Guizhou, China. *Sci. China: Earth Sci.* **2014**, *57*, 2781–2790. [CrossRef]
4. Conway-Morris, S. The Cambrian "explosion": Slow-fuse or megatonnage? *Proc. Natl. Acad. Sci. USA* **2000**, *97*, 4426–4429. [CrossRef]
5. Li, G.X.; Steiner, M.; Zhu, X.J.; Yang, A.H.; Wang, H.F.; Erdtmann, B.D. Early Cambrian metazoan fossil record of South China: Generic diversity and radiation patterns. *Palaeogeogr. Palaeoclimatol. Palaeoecol.* **2007**, *254*, 229–249. [CrossRef]
6. Shu, D. Cambrian explosion: Birth of tree of animals. *Gondwana Res.* **2008**, *14*, 219–240. [CrossRef]
7. Brasier, M.D.; Corfield, R.M.; Derry, L.A.; Rozanov, A.Y.; Zhuravlev, A.Y. Multiple $\delta^{13}$C excursions spanning the Cambrian explosion to the Botomian crisis in Siberia. *Geology* **1994**, *22*, 455–458. [CrossRef]
8. Guo, J.F.; Li, Y.; Han, J.; Zhang, X.L.; Zhang, Z.F.; Ou, Q.; Liu, J.N.; Shu, D.G.; Maruyama, S.; Komiya, T. Fossil association from the Lower Cambrian Yanjiahe Formation in the Yangtze Gorges Area, Hubei, South China. *Acta Geol. Sin.-Engl. Ed.* **2008**, *82*, 1124–1132.
9. Chen, D.Z.; Wang, J.G.; Qing, H.R.; Yan, D.T.; Li, R.W. Hydrothermal venting activities in the Early Cambrian, South China: Petrological, geochronological and stable isotopic constraints. *Chem. Geol.* **2009**, *258*, 168–181. [CrossRef]
10. Zhou, C.M.; Zhang, J.M.; Li, G.X.; Yu, Z.Y. Carbon and oxygen isotopic record of the Early Cambrian from the Xiaotan section, Yunnan, South China. *Sci. Geol. Sin.* **1997**, *32*, 201–211 (In Chinese with English abstract)
11. Shen, Y.A.; Schidlowski, M. New C isotope stratigraphy from southwest China: Implications for the placement of the Precambrian-Cambrian boundary on the Yangtze Platform and global correlations. *Geology* **2000**, *28*, 623–626. [CrossRef]
12. Brasier, M.D.; Shields, G.; Kuleshov, V.N.; Zhegallo, E.A. Integrated chemo-and biostratigraphic calibration of early animal evolution: Neoproterozoic-early Cambrian of southwest Mongolia. *Geol. Mag.* **1996**, *133*, 445–485. [CrossRef]
13. Yu, B.S.; Dong, H.L.; Widom, E.; Chen, J.Q.; Lin, C.S. Geochemistry of basal Cambrian black shales and cherts from the Northern Tarim Basin, Northwest China: Implications for depositional setting and tectonic history. *J. Asian Earth Sci.* **2009**, *34*, 418–436. [CrossRef]
14. Li, M.J.; Wang, T.G.; Lillis, P.G.; Wang, C.J.; Shi, S.B. The significance of 24-norcholestanes, triaromatic steroids and dinosteroids in oils and Cambrian-Ordovician source rocks from the cratonic region of the Tarim Basin, NW China. *Appl. Geochem.* **2012**, *27*, 1643–1654. [CrossRef]
15. Yao, C.Y.; Ding, H.F.; Ma, D.S.; Li, G.X. Carbon isotope features of the Sugetbrak Section in the Aksu-Wushi Area, Northwest China: Implications for the Precambrian/Cambrian stratigraphic correlations. *Acta Geol. Sin.* **2014**, *88*, 1535–1546. [CrossRef]
16. Zhou, X.Q.; Chen, D.Z.; Dong, S.F.; Zhang, Y.Q.; Guo, Z.H.; Wei, H.Y.; Yu, H. Diagenetic barite deposits in the Yurtus Formation in Tarim Basin, NW China: Implications for barium and sulfur cycling in the earliest Cambrian. *Precambr. Res.* **2015**, *263*, 79–87. [CrossRef]
17. Cai, C.F.; Li, K.K.; Ma, A.L.; Zhang, C.M.; Xu, Z.M.; Worden, R.H.; Wu, G.H.; Zhang, B.S.; Chen, L.X. Distinguishing Cambrian from Upper Ordovician source rocks: Evidence from sulfur isotopes and biomarkers in the Tarim Basin. *Org. Geochem.* **2009**, *40*, 755–768. [CrossRef]
18. He, T.H.; Zeng, Q.H.; Lu, S.F.; Li, W.H.; Li, M.Q.; Wen, Z.G.; Yang, E.Q.; Jing, T.T.; Ying, J.F.; Zhu, P.F.; et al. Aryl isoprenoids from the Lower Paleozoic in the Tarim Basin, NW China: Insight into deep ancient hydrocarbon exploration. *Geoenergy Sci. Eng.* **2023**, *225*, 211666. [CrossRef]
19. Zhu, B.; Yang, T.; Wang, J.; Chen, X.; Pan, W.Q.; Chen, Y.Q. Multiple controls on the paleoenvironment of the early Cambrian black shale-chert in the northwest Tarim Basin, NW China: Trace element, iron speciation and Mo isotopic evidence. *Mar. Pet. Geol.* **2022**, *136*, 105434. [CrossRef]
20. Zhang, C.Y.; Guan, S.W.; Wu, L.; Ren, R.; Wang, L.N.; Wu, X.Q. Depositional environments of early Cambrian marine shale, northwestern Tarim Basin, China: Implications for organic matter accumulation. *J. Pet. Sci. Eng.* **2020**, *194*, 107497. [CrossRef]
21. Zhang, Y.G.; Yang, T.; Hohl, S.V.; Zhu, B.; He, T.C.; Pan, W.Q.; Chen, Y.Q.; Yao, X.Z.; Jiang, S.Y. Seawater carbon and strontium isotope variations through the late Ediacaran to late Cambrian in the Tarim Basin. *Precambr. Res.* **2020**, *345*, 105769. [CrossRef]
22. Zhu, G.Y.; Li, T.T.; Zhao, K.; Li, C.; Cheng, M.; Chen, W.Y.; Yan, H.H.; Zhang, Z.Y.; Algeo, T.J. Mo isotope records from Lower Cambrian black shales, northwestern Tarim Basin (China): Implications for the early Cambrian ocean. *Geol. Soc. Am. Bull.* **2021**, *134*, 3–14. [CrossRef]

23. Grantham, P.J.; Wakefield, L.L. Variations in the sterane carbon number distributions of marine source rock derived crude oils through geological times. *Org. Geochem.* **1988**, *12*, 61–77. [CrossRef]
24. Li, Z.X.; Bogdanova, S.V.; Collins, A.S.; Davidson, A.; De Waele, B.; Ernst, R.E.; Fitzsimons, I.C.W.; Fuck, R.A.; Gladkochub, D.P.; Jacobs, J.; et al. Assembly, configuration, and break-up history of Rodinia: A synthesis. *Percambrian Res.* **2008**, *160*, 179–210. [CrossRef]
25. Zhang, C.L.; Zhou, H.B.; Li, H.K.; Wang, H.Y. Tectonic framework and evolution of the Tarim block in NW China. *Gondwana Res.* **2013**, *23*, 1306–1315. [CrossRef]
26. Zhang, C.L.; Zhou, H.B.; Wang, H.Y.; Li, H.K.; Ye, H.M. Multiple phases of the Neoproterozoic igneous activity in Quruqtagh of the northeastern Tarim Block, NW China: Interaction between plate subduction and mantle plume? *Percambrian Res.* **2012**, *222–223*, 488–502. [CrossRef]
27. Li, Z.X.; Powell, C.M. An outline of the palaeogeographic evolution of the Australasian region since the beginning of the Neoproterozoic. *Earth Sci. Rev.* **2001**, *53*, 237–277. [CrossRef]
28. Turner, S.A. Sedimentary record of Late Neoproterozoic rifting in the NW Tarim Basin, China. *Precambr. Res.* **2010**, *181*, 85–96. [CrossRef]
29. Dong, L.; Xiao, S.H.; Shen, B.; Zhou, C.M.; Li, G.X.; Yao, J.X. Basal Cambrian microfossils from the Yangtze Gorges area (South China) and the Aksu area (Tarim block, northwestern China). *J. Paleontol.* **2009**, *83*, 30–44. [CrossRef]
30. Li, W.H.; Zhang, Z.H.; Li, Y.C.; Fu, N. The effect of river-delta system on the formation of the source rocks in the Baiyun Sag, Pearl River Mouth Basin. *Mar. Pet. Geol.* **2016**, *76*, 279–289. [CrossRef]
31. Zhan, Z.W.; Zou, Y.R.; Shi, J.T.; Sun, J.N.; Peng, P.A. Unmixing of mixed oil using chemometrics. *Org. Geochem.* **2016**, *92*, 1–15. [CrossRef]
32. Joachimski, M.M. Water column anoxia, enhanced productivity and concomitant changes in $\delta^{13}C$ and $\delta^{34}S$ across the Frasnian-Famennian boundary (Kowala—Holy Cross Mountains/Poland). *Chem. Geol.* **2001**, *175*, 109–131. [CrossRef]
33. Rimmer, S.M.; Thompson, J.A.; Goodnight, S.A.; Robl, T.L. Multiple controls on the preservation of organic matter in Devonian–Mississippian marine black shales: Geochemical and petrographic evidence. *Palaeogeogr. Palaeoclimatol. Palaeoecol.* **2004**, *215*, 125–154. [CrossRef]
34. Zhang, T.G.; Trela, W.; Jiang, S.Y.; Nielsen, J.K.; Shen, Y.A. Major oceanic redox condition change correlated with the rebound of marine animal diversity during the Late Ordovician. *Geology* **2011**, *39*, 675–678. [CrossRef]
35. Brocks, J.J.; Jarrett, A.J.M.; Sirantoine, E.; Kenig, F.; Moczydlowska, M.; Porter, S.; Hope, J. Early sponges and toxic protists: Possible source of cryostane, an age diagnostic biomarker antedating Sturtian Snowball Earth. *Geobiology* **2016**, *14*, 129–149. [CrossRef] [PubMed]
36. Brocks, J.J.; Jarrett, A.J.M.; Sirantoine, E.; Hallmann, C.; Hoshino, Y.; Liyanage, T. The rise of algae in Cryogenian oceans and the emergence of animals. *Nature* **2017**, *548*, 578–581. [CrossRef]
37. Schwark, L.; Empt, P. Sterane biomarkers as indicators of palaeozoic algal evolution and extinction events. *Palaeogeogr. Palaeoclimatol. Palaeoecol.* **2006**, *240*, 225–236. [CrossRef]
38. Jiang, G.Q.; Wang, X.Q.; Shi, X.Y.; Xiao, S.H.; Zhang, S.H.; Dong, J. The origin of decoupled carbonate and organic carbon isotope signatures in the early Cambrian (ca. 542–520Ma) Yangtze platform. *Earth Planet. Sci. Lett.* **2012**, *317–318*, 96–110. [CrossRef]
39. Wang, J.G.; Chen, D.Z.; Yan, D.T.; Wei, H.Y.; Xiang, L. Evolution from an anoxic to oxic deep ocean during the Ediacaran-Cambrian transition and implications for bioradiation. *Chem. Geol.* **2012**, *306–307*, 129–138. [CrossRef]

processes

MDPI

Article

# Experimental Study on the Acid Fracturing Fracture Propagation Law of a Fractured Carbonate Reservoir in the Majiagou Formation

**Yongchun Zhang [1,2], Jianchao Kuang [1], Hao Zhang [1,*], Ying Zhong [1] and Shijie Dong [3]**

[1] College of Energy Resource, Chengdu University of Technology, Chengdu 610059, China; hbjzyc@126.com (Y.Z.)
[2] Sinopec North China Oil & Gas Company, Zhengzhou 450006, China
[3] College of Petroleum Engineering, China University of Petroleum (Beijing), Beijing 102249, China
* Correspondence: zhanghao@cdut.edu.cn

**Abstract:** Acid fracturing is a crucial method for reservoir reconstruction in carbonate reservoirs, and the propagation pattern of acid-etched fractures plays a key role in determining the scope of reservoir enhancement and post-fracturing productivity. However, large-scale physical simulations directly using acid solutions in fracturing experiments are limited, and the fracture propagation patterns under acid fracturing remain unclear. To address this gap, in this study, we collected carbonate rock samples from the Majiagou Formation in the Daniudi area, preparing large-scale fracturing specimens with side lengths of 30 cm. The propagation of acid fracturing fractures was investigated using self-developed true-triaxial acid fracturing equipment. Based on post-fracturing fracture morphology and pressure curves, the effects of fracturing fluid type, injection rate, injection mode, and natural fractures (NFs) on acid fracturing fracture propagation were analyzed. The experimental results showed that the acid solution effectively weakens the mechanical properties of the open-hole section, creating multiple mechanical weak points and promoting the initiation of fractures. Pre-fracturing treatment with low-viscosity acid can significantly enhance fracture complexity near the wellbore and expand the near-well stimulation zone. Lowering the injection rate increases the acid solution's filtration loss into natural fractures, weakening the cementation strength of these fractures and encouraging the formation of complex fracture networks. Furthermore, employing a multi-stage alternating injection of high-viscosity and low-viscosity acids can reduce fracture temperature and acid filtration loss while also enhancing differential etching through viscous fingering. This approach improves the conductivity and conductivity retention of the acid-etched fractures. The results of this study can provide a reference for the acid fracturing stimulation of fractured carbonate reservoirs.

**Keywords:** acid fracturing; experimental research; natural fracture; carbonate rock; fracture propagation; Majiagou Formation

## 1. Introduction

Common techniques for stimulating carbonate reservoirs include matrix acidizing and acid fracturing. Matrix acidizing involves injecting an acid solution at a pressure lower than the formation fracture pressure. The acid then flows through the formation, reacting with the rock to form wormholes that bypass the damaged zone, thereby restoring or enhancing productivity by improving access to the near-wellbore region. This technique is typically employed in reservoirs with favorable formation properties [1,2]. However, for deep carbonate reservoirs, the physical properties of the formation are typically poor due to sedimentary compaction [3]. Therefore, acid fracturing is required to create high-conductivity flow channels in the formation, increasing the seepage area and ultimately enhancing production. Unlike hydraulic fracturing, acid fracturing does not rely on proppants. Instead, it generates fractures with high conductivity through the chemical reaction between the acidic fracturing fluid and the minerals on the surface of carbonate rock fractures [4,5]. The effectiveness of acid fracturing stimulation depends on the length, coverage area, and conductivity of the acid-etched fractures [6–8]. As the primary pathways for acid flow and subsequent oil and gas production, studying the propagation process and morphology of acid-etched fractures is crucial for the design and optimization of acid fracturing treatments.

The propagation of acid fracturing fractures is a coupled process involving multiple fields, including the flow and reaction of the acid solution, as well as the compressive deformation and fracturing of the rock. When natural fractures are present in the formation, the process also includes the weakening of the mechanical properties of these fractures due to the dissolution effects of the acid solution [9,10]. The numerical simulation method must account for the coupling of fluid, solid, and chemical interactions, resulting in a substantial computational burden and challenges in achieving convergence [11]. Currently, numerical simulations typically employ equivalent methods or simulate the acid fracturing process after decoupling it. Despite these complexities, the simulation results reflect the propagation process of acid fracturing fractures to a certain extent [9,12]. Indoor true triaxial physical simulation experiments can effectively replicate the propagation of acid fracturing fractures under realistic formation conditions, allowing for the coupling of fluid, solid, and chemical interactions. This makes it possible to evaluate the impact of various factors on the fracture propagation process. Experimental results indicate that acid concentration significantly influences fracture formation: at low acid concentrations, the stress field primarily governs fracture formation, while at high acid concentrations, chemical reactions become the dominant factor [13]. The dissolution rates of different acid solutions vary, leading to different degrees of weakening in the rock's mechanical properties [14]. When using low-viscosity acid fluid, the weakening of rock mechanical properties is more intense, and the formation fracture pressure during acid fracturing is significantly lower than that when using high-viscosity acid fluid [7,15]. In addition to the acid solution, the lithological composition of the formation plays a critical role in the propagation of acid-induced fractures. The natural fracture network within the formation serves as a conduit for acid fluid filtration. As these natural fractures evolve, they contribute to increased complexity in the fracture network during the acid fracturing process [16–18]. CT scans of post-fracturing samples indicate that natural fractures significantly impact fracture propagation. Whether in hydraulic fracturing or acid fracturing, fractures typically extend in the direction of the maximum horizontal stress. However, the presence of voids and natural fractures can disrupt the fracture path [19,20]. Further research has revealed that when the horizontal stress difference is significant, fractures tend to propagate directly through voids and natural fractures. However, when the stress difference is smaller and the approach angle between natural and artificial fractures is low, natural fractures are more likely to be activated, thereby influencing the direction of fracture propagation [17]. The reaction between acid and rock also exacerbates fracture complexity. The dissolution of the natural fracture cement by the acid solution will reduce the failure strength of the natural fracture. Under the combined action of the dissolution of the acid solution and the hydraulic force, the natural fracture is more likely to be activated, further increasing the complexity of the fracture [21,22]. Due to the involvement of post-fracturing acid treatment in acid fracturing experiments and the potential for acid to cause some degree of corrosion to fracturing equipment, there are currently few studies on acid fracturing physical simulations based on true triaxial conditions [23].

To reveal the fracture propagation patterns during acid fracturing, this study used large-scale fracturing samples made from outcrops of the Majiagou Formation. Experiments on acid fracture propagation were conducted using a self-developed true triaxial acid fracturing apparatus. Based on the post-fracturing fracture morphology and pressure curves, the study examined the effects of fracturing fluid type, pump injection rate, injection mode, and natural fractures on acid fracture propagation. The experimental results can provide a reference for acid fracturing in fractured carbonate reservoirs.

## 2. Experimental Method

### 2.1. Experimental Devices

This study utilizes a self-developed true triaxial fracturing physical simulation apparatus to conduct fracturing experiments. The apparatus, in conjunction with a multi-stage alternating injection system, can achieve the alternating injection of up to five different fluids. The experimental apparatus consists of four main components: the control system, the injection system, the triaxial loading system, and the multi-channel acid alternating injection system. A schematic diagram of the apparatus is shown in Figure 1.

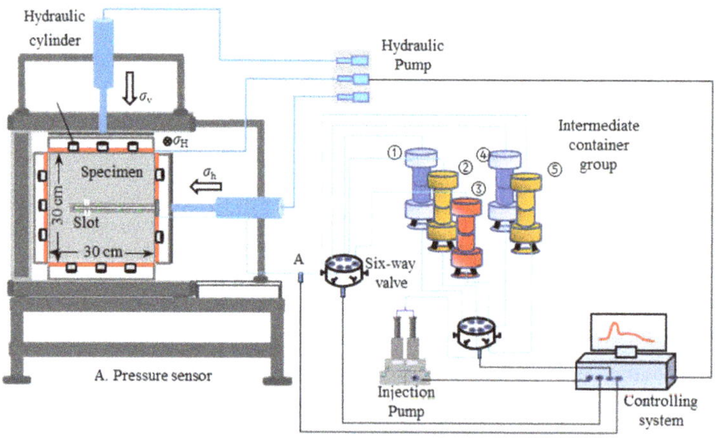


**Figure 1.** Schematic diagram of the true triaxial fracturing physical simulation apparatus.

#### 2.1.1. Control System

The control system primarily consists of a computer and fracturing system controlling software, serving two main functions. First, it controls the injection system and the true triaxial pressure system. By setting injection parameters and stress parameters in the fracturing simulation software, it manages the injection rate and the loading and control of triaxial pressure. Second, it provides data acquisition and recording functions, mainly used to record changes in wellhead pressure over the injection time during the fracturing process.

#### 2.1.2. Injection System

The injection system mainly consists of an injection pump and various valves and pipelines. Its primary function is to provide pressure for injecting the fracturing fluid. This system can precisely control the injection rate of both fracturing and acid fluids. The main technical parameters are as follows: The maximum output pressure of the booster piston is

140 MPa, with a resolution of 0.05 MPa; the effective volume of the booster is 600 mL; the minimum injection rate is 1 mL/min, and the maximum injection rate is 200 mL/min.

#### 2.1.3. Stress Loading System

The triaxial stress loading system consists mainly of a hydraulic servo pump and a hydraulic chamber. It provides confining pressure for large true triaxial physical models, simulating the triaxial stress environment of the formation, with independent pressure application in all three directions. In Figure 1, $\sigma_H$ means the maximum horizontal stress, $\sigma_h$ means the minimum horizontal stress, and $\sigma_V$ means the vertical stress. The system can accommodate specimens with a maximum size of 500 mm × 500 mm × 500 mm, and the maximum load in each direction can reach 8000 kN. The system ensures uniform force distribution, with axial forces evenly transmitted to the specimen through connecting plates and force transfer plates. During loading, the central position of the specimen can be maintained through program control, effectively preventing eccentric loading and the generation of bending moments.

#### 2.1.4. Multi-Channel Acid Alternating Injection System

The multi-channel acid alternating injection system comprises acid piston containers, connecting pipelines, valves, and a sophisticated valve control system. This setup allows for the alternating injection of up to five different acid solutions. By utilizing solenoid valves to manage the connection and disconnection of multiple intermediate containers, the system efficiently facilitates the alternating injection of various fluids.

### *2.2. Sample Preparation*

The dolomitic limestone outcrop of the Ma5 Member of the Majiagou Formation in Daniudi Gas Field, Ordos Basin, was used in the experiment. Natural fractures are developed in the outcrop, and almost all natural fractures are filled with cement. The mechanical properties of the outcrop matrix were tested with the triaxial rock mechanics comprehensive test system (TAW2000) manufactured by Changchun Chaoyang machinery works, Changchun, China. Under the confining pressure of 60 MPa, the average Young's modulus is 38.3 GPa, the average Poisson's ratio is 0.24, and the average compressive strength is 148.5 MPa. The tensile strength is 6.6 MPa. It is similar to the mechanical properties of the downhole core of the Majiagou Formation in the Daniudi gas field. The outcrop specimen was cut into a cubic sample with dimensions of 300 × 300 × 300 mm. To ensure uniform loading of triaxial stress on the specimen, the cutting error of the cube's edge length must not exceed 2 mm. After forming the cubic specimen, a blind hole is drilled in the center of one face. The hole has a diameter of 28 mm and a depth of 250 mm. At a depth of 240 mm within the blind hole, an initial fracture with a cutting depth of 3 mm is created using a grooving tool. Once drilling is complete, the rock debris inside the hole is cleaned out and allowed to dry. A steel tube, 200 mm in length with an inner diameter of 9 mm and an outer diameter of 20 mm, is used as an artificial wellbore. One end of the tube is threaded internally, and this threaded end is inserted into the drilled hole in the specimen, facing outward. The artificial wellbore is fixed in place using epoxy resin adhesive, leaving a 50 mm open hole section at the bottom of the specimen, as shown in Figure 2.

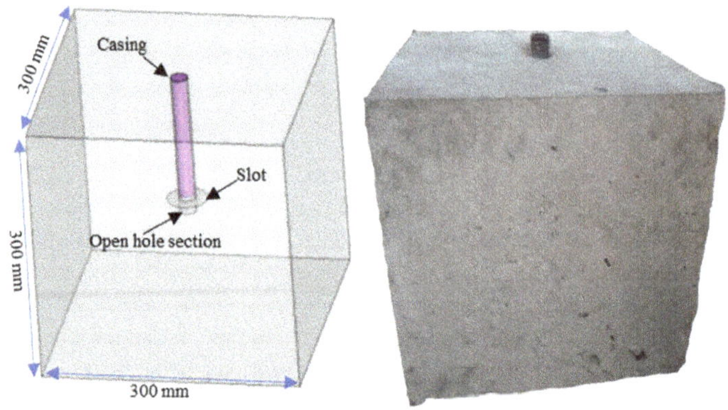


**Figure 2.** Schematic diagram of the structure of the fracturing sample (**left**) and the physical object (**right**).

*2.3. Experimental Scheme*

The experiment primarily considers the effects of different injection rates, types of fracturing fluids, injection methods, and natural fractures on the fracture propagation patterns during acid fracturing. Triaxial stress loading simulates the stress state of a horizontal well, with the direction parallel to the wellbore being the horizontal minimum principal stress ($\sigma_h$). The other two directions are the vertical stress ($\sigma_V$) and the horizontal maximum principal stress ($\sigma_H$), respectively. According to the in situ stress test results from the Daniudi gas field, the horizontal maximum principal stress ranges from 48 to 55 MPa, and the horizontal minimum principal stress ranges from 35 to 40 MPa, with a horizontal stress difference of 13 to 15 MPa. Since the main factor affecting fracture propagation and the interaction between artificial and natural fractures is the horizontal stress difference, and considering the limitations of the stress loading equipment, the experiment reduces the magnitude of the triaxial stresses based on the actual stress conditions. The vertical stress is set at 25 MPa, the horizontal maximum principal stress is set at 20 MPa, and the horizontal minimum principal stress is set at 5 MPa. The injection rates in the experiment are converted using similarity criteria. The field injection rates range from 6 to 12 $m^3/min$, and in the experiment, the rates are set at 60 mL/min, 90 mL/min, and 120 mL/min to simulate field rates of 6 $m^3/min$, 9 $m^3/min$, and 12 $m^3/min$, respectively. It should be noted that due to the limitations of laboratory equipment and sample size, these rate conversions can only partially reflect the impact of rate changes on fracture initiation and near-wellbore propagation. Fracturing fluid mainly includes high viscosity fracturing fluid, gelled acid, and crosslinked acid used in field fracturing. Since the experiments were conducted at room temperature, the viscosities of the fluids used were based on measurements of the field fracturing fluid viscosity at formation temperature. For instance, in the Majiagou Formation, the field fracturing fluid has a viscosity of approximately 100 mPa·s at formation temperature. In our experiments, the thickener concentration was adjusted to match the viscosity of the field fracturing fluid under formation conditions. The high-viscosity fracturing fluid was formulated with 0.6% polymer thickener BFC-200 and 1.2% crosslinker JX-JL-1, achieving a viscosity of up to 100 mPa·s at room temperature. The gelled acid is composed of 20% HCl and 0.7% thickener GDCH-1, with a viscosity of 40 mPa·s at room temperature. The crosslinked acid consists of 20% HCl, 0.8% thickener GDCH-2, and 0.7% crosslinker GDJL-1, with a viscosity of 70 mPa·s at room temperature. For Sample #6, the crosslinker content was increased to 1%, resulting in a viscosity of up to 100 mPa·s. A red dye was added to the fracturing and acid fluids to trace the fracture propagation paths during the acid fracturing process. The specific experimental scheme is detailed in Table 1.

**Table 1.** Acid fracturing experiment scheme.

| Specimen No. | Stress (MPa) $\sigma_h$-$\sigma_H$-$\sigma_v$ | Flowrate (mL·min$^{-1}$) | Viscosity (mPa·s) | Fluid Type | Purpose |
|---|---|---|---|---|---|
| 1# | 5-20-25 | 120 | 100 | Fracturing fluid | Effect of fluid types |
| 2# | | 120 | 40 | Gelled acid | |
| 3# | | 120 | 70 | Crosslinked acid | |
| 4# | | 90 | 70 | Crosslinked acid | Effect of pumping rate |
| 5# | | 60 | 70 | Crosslinked acid | |
| 6# | | 120 | 100 | Crosslinked acid | Effect of NFs |
| 7# | | 120 | 40/70 | Gelled acid/ Crosslinked acid | Injection method |

### 2.4. Experimental Procedures

The main experimental steps of the acid fracturing physical simulation experiment include stress loading of the samples, preparation of the fracturing fluid, fluid injection and fracturing, and comprehensive post-fracturing analysis. The key experimental stages are as follows.

(1) Sample Loading: Before installation, the samples need to be wrapped in plastic wrap to prevent the acid from damaging the experimental equipment after the samples are fractured. The wrapped samples are placed in the rock sample pressure chamber, where a hydraulic stress loading system is used to apply triaxial pressure simultaneously. Considering horizontal well fracturing, the $\sigma_h$ direction is parallel to the wellbore, $\sigma_V$ is perpendicular to the bedding plane, and $\sigma_H$ is perpendicular to both of the aforementioned directions. The pressure is gradually increased in all three directions until the set values are reached.

(2) Fracturing Fluid Preparation: The high-viscosity fracturing fluid, gelled acid, and crosslinked acid are prepared according to the specified formulations and added to the alternating injection system. When using a single-fluid fracturing method, only one intermediate container is used. For alternating injection of multiple fluids, the fluids are placed in separate intermediate containers as needed. The intermediate containers containing the fracturing fluids are connected to the sample and the injection pump.

(3) Acid Fracturing: The corresponding valves are opened and the injection pump is started according to the experimental scheme. When the pressure reaches 1 MPa, the pump is stopped to hold the pressure and leaks are checked. If there is no pressure change within 3 min, the pump is restarted and the pressure is further increased. During this process, pressure and other relevant data are collected by the data acquisition system.

(4) Post-Fracturing Analysis: The experiment is finished when fracturing fluid seeps from the sample or when there is a sudden drop in pressure without significant recovery. At this point, the injection pump is turned off to release the pressure and the confining pressure system is shut down. The post-fracturing sample is then removed and cleaned and then a comprehensive analysis of the internal and external fracture conditions is conducted using pressure curves, dye distribution, rock sample sectioning, and laser scanning.

## 3. Results and Analysis

After the experiment, the fracture morphology, breakdown pressure, and fracture surface erosion patterns were used to evaluate the impact of fracturing fluid type, injection rate, and injection method on the propagation behavior of acid fractures. In the analysis below, blue and yellow lines are used to outline the fractures on the surface of the samples to aid in observing fracture morphology. The blue solid lines represent artificial fractures, the yellow solid lines indicate unopened natural fractures, and the yellow dashed lines represent opened natural fractures.

### 3.1. Effect of Fracturing Fluid Types

To study the influence of fracturing fluid type on the propagation of acid fracturing fractures, high-viscosity fracturing fluid (non-reactive fluid), gelled acid, and crosslinked acid were used to fracture Samples 1#, 2#, and 3#, respectively. In addition to differences in viscosity, these fluids also vary in their reactivity with the rock. All other experimental parameters were kept constant, with an injection rate of 120 mL/min for each test.

The fracture morphologies of the three samples after fracturing are shown in Figure 3. Since horizontal fracturing was simulated, all three samples developed transverse fractures perpendicular to the wellbore direction. The fractures initiated at the open hole section at the bottom of the wellbore and extended to the surface of the samples. The differences are as follows: Sample 1#, using high-viscosity non-reactive fluid, resulted in simple, straight double-wing cracks; Sample 2#, using gelled acid, exhibited more complex branching fractures due to the lower viscosity, higher leak-off capability, and faster acid–rock reaction rate of the gelled acid, which created multiple mechanical weak points in the open hole section, leading to the initiation of multiple fractures; and Sample 3#, using crosslinked acid for fracturing, resulted in relatively simple fractures. The crosslinked acid has a viscosity lower than the fracturing fluid but higher than the gelled acid, making its leak-off stronger than the fracturing fluid but weaker than the gelled acid. Its acid–rock reaction capability is also weaker than that of the gelled acid. Under these conditions, the fractures formed in Sample 3# were relatively simple. During the extension process, the fracture encountered a large-angle intersecting natural fracture, which was opened on one side due to acid leak-off and dissolution.

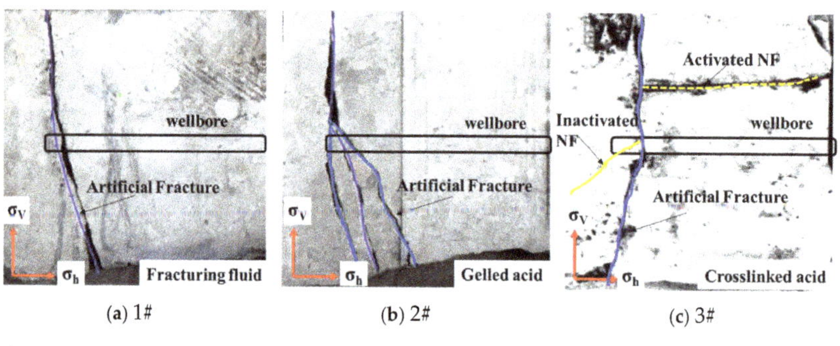


**Figure 3.** Fracture morphology formed after the injection of different types of fracturing fluids: (**a**) a straight, double-wing crack formed in Sample 1# with high-viscosity non-reactive fluid; (**b**) multiple fractures were generated with gelled acid in Sample 2#; and (**c**) relatively simple fractures formed with crosslinked acid in Sample 3#.

The variations in fracture morphologies resulting from different fluid properties are also evident in the pressure curves, as illustrated in Figure 4. Three distinct characteristics differentiate the pressure curves of the samples. Firstly, the most notable feature is that the breakdown pressure increases with the increase in viscosity. Sample 1#, which used the highest viscosity fracturing fluid, had the highest breakdown pressure at 18.5 MPa. Sample 2#, using gelled acid with the lowest viscosity of only 40 mPa, had a breakdown pressure of 14.6 MPa. Sample 3#, using crosslinked acid with intermediate viscosity, had a breakdown pressure of 17.2 MPa. The second characteristic difference is the rate of pressure increase at the wellhead before the samples fracture, as indicated by the red dashed lines in the figure. Sample 1#, having no reactivity and the highest viscosity, had the weakest leak-off performance, resulting in the fastest pressure increase. Sample 3#, with intermediate viscosity and reactivity, had a pressure increase rate second only to Sample 1#. Sample 2#, with the lowest viscosity and strongest reactivity, had the strongest leak-off capability, resulting in the slowest pressure increase. The final characteristic is the distinct differences in the shapes of the three curves. Sample 1# exhibits typical homogeneous rock hydraulic fracturing characteristics, with a sudden pressure drop after the bi-wing fractures extend to the boundary. Sample 2# shows two peaks after reaching the breakdown pressure, indicating the initiation of multiple fractures. The curve shape of Sample 3# is similar to that of Sample 1#, corresponding to their similar fracture morphologies.

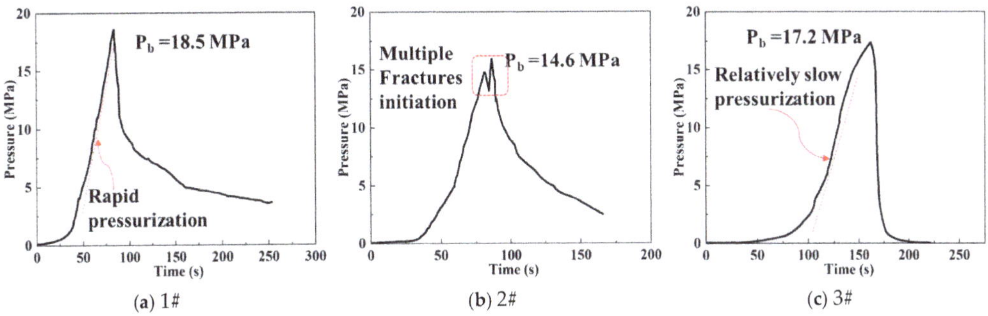


**Figure 4.** Wellhead pressure change curves during fracturing with different types of fracturing fluids: (**a**) Sample 1# had the highest breakdown pressure at 18.5 MPa with the highest viscosity fracturing fluid; (**b**) Sample 2#, using gelled acid with the lowest viscosity, had a breakdown pressure of 14.6 MPa; (**c**) Sample 3#, using crosslinked acid with intermediate viscosity, had a breakdown pressure of 17.2 MPa.

### 3.2. Effect of Injection Rates

Injection rate is another important factor affecting fracture propagation. To study the influence of injection rate on acid fracturing crack propagation, Samples 3#, 4#, and 5# were fractured at rates of 120 mL/min, 90 mL/min, and 60 mL/min, respectively. In the Majiagou Formation, crosslinked acid is the primary acid system used to etch the fracture surface, as its higher viscosity, compared to gelled acid, allows for greater acid penetration. Thus, crosslinked acid was used as the fracturing fluid, and all other experimental parameters were kept constant.

After acid fracturing at different injection rates, the fracture morphology shows significant differences, as illustrated in Figure 5. The primary difference is the change in fracture complexity due to the activation of natural fractures. At higher injection rates, the pressure rises quickly, and the acid fluid has a shorter leak-off time before the sample breaks, resulting in reduced leak-off volume. This leads to the acid having weaker dissolution and reaction effects. Natural fractures intersecting with the wellbore do not have sufficient time to dissolve before reaching the breakdown pressure. Once the fractures break through the boundary, the pressure within the fractures decreases, preventing the activation of natural fractures. As the injection rate decreases, the acid leak-off time and dissolution time increase along with the dissolution volume, creating multiple weak points within the wellbore and promoting the formation of multiple fractures. When the injection rate is reduced to 60 mL/min, a natural fracture develops along the wellbore and connects with other natural fractures, causing a large amount of acid to leak into the natural fracture network, ultimately forming an intricate fracture system.

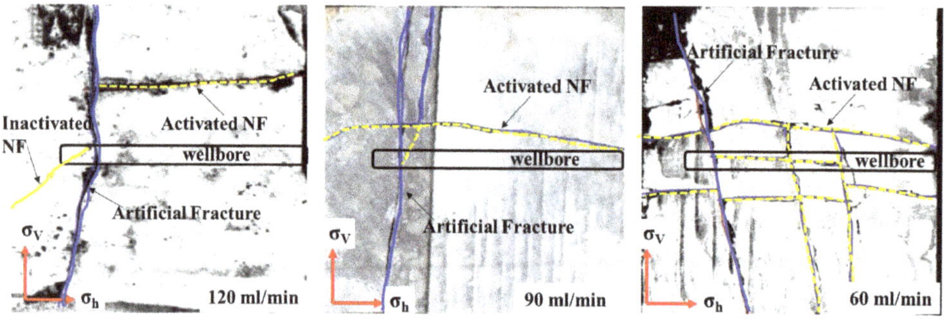


(**a**) 3# (**b**) 4# (**c**) 5#

**Figure 5.** Fracture morphologies formed after acid fracturing at different injection rates: (**a**) one NF was activated with an injection rate of 120 mL/min; (**b**) two NFs were activated with an injection rate of 90 mL/min; and (**c**) complex fracture network formed with an injection rate of 60 mL/min due to NF activation.

The differences in fracture morphology can be indirectly reflected in the pressure curves. The injection pressure curves for the three experiments at different injection rates are shown in Figure 6. A comparative analysis was conducted from three aspects: breakdown pressure, pressure increase rate, and curve fluctuations. As the injection rate increases, the breakdown pressure rises, and the pressure increase rate also significantly accelerates, both of which are related to acid leak-off and dissolution. Additionally, for Samples 4# and 5#, noticeable fluctuations in the pressure curve can be observed before and after breakdown. This is due to the partial activation of natural fractures after dissolution. These analytical results correspond to the observed fracture morphologies.

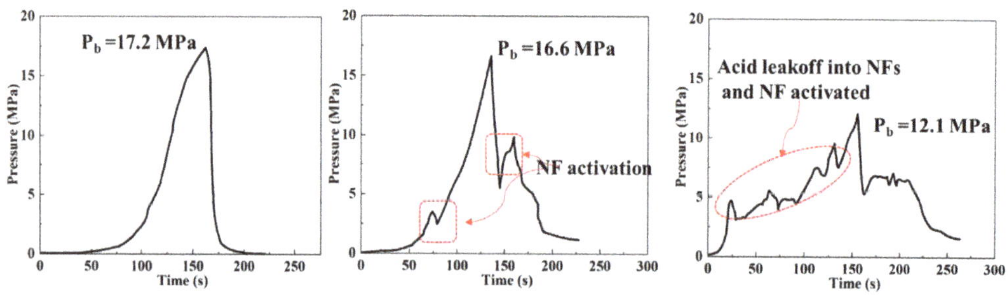


(**a**) 3# (**b**) 4# (**c**) 5#

**Figure 6.** Wellhead pressure variation curves during the acid fracturing process at different injection rates: (**a**) the highest breakdown pressure was observed among three samples with different injection rates; (**b**) intermediate breakdown pressure along with pressure fluctuations was observed when fractured with the intermediate injection rate; and (**c**) noticeable fluctuations in the pressure curve were recorded with the low injection rate.

### 3.3. *Effect of Injection Method*

In acid fracturing, a multi-stage alternating injection of high- and low-viscosity fluids is often used to enhance the differential etching of the fracture surfaces. However, there has been limited research on its impact on fracture propagation patterns. In the experiment, Samples 7# and 3# were compared to study the effect of injection patterns on the morphology of acid fractures. After fracturing, both samples were cut open along the surface fractures, and a 3D laser profilometer was used to capture the dissolution morphology of the rock. Sample 3# underwent crosslinked acid fracturing, while Sample 7# used a combination of crosslinked acid and gelled acid fracturing with a volume ratio of 7:3. Acid was injected for an additional 10 min after the fractures reached the boundary to observe the etching patterns on the fracture surfaces. The post-fracturing morphology and surface etching patterns of the fractures in both samples are shown in Figure 7.

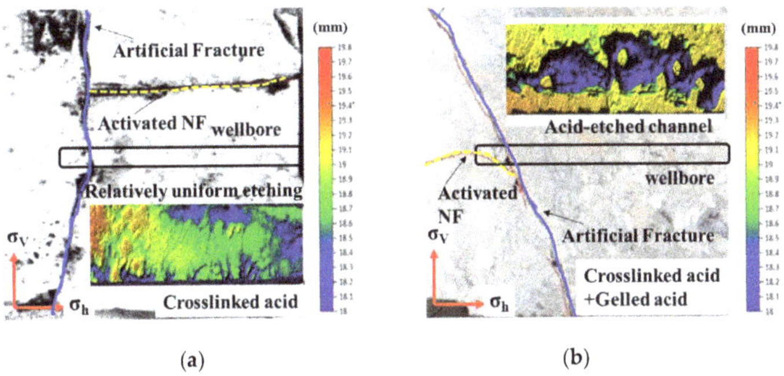


**Figure 7.** Comparison of fracture morphology and surface etching morphology formed through single-acid hydraulic fracturing and multi-stage alternating injection fracturing: (**a**) fracture morphology and surface etching morphology formed through single-acid hydraulic fracturing. (**b**) fracture morphology and surface etching morphology formed through alternating injection fracturing.

Comparing the fracture morphologies formed by the two fracturing modes, the differences are minimal. Both consist of a transverse fracture perpendicular to the wellbore and a natural fracture intersecting the main fracture at a high angle. The rock samples were cut along the main fracture marked in blue, and a laser scanner was used to scan the local acid etching in representative areas, resulting in the color contour maps shown in the figure. In these maps, blue indicates deeper etched areas, while red represents raised areas. The contour map in Figure 7a shows that when a single crosslinked acid is injected, the etching is relatively uniform, with local pits that do not connect to form continuous channels. However, with alternating injection, the viscosity difference between the two acids causes the gelled acid to form connected grooves within the crosslinked acid. This increases the degree of non-uniform etching on the fracture surfaces. This fracturing method enhances the ability of acid-etched fractures to maintain conductivity under closure stress.

### 3.4. Effect of NFs

The above section analyzes the influence of engineering factors on acid fracturing crack propagation. Next, the influence of natural fractures on crack propagation under different injection conditions is analyzed from the aspect of geological factors. Samples 1# and 6# were selected to analyze the influence of different types of fracturing fluids on the interaction between natural and artificial fractures. Samples 6# and 3# were selected to compare the influence of natural fracture inclination angles on the interaction between artificial fractures and natural fractures. All three samples have a flow rate of 120 mL/min, with the other experimental parameters remaining constant.

In comparing Sample 1# and Sample 6#, both the artificial fractures (blue) and natural fractures (yellow) intersect at an angle of approximately 50 degrees, as shown in Figure 8a,b. In Sample 1#, the artificial fracture passes directly through the natural fracture without activating it. However, in Sample 6#, the artificial fracture passes through and activates the natural fracture. The role of acid in opening natural fractures is significant. Acid can infiltrate natural fractures through filtration, reducing the cementation strength within these fractures and making them more prone to activation. In contrast, non-reactive fracturing fluids do not have the same effect, making natural fractures less likely to open at high approach angles. Figure 8b,c compare the opening of natural fractures under the same acid solution but with different approach angles of the natural fractures. In Sample 3#, the artificial fracture intersects the natural fracture at an angle close to 90 degrees. The artificial fracture passes through the natural fracture and opens one side of it. This occurs because when the natural fracture is horizontally developed, it is subjected to vertical stress. Since the vertical stress is much greater than the minimum horizontal principal stress, the difficulty of opening the fracture increases.

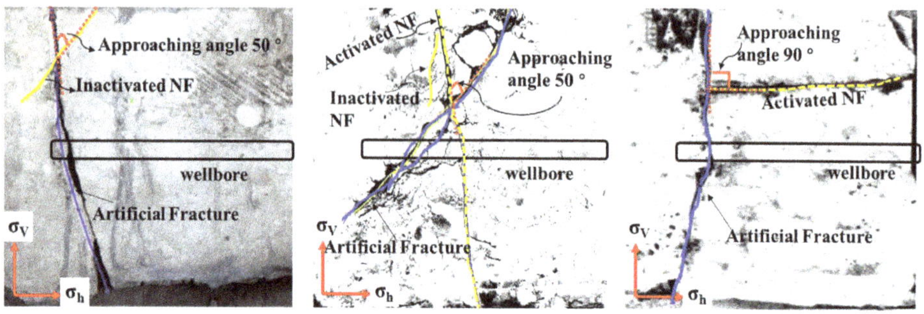


(**a**) 1# Non-reactive fracturing fluid (**b**) 6# Crosslinked acid (**c**) 3# Crosslinked acid

**Figure 8.** Impact of nature fractures on fracture propagation under different injection conditions: (**a**) and (**b**) compare the influence of different types of fracturing fluids on the interaction between natural and artificial fractures; (**b**,**c**) compare the influence of natural fracture inclination angles on the interaction between artificial fractures and natural fractures.

## 4. Discussion

Acid fracturing is an important technique for the stimulation of carbonate reservoirs. This is particularly true for tight fractured carbonate formations where maximizing the acid action distance and expanding the stimulated volume are critical for successful acid fracturing. The use of pre-pad fluid in acid fracturing can help reduce filtration loss of subsequent acid and lower the temperature within the fracture. However, because the pre-pad fluid is non-reactive, its activation effect on natural fractures is relatively weak, often resulting in the formation of a single dominant fracture in the open hole. Acid can weaken the mechanical properties of natural fractures in the open hole, creating multiple mechanical weak points. This can ultimately lead to the simultaneous initiation of multiple fractures within the wellbore, increasing the complexity of fractures in the near-wellbore area. Therefore, before acid fracturing, gelled acid can be used to pre-treat the open hole section, reducing the fracture pressure while also promoting the initiation of multiple fractures.

Elevating the pumping displacement results in a decrease in filtration time and an augmentation in the height and length of the primary fracture. The infiltration of acid into natural fractures fosters the weakening of their mechanical properties, which in turn aids in the activation of these fractures, thereby expanding the stimulation volume. Consequently, a strategy involving alternating injections of low-viscosity acid at lower displacement levels and high-viscosity acid at higher displacement levels should be employed to enhance the overall stimulation volume. Furthermore, the application of a multi-stage alternating injection technique during acid fracturing is advisable. This method not only mitigates filtration loss but also promotes non-uniform etching patterns on acid-etched fracture surfaces via viscous fingering, ultimately yielding improved flow conductivity maintenance capabilities.

## 5. Conclusions

In this paper, a series of large-scale acid fracturing physical simulation experiments were carried out using true triaxial acid fracturing equipment to study the influence of fracturing fluid types, injection displacements, injection methods, and natural fractures on the propagation laws of acid fracturing fractures. The main conclusions obtained are as follows:

(1) After fracturing with non-reactive fluid, the number of fracture initiations is lower; the acid can form multiple weaknesses in the wellbore through dissolution, which promotes the initiation of multiple cracks. Using low-viscosity acid as pre-treatment before fracturing can effectively improve the complexity of fractures near the well and improve the range of near-well stimulation.

(2) Under conditions of high displacement and high viscosity, the fracture morphology tends to be straightforward. In contrast, injecting acid at low viscosity and low displacement promotes the initiation and extension of natural fractures, thereby increasing the overall complexity of the fracture network.

(3) The multi-stage alternating injection method can reduce acid fluid loss while simultaneously increasing the degree of differential etching through viscous fingering. This approach enables acid-etched fractures to attain a high level of flow conductivity and effectively maintain that conductivity over time.

**Author Contributions:** Conceptualization, Y.Z. (Yongchun Zhang) and H.Z.; methodology, Y.Z. (Yongchun Zhang) and H.Z.; validation, Y.Z. (Yongchun Zhang), H.Z., Y.Z. (Ying Zhong) and J.K.; investigation, H.Z.; resources, Y.Z. (Yongchun Zhang); data curation, Y.Z. (Ying Zhong); writing—original draft preparation, Y.Z. (Yongchun Zhang); writing—review and editing, S.D.; visualization, J.K.; supervision, Y.Z. (Yongchun Zhang); project administration, Y.Z. (Yongchun Zhang). All authors have read and agreed to the published version of the manuscript.

**Funding:** This research was funded by The Basic Science Center Project of the National Natural Science Foundation of China, "Flow Regulation in Ultra-Deep Oil and Gas Drilling and Production" (Project No.: 52288101).

**Data Availability Statement:** The raw data supporting the conclusions of this article will be made available by the authors upon request.

**Conflicts of Interest:** Author Yongchun Zhang is employed by the company Sinopec North China Oil & Gas Company. The remaining authors declare that the research was conducted in the absence of any commercial or financial relationships that could be construed as a potential conflict of interest.

## References

1. Mou, J. *Modeling Acid Transport and Non-Uniform Etching in a Stochastic Domain in Acid Fracturing*; Texas A&M University: College Station, TX, USA, 2010.
2. Mou, J. *Principle of Acidizing Stimulation Technology*; Petroleum Industry Press: Beijing, China, 2023.
3. Zhang, X.; Zhang, S.; Mou, J. Numerical investigation of hydraulic fracture propagation behavior in vuggy, naturally fractured carbonate reservoir—A case study of shunbei oil field. *Geoenergy Sci. Eng.* **2024**, *242*, 213257. [CrossRef]
4. Deng, J.; Hill, A.; Zhu, D. A theoretical study of acid-fracture conductivity under closure stress. *SPE Prod. Oper.* **2011**, *26*, 9–17. [CrossRef]
5. Salimi, S.; Ghalambor, A.; Hayer, H. Insights Into the Process of Effectively Acidizing Naturally Fractured Reservoirs. In Proceedings of the SPE International Symposium and Exhibition on Formation Damage Control, Lafayette, LA, USA, 26–28 February 2014. SPE-168126-MS.
6. Panga, M.; Ziauddin, M.; Balakotaiah, V. Two-scale continuum model for simulation of wormholes in carbonate acidization. *AIChE J.* **2010**, *51*, 3231–3248. [CrossRef]
7. Taylor, K.; Nasr-El-Din, H. Laboratory evaluation of in-situ gelled acids for carbonate reservoirs. *SPE J.* **2003**, *8*, 426–434. [CrossRef]

8. Ugursal, A.; Zhu, D.; Hill, A. Development of Acid Fracturing Model for Naturally Fractured Reservoirs. *SPE Prod. Oper.* **2019**, *34*, 735–748. [CrossRef]
9. Gou, B.; Ma, H.; Liu, Z. Research progress and prospect of numerical modeling for acid fracturing of heterogeneous carbonate reservoirs. *Nat. Gas Ind.* **2019**, *39*, 87–98.
10. Zeng, Q.; Li, T.; Bo, L. Effect of acid pre-treatment on hydraulic fracturing in heterogeneous porous media. *Geoenergy Sci. Eng.* **2025**, *246*, 213601. [CrossRef]
11. Budong, G.; Jianye, M.; Panpan, L. Numerical investigation into the acid flow and reaction behavior in the tight, naturally fractured carbonate reservoir during acid fracturing. *Phys. Fluids* **2024**, *36*, 113614.
12. Guo, J.; Bo, G.; Dengyun, L. Advance and prospect of acid fracturing in deep carbonate reservoirs. *Drill. Prod. Technol.* **2024**, *47*, 121–129.
13. Zheng, H.; Liao, R.; Cheng, N. Study of mechano-chemical effects on the morphology of hydraulic fractures. *J. Pet. Sci. Eng.* **2021**, *206*, 109031. [CrossRef]
14. Li, C.; Wang, Y.; Deng, P. Influences of gelled acid on mechanical properties of the ordovician carbonate rocks in block shunbei, Tahe oilfield. *Xinjiang Pet. Geol.* **2018**, *39*, 689–695.
15. Guo, J.; Ren, J.; Wang, S. Numerical simulation and application of multi-field coupling of acid fracturing in farctured tight carbonate reservoirs. *Acta Pet. Sin.* **2020**, *41*, 1219–1228.
16. Aljawad, M.; Schwalbert, M.; Mahmoud, M. Impacts of natural fractures on acid fracture design: A modeling study. *Energy Rep.* **2020**, *6*, 1073–1082. [CrossRef]
17. Gou, B.; Zhan, L.; Guo, J. Effect of different types of stimulation fluids on fracture propagation behavior in naturally fractured carbonate rock through CT scan. *J. Pet. Sci. Eng.* **2021**, *201*, 108529. [CrossRef]
18. Wang, Y.; Hou, B.; Zhang, K. Laboratory true triaxial acid fracturing experiments for carbonate reservoirs. *Pet. Sci. Bull.* **2020**, *3*, 412–419.
19. Mohtarami, E.; Baghbanan, A.; Eftekhari, M. Investigating of chemical effects on rock fracturing using extended finite element method. *Theor. Appl. Fract. Mech.* **2017**, *89*, 110–126. [CrossRef]
20. Zhang, N.; Chen, X.; Luo, Z. Experimental study of fracture conductivity in dolomite reservoirs treated with different acid fracturing technologies. *Geoenergy Sci. Eng.* **2023**, *227*, 211914. [CrossRef]
21. Hou, B.; Zhang, R.; Chen, M. Investigation on acid fracturing treatment in limestone formation based on true tri-axial experiment. *Fuel* **2019**, *235*, 473–484. [CrossRef]
22. Zhang, K.; Chen, M.; Zhou, C. Study of alternating acid fracturing treatment in carbonate formation based on true tri-axial experiment. *J. Pet. Sci. Eng.* **2020**, *192*, 107268. [CrossRef]
23. Su, X.; Qi, N.; Lu, Y. Experimental study on acid etching mechanism of acid fracturing in carbonate reservoirs. *Geoenergy Sci. Eng.* **2024**, *241*, 213079. [CrossRef]

processes

MDPI

Article

# Experimental and CFD-Integrated Investigation into the Intricate Oil–Water Two-Phase Flow Dynamics Within Blind Tees: Uncovering Flow Behaviors for Advanced Process Engineering Applications

Chen Cheng [1,*], Qingming Gan [2], Yubin Su [2], Yang Cheng [3] and You Zhang [4,*]

[1] Sinopec Natural Gas Branch, Beijing 100020, China
[2] Oil & Gas Technology Research Institute, Changqing Oilfield Branch Company, PetroChina, Xi'an 710021, China; ganqm_cq@petrochina.com.cn (Q.G.); suyb_cq@petrochina.com.cn (Y.S.)
[3] Petroleum Engineering Technology Research Institute, Jianghan Oilfield Branch Company, Sinopec, Wuhan 430000, China; 18389599309@163.com
[4] Petroleum Engineering College, Yangtze University, Wuhan 430100, China
* Correspondence: chengchen_sinopec@163.com (C.C.); 15272424010@163.com (Y.Z.)

**Abstract:** Blind tees are widely used in offshore oil systems. To enhance their rectification performance, an indoor experimental platform was established to study the flow field behavior within blind tees and bend pipes. Fluent2021 analyzed the significant effects of various structural parameters and operating conditions on the oil–water rectification performance of blind tees. The study evaluates the device's effectiveness by establishing a rectification coefficient $S$, investigating the impacts of blind end length, vertical pipe diameter, inlet flow velocity, and inlet oil content. Results indicate that internal vortices within the blind tee primarily cause disturbances leading to uniform distribution of oil–water phases. The size and intensity of the vortices determine the oil–water rectification efficiency of the blind tee. Blind tees of different lengths correspond to a specific velocity at which the blind end becomes filled with vortices, causing fluid stagnation. This velocity is defined as the blind end closure velocity. Avoiding this closure velocity, shorter blind ends, smaller vertical pipe diameters, and higher inlet velocities significantly enhance the oil–water flow straightening effectiveness. The rectification efficiency of the blind tee was improved by up to76.68% compared to the bend.

**Keywords:** offshore oil pipelines; rectifiers; oil–water two-phase flow; CFD; vortex detection

## 1. Introduction

Multiphase flow of oil and water is ubiquitous in the offshore oil industry, where crude oil often coexists with water, natural gas, and other components in many offshore oil fields [1]. Following gas–liquid separation, thorough research and precise measurement of oil–water two-phase flow are crucial for better understanding fluid behavior and flow characteristics. This understanding holds significant importance for assessing crude oil production, predicting the lifespan of oil well developments, monitoring well conditions, and enhancing energy efficiency [2]. However, the presence of significant phase slip within oil–water two-phase flow complicates measurements, necessitating the use of flow rectifiers before measurement instruments to improve accuracy [3]. Due to the scarcity of space on offshore drilling platforms, complex and massive rectifier devices are not suitable for installation, operation, and maintenance. The design and implementation of effective and simple flow pattern rectifiers are, therefore, of paramount practical significance for multiphase flow measurements [4].

Bend configurations are the most common structural elements in subsea pipelines. Patankar et al. [5] successfully predicted the flow characteristics in curved pipes, encompassing both developing and fully developed flow regimes, by incorporating a two-equation turbulence model into a numerical solution method. Their predictions showed good agreement with experimental data, and the computational time remained within a reasonable range. The research of Hidesato et al. [6] demonstrated that varying hydrodynamic conditions at the inlet significantly influence flow behavior in bends; for instance, two serially connected 90° bends generate intense swirling flows downstream, depending on the combination angles. Liberto et al. [7] investigated turbulent flow in curved pipes with three curvature parameters ($\delta$) using direct numerical simulation (DNS). Dutta [8] conducted a numerical study on turbulent single-phase flow in a circular 90° bend, focusing on the effects of different curvature ratios on flow characteristics. Noorani [9] investigated the influence of pipe curvature on the transport and accumulation of micro-particles. Ayala [10] performed a numerical study on turbulent flow in a smooth-walled circular 90° bend, aiming not only to estimate the resistance coefficient of the bend through pressure drop but also to incorporate information on downstream flow behavior in the outlet section. Additionally, Jarrahi [11] and Kim [12] systematically examined the swirling flow characteristics and mixing effects induced by 90° elbows.

The blind tee is a specialized pipe connection structure widely used in the petroleum and natural gas industries, distinguished from standard tees by having one closed end. Numerous scholars have conducted research on this structure. Pouraria found that blind tees exhibit greater erosion resistance for small diameter pipes; however, as pipe diameter increases, the erosion caused by blind tees can be comparable to or even more severe than that caused by standard bends [13]. Chen applied a CFD erosion prediction model to predict erosion in blind tees and bends under air/sand flow conditions, revealing significant influences of sand load on relative erosion severity. Tests under high sand concentrations showed that erosion in blind tees was two orders of magnitude less than that in bends [14]. Other studies have highlighted the blind tee's special promotion of heat and mass transfer within the fluid. Han used three-dimensional numerical simulations to study laminar flow within blind tees at Re = 500, 1000, 1500, and 2000, finding that higher flow velocities and blind tee lengths up to 3D facilitate good fluid mixing within the tee [15]. Ma installed blind tees upstream of Venturi meters and gamma-ray segments to measure oil, gas, and water three-phase flows, significantly improving measurement accuracy [16]. Bhakta observed that downstream of blind tee outlets, gas–liquid two-phase flows exhibit more symmetrical and uniform characteristics [17]. Millington described the variation in phase distribution at the outlet of the blind tee and the impact of inlet flow structures on mixing performance. Before the flow accelerated through a transparent plexiglass Venturi tube, phase area distribution was captured using an Electrical Capacitance Tomography (ECT) system in parallel with a high-speed video recorder. The results indicated that the mixing capability of the blind tee depends on the inlet flow patterns and oil viscosity, thereby affecting phase dispersion and slip characteristics through the Venturi section [18]. Experiments conducted by Hjertaker using a high-speed gamma-ray tomography instrument in the multiphase flow loop at the Christian Michelsen Research Institute demonstrated that near the blind tee, the flow is well mixed but significantly deviates from axisymmetry. Further up along the vertical pipe section, the flow develops towards an axisymmetric profile with increased slip. At lower flow rates, the vertical flow pattern develops more rapidly, resulting in a higher degree of axisymmetry, increased gas holdup in the pipe center, and greater slip between the gas and liquid phases [19]. In multiphase flow measurement systems, blind tees are often used as flow mixers to homogenize the mixture when installed upstream of multiphase flow meters.

Based on a review of international research, it was found that studies on blind tees predominantly focus on erosion, single-phase flow, and the mixing characteristics of gas–liquid multiphase flow. However, there is limited research on the actual changes in the internal flow field of blind tees, their rectification effects on oil–water two-phase flow, and mixing mechanisms. Therefore, this study first investigated the actual changes in the internal flow field of blind tees through indoor experiments. Subsequently, different blind tee models with varying blind end lengths and vertical pipe diameters were established, and numerical simulations were used to explore the effects of different flow rates, blind end lengths, vertical pipe diameters, and inlet oil contents on the internal oil–water flow characteristics and rectification performance of blind tees. This research aims to provide references for the field of oil–water two-phase flow measurement and rectification.

## 2. Experimental Apparatus and Procedure

### 2.1. Experiment Set up and Measurement Techniques

To investigate the internal flow field patterns of blind tees and bend pipes, Particle Image Velocimetry (PIV) technology was used to conduct experiments on both structures. A flow pattern investigation experimental system was established at the Key Laboratory of Multiphase Flow in Yangtze University, which is shown in Figure 1. The system mainly consists of a liquid supply system, pipeline instrumentation, PIV measurement device (Hefei Zhongke Junda Vision Technology Co., Ltd., Hefei, China), and test sections. The experimental site is shown in Figure 2. In the experimental section, the bent pipe is a standard 90° bend with a diameter of 20 mm, while the blind tee has a main pipe diameter of 20 mm and a blind end length of 40 mm.

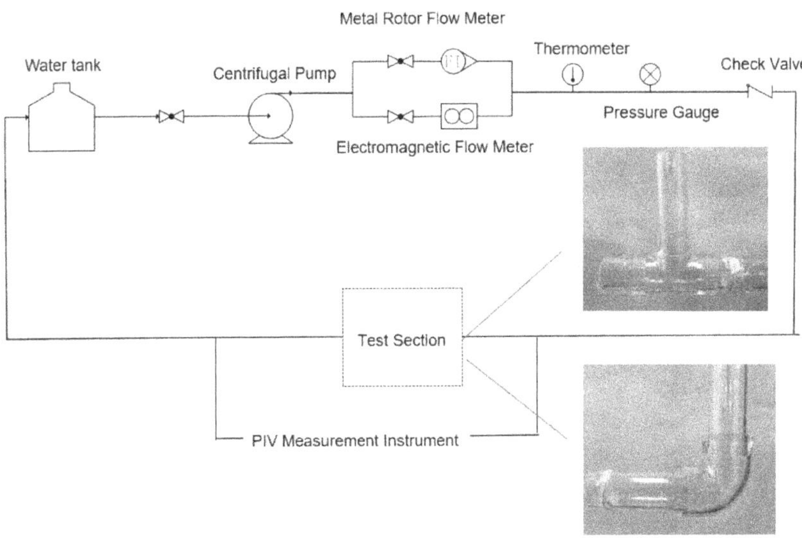


**Figure 1.** Schematic diagram and testing apparatus of experiment.

### 2.2. PIV Data Processing Techniques

The PIV data processing was performed using the manufacturer's proprietary software RFlow4 (Hefei Zhongke Junda Vision Technology Co., Ltd., Hefei, China), with a unique grid refinement function applied near the boundary wall to ensure that the closest measurement point to the boundary was half of the minimum diagnostic window size.

The calculation accuracy for both the velocity field and solid-phase particle motion was 0.1 pixels, with validation based on actual data from the PIV Challenge. Additionally, AI-based automatic masking, gamma correction, and high-pass filtering techniques were utilized in the data processing to reduce image noise.

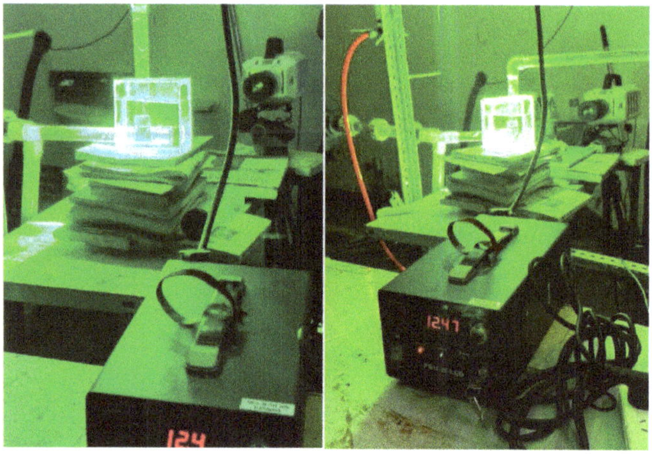

**Figure 2.** Experimental test site.

### 2.3. *Experimental Conditions and Control*

Since this experiment primarily focuses on the hydrodynamic behavior and vortex evolution mechanisms within blind tee and elbow structures, pure water was used as the working fluid to reduce experimental costs and enhance visualization of the internal flow characteristics. At 25 °C, the density of water is 998 kg/m$^3$, and the dynamic viscosity is 0.89 mPa·s. The volumetric flow rate was controlled by a variable-frequency pump set at 0.4 m$^3$/h. The experiment was conducted at 25 °C, with both inlet and outlet pressures maintained at atmospheric pressure to ensure consistency in boundary conditions.

The experimental apparatus was designed with a geometric scale ratio of 1:25 (model pipe diameter D = 20 mm, corresponding to a typical offshore pipeline prototype diameter of $D_0$ = 500 mm). The Reynolds number of the model was Re = 6995.88, and the calculated prototype Reynolds number was $Re_0$ – 70,382 (Q =100 m$^3$/h), both within the turbulent regime (Re > 4000), ensuring flow similarity.

To ensure the measurement accuracy of the PIV system, rigorous calibration and validation were performed prior to the experiment, including laser source adjustment, camera focal correction, and optimization of particle imaging quality. Additionally, three repeated measurements were conducted to evaluate data stability and minimize the influence of random errors. These experimental conditions ensured the stability of the working fluid, the consistency between the experimental environment and practical engineering conditions, and provided reliable visualization data for investigating the internal flow field evolution and vortex structures within blind tees and elbows.

### 2.4. *Analysis of Experimental Results*

In the design of flow rectifiers, the types and characteristics of vortices generated within the device serve as key indicators of rectification effectiveness [20–22]. Therefore, this study employed Particle Image Velocimetry (PIV) technology to detect vortices within the internal flow field. Figures 3 and 4, obtained using PIV post-processing software RFlow4, present the vector field and streamline plot of the blind end region of the blind tee, providing detailed insights into the flow characteristics and vortex generation mechanisms of fluid within the blind end. The vector field arrows indicate that the rotational motion of the fluid within the blind end leads to the formation of lateral vortices, where the vortex core region exhibits lower velocity, while the outer region has higher velocity. The formation of these vortices is primarily due to the impact of the incoming fluid on the structural walls of the blind end, inducing localized rotational motion. This process facilitates the transfer of kinetic energy from the main flow into the vortex, establishing a primary energy dissipation region. The streamline plot further reveals the fluid motion paths and flow states within the blind end. Fluid particles within this region follow rotational streamlines, forming sustained recirculation and circular motion, significantly increasing fluid residence time in the blind end. This enhances energy exchange between fluid particles, thereby improving the rectification performance of the blind tee. Observing the vector arrows, it is evident that the vortex core region exhibits lower velocity, whereas velocity gradually increases at the interface between the vortex periphery and the main flow, forming a velocity gradient. This velocity distribution suggests that vortex formation is a primary contributor to kinetic energy dissipation, and that velocity gradients facilitate localized momentum balance and rectification effects. In the high-velocity regions near the blind end, fluid transfers energy from the main flow into the blind end, maintaining vortex stability. These analyses collectively highlight that the blind end structure plays a crucial role in regulating flow behavior, enhancing rectification performance, and promoting fluid mixing.

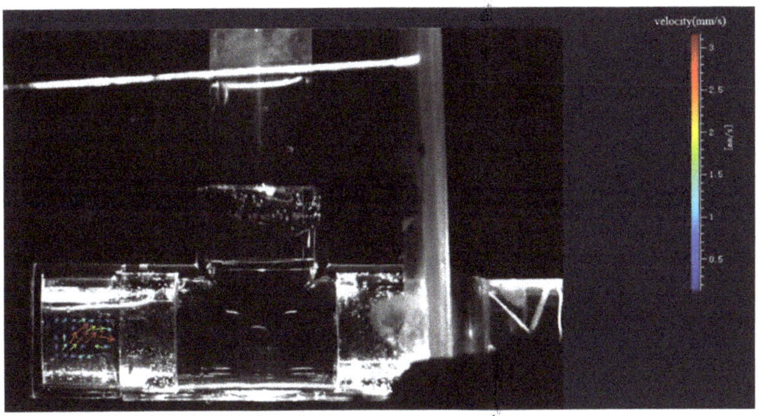

**Figure 3.** Flow vector diagram at the blind end of the blind tee.

**Figure 4.** Streamline diagram at the blind end of the blind tee.

Figures 5 and 6 present the streamline plots of the turning regions in the blind tee and elbow. The streamline plots reveal that fluid flow in the turning regions of both the blind tee and elbow is influenced by centrifugal forces and geometric constraints, leading to the formation of a typical secondary flow structure. Fluid particles near the inner wall of the elbow exhibit recirculation, generating vortices in this region. The streamline distribution in the blind tee appears more disordered, indicating stronger mixing effects in two-phase flow. In contrast, the vortices in the elbow are smaller, but they still play a positive role in promoting fluid mixing. These findings demonstrate that the blind tee and elbow each exhibit unique flow dynamics in the turning regions, yet both induce vortex formation, which is critical in enhancing fluid mixing and rectification performance.

**Figure 5.** Streamline diagram at the elbow of the blind tee.

**Figure 6.** Streamline diagram at the elbow of the bend pipe.

This experiment provided deeper insights into the internal flow characteristics of the blind tee and elbow, which can also serve as a validation basis for subsequent numerical simulation studies.

## 3. Numerical Simulation

### 3.1. Geometric Model

A model of a blind tee with a blind end length of 1D and a vertical inner diameter of 40 mm is shown in Figure 7, where D is 40 mm, C is 20 mm, and L is 190 mm. In the subsequent simulations, D was set to 20 mm, 25 mm, 30 mm, 35 mm, and 40 mm; L was set to 190 mm, 210 mm, 230 mm, 250 mm, and 270 mm; and C was set to 20 mm, 40 mm, 60 mm, 80 mm, and 100 mm. To analyze the rectification effect on the oil–water two-phase flow, it was essential to ensure that the oil and water phases were in a non-homogeneous mixed state upon entering the device. Therefore, two inlets were configured: a pure water inlet for the water phase and an oil–water mixture inlet for the oil phase.

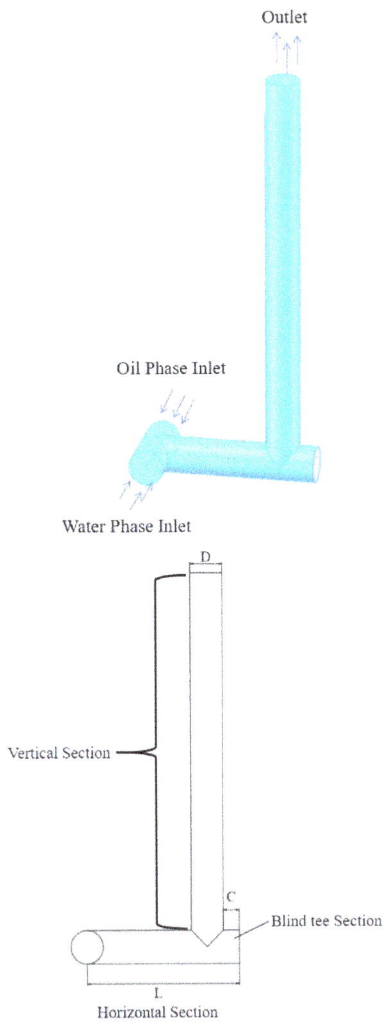


**Figure 7.** Blind tee mode.

### 3.2. Methodology

#### 3.2.1. Multiphase Flow Model and Boundary Conditions

The multiphase flow model adopts the Eulerian model.

Continuity equation:

$$\frac{\partial}{\partial t}(\alpha_q \rho_q) + \nabla \cdot (\alpha_q \rho_q v_q) = \sum_p \dot{m}_{pq} \tag{1}$$

where $\alpha_q$ is the volume fraction of phase $q$, $\rho_q$ is the density of phase $q$, $v_q$ is the velocity of phase $q$, $\dot{m}_{pq}$ is the mass source term transferred from phase $p$ to phase $q$, representing interphase mass exchange.

The momentum equation is expressed as follows:

$$\frac{\partial}{\partial t}(\alpha_q \rho_q v_q) + \nabla \cdot (\alpha_q \rho_q v_q v_q) = \alpha_q \nabla P + \nabla \cdot (\tau_q) + \alpha_q \rho_q g + \sum_p M_{pq} \tag{2}$$

where $P$ is the pressure, $\tau_q$ is the stress tensor of the phase $q$, $g$ is the gravitational acceleration, $M_{pq}$ is the interphase momentum exchange term.

The RNG (Renormalization Group) k-ε turbulence model was chosen for this simulation, as it accounted for the effects of swirl on turbulence viscosity, making it more effective in handling highly swirling and curved flows. A three-dimensional steady-state model was employed, and the SIMPLE (Semi-Implicit Method for Pressure-Linked Equations) algorithm was used for iterative calculations. In the numerical simulation, the turbulence intensity was set to 5%, which is a commonly used initial turbulence intensity value for industrial pipeline flows. A first-order discretization scheme was adopted, offering high computational efficiency and fast convergence, typically used for initial solutions or when high stability is required. The boundary layer thickness was set to 1.9 mm.

For the boundary conditions, the water phase inlet consisted of pure water, while the oil phase inlet consisted of oil–water mixtures with oil concentrations of 10%, 20%, 30%, 40%, and 50%. The inlet velocities were set at 1 m/s, 2 m/s, 3 m/s, 4 m/s, and 5 m/s, respectively.

#### 3.2.2. Turbulence Model

In the RNG k-epsilon model, the influence of small-scale motions is accounted for through modified viscosity terms in the large-scale motion equations, systematically eliminating these small-scale effects from the governing equations [23]. The equations are as follows:

$$\frac{\partial(\rho K)}{\partial t} + \frac{\partial(\rho \overline{u_j} K)}{\partial x_j} = \frac{\partial}{\partial x_j}\left[\left(\mu + \frac{\mu_t}{Pr_K}\right)\frac{\partial K}{\partial x_j}\right] + P_K + G_b - \rho\varepsilon - Y_M \tag{3}$$

$$\frac{\partial(\rho\varepsilon)}{\partial t} + \frac{\partial(\rho \overline{u_j}\varepsilon)}{\partial x_j} = \frac{\partial}{\partial x_j}\left[\alpha_\varepsilon\left(\mu + \frac{\mu_t}{\mathrm{Pr}_\varepsilon}\right)\frac{\partial \varepsilon}{\partial x_j}\right] + C_{\varepsilon 1}\frac{\varepsilon}{K}(P_K + C_{\varepsilon 3} G_b) - C^*_{\varepsilon 2}\rho\frac{\varepsilon^2}{K} \tag{4}$$

In these equations:

$$\mu_t = C_\mu \rho \frac{K^2}{\varepsilon} \tag{5}$$

$$C^*_{\varepsilon 2} = C_{\varepsilon 2} + \frac{C_\mu \rho \eta^3 \left(1 - \frac{\eta}{\eta 0}\right)}{1 + \beta \eta^3} \tag{6}$$

$$\eta = \frac{K}{\varepsilon}\overline{S} \tag{7}$$

$$\overline{S} = \sqrt{2\overline{S}_{ij}\overline{S}_{ij}} \tag{8}$$

$$\overline{S}_{ij} = \frac{1}{2}\left(\frac{\partial \overline{u}_i}{\partial x_j} + \frac{\partial \overline{u}_j}{\partial x_i}\right) \tag{9}$$

where $\rho$ is the fluid density, $K$ is the turbulence kinetic energy, $t$ is time, $\overline{u_j}$ is the mean velocity component in the $j$-th direction, $x_j$ is the spatial coordinate in the $j$-direction, $\mu$ is dynamic viscosity, $\mu_t$ is turbulent viscosity, $\mathrm{Pr}_K$ is the Prandtl number for turbulence kinetic energy, $P_K$ is the generation rate of turbulence kinetic energy, $G_b$ is the turbulence kinetic energy generated by buoyancy, $\varepsilon$ is the dissipation rate of turbulence kinetic energy, and $Y_M$ is the influence of fluctuating dilation on the turbulence dissipation rate.

### 3.2.3. Grid Resolution Testing

The model was meshed using Fluent meshing, with refined meshes applied to the corner regions and 10 layers of boundary layers set, as shown in Figure 8. To validate the mesh accuracy, five different mesh models were established: mesh1 with 278,256 elements, mesh2 with 327,506 elements, mesh3 with 453,919 elements, mesh4 with 571,184 elements, and mesh5 with 678,590 elements for simulation. Figure 9 shows the velocity distribution along the $x$-axis at the outlet for each mesh model. It can be observed that the simulation results of the five models differ only slightly, with only one sampling point having an error of around 10%, while the errors at other sampling points remain within single digits. The reason is that the model is relatively simple, with no small regions requiring fine meshes for high-precision calculations. To ensure simulation accuracy and reduce simulation time, the mesh3 model with 453,919 elements was selected for the simulation study.

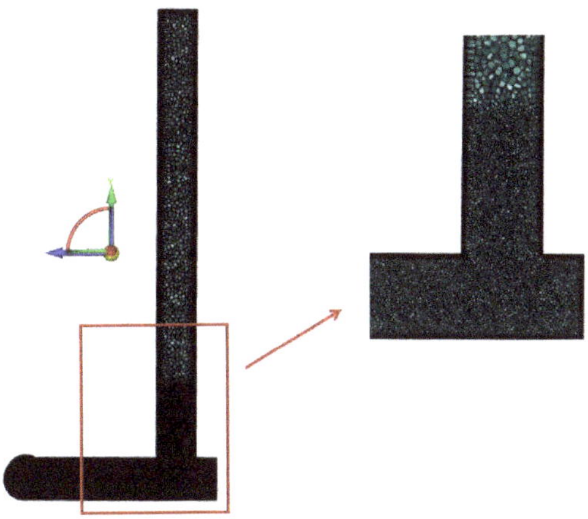

**Figure 8.** Mesh partitioning.

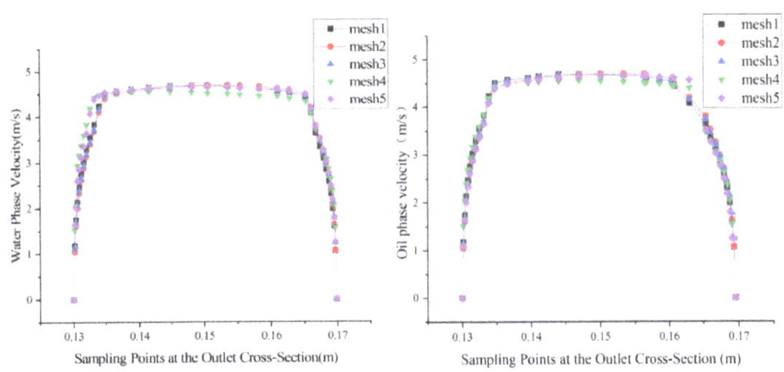


**Figure 9.** The velocity distributions of the water phase (**left**) and the oil phase (**right**) at the outlet under different mesh models.

### 3.3. Evaluation Mathematical Model

To quickly assess the rectification efficiency of the model, coefficient S was defined as the rectification coefficient, where its magnitude reflected the uniformity of flow patterns within the pipeline. An S value of 0 indicated perfect mixing, while increasing S values greater than 0 indicated increasing flow non-uniformity. The expression was given as follows:

$$S = \sqrt{\frac{1}{N}\sum_{i=1}^{N}\left(\rho_{x_iy_j} - \overline{\rho}\right)^2} \tag{10}$$

where $N$ is the number of sampling points. A larger $N$ indicates a more comprehensive evaluation of the rectification efficiency of the device by the mathematical model. $\rho_{x_iy_j}$ is the density of sampling points at different locations (in this study, points at the cross section y = 400 mm are selected to evaluate the rectification efficiency of the device) and $\overline{\rho}$ is the average density of the fluid entering the device.

### 3.4. Analysis of Simulation Results

#### 3.4.1. Flow Characteristics of Oil–Water Two-Phase Flow Inside a Bent Pipe

To verify the superiority of the blind tee over traditional bent pipes in adjusting multiphase flow patterns, an internal flow field analysis was first conducted on a commonly used bent pipe in the petroleum industry. Numerical simulations were performed on a 90-degree bent pipe with a curvature radius of 1.5D, a diameter of 40 mm, a vertical length of 420 mm, and a horizontal length of 190 mm, followed by a detailed analysis of the internal flow field.

Figure 10 shows the schematic diagram of the bent pipe. Pure water was introduced at the water phase inlet, while an oil–water mixture with 40% oil content was introduced at the oil phase inlet. Figure 11 presents the phase distribution contour plot at the y = 0.002 m cross section. It can be observed that the fluids from both inlets collide in the middle of the pipe and flow into the bent section. In the bent section, the oil phase exhibits a trapezoidal distribution on the zx plane, indicating a highly turbulent state of oil–water two-phase flow. The inlet design of the model meets the requirement for non-uniform mixing of the two phases upon entry into the pipe.

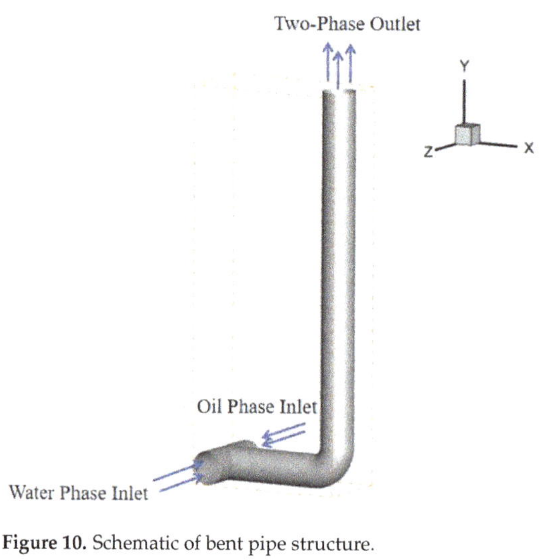


**Figure 10.** Schematic of bent pipe structure.

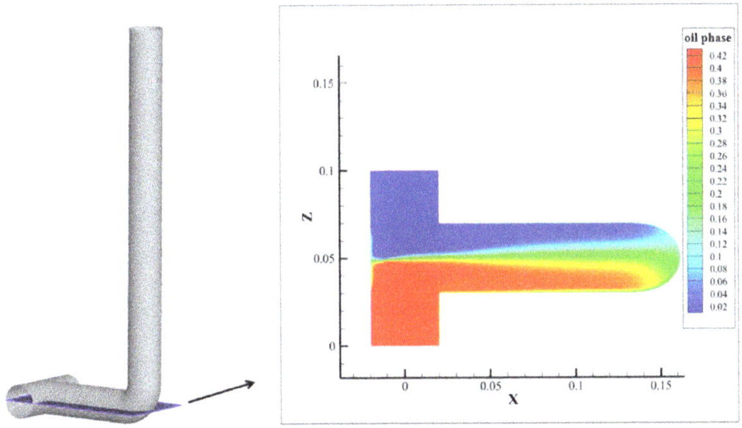


**Figure 11.** Phase distribution contour plot at cross section y = 0.002 m of bent pipe.

Figure 12 depicts the velocity and phase distribution contour plots of the oil–water two-phase flow at the z = 0.05 m cross section. The fluids from both inlets collide in the middle of the pipe, causing velocity superposition, which increases the fluid velocity in the horizontal pipe section from 1–2 m/s to approximately 5 m/s. As the oil–water flow enters the bend area, the fluid velocity near the inner side of the bend remains around 5 m/s, whereas the velocity near the outer side is noticeably lower. After exiting the bend, the high-speed fluid near the inner side flows towards the outer side of the vertical pipe section, where the fluid velocity is lower near the inner wall, forming a low-speed elliptical region. Comparison of the velocity and phase distribution contour plots at the z = 0.05 m cross section reveals significant overlap between the high-velocity region and the region with a high oil phase volume fraction near the bend. This indicates that the oil phase velocity exceeds the water phase velocity, suggesting significant velocity slip between the oil and water phases. These phenomena arise because the kinetic energy generated by the collision of inlet fluids results in velocity superposition. Due to the lower density of oil compared to water, the oil accelerates more rapidly. As the oil moves through the horizontal pipe section, buoyancy causes it to rise towards the inner side of the bend, resulting in higher fluid velocities near the inner side.

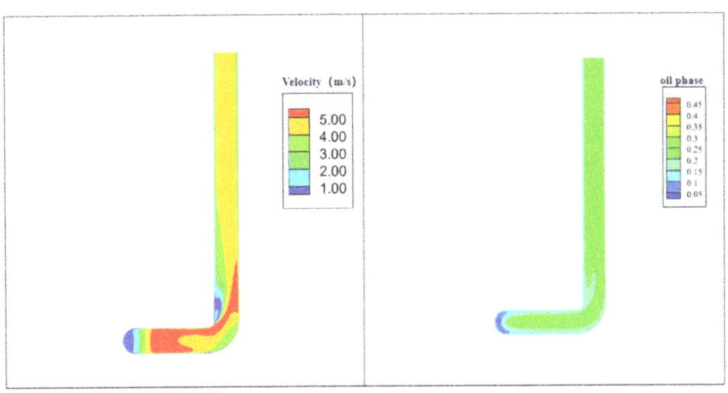


**Figure 12.** Velocity distribution contour plot (left) and phase distribution plot (right) of oil–water two-phase flow at cross section z = 0.05 m in the bent pipe.

Figures 13 and 14 present the velocity and phase distribution contour maps at different y-positions at the outlet of the bent pipe. After passing through the bend, the oil–water velocity gradually increases from the inner side to the outer side, forming a trapezoidal distribution, where high-speed fluid accumulates near the outer wall, while low-speed fluid is concentrated near the inner wall. This velocity gradient enhances shear interaction between the fluids, promoting interphase momentum exchange and improving mixing efficiency. However, excessive velocity differences may lead to flow instability and even phase stratification, which, in severe cases, can reduce mixing uniformity. At the initial entrance into the curved section, the phase distribution trend is similar to that in the horizontal pipe, with oil content gradually decreasing along the *z*-axis. The presence of vortex structures affects the phase distribution, which, in turn, determines the configuration of the oil–water interface. When the phase distribution is uniform, vortices can disrupt the two-phase interface, allowing the oil phase to disperse more easily into the water phase, thereby improving mixing efficiency. However, if the phase distribution is uneven, the oil phase may accumulate in specific regions, leading to localized oil–water separation, which reduces both rectification and mixing performance. As the fluid flows downstream, the low-speed region on the inner side gradually shrinks. Due to the mixing of the water and oil phases, the velocity distribution becomes more uniform. The superposition of high-speed water flow and low-speed oil flow contributes to this uniformization phenomenon.

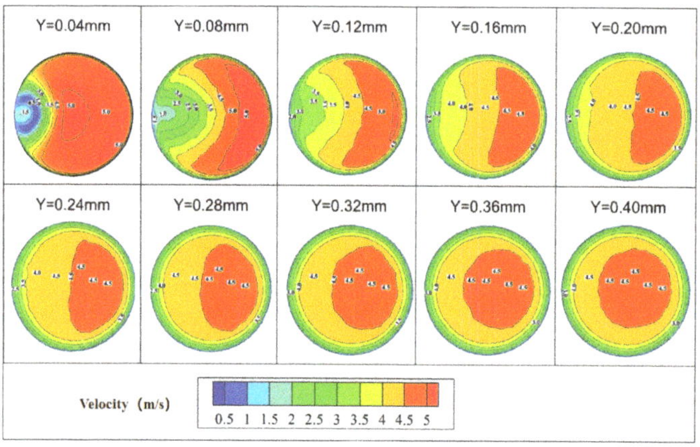


**Figure 13.** Velocity contour plots at cross sections of the outlet sections at different positions in the y-direction of the bent pipe.

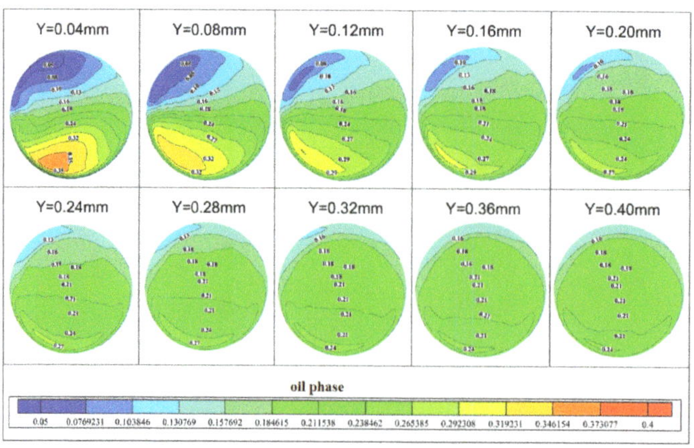


**Figure 14.** Phase distribution contour plots at cross sections of the outlet sections at different positions in the y-direction of the bent pipe.

Figure 15 shows that a longitudinal vortex generated at the bend causes streamlines from the outer side of the horizontal pipe to reach the inner side, forming secondary flow. This vortex disturbs the inner boundary layer, enhancing energy transfer and mixing efficiency, thereby further uniformizing the velocity and phase distribution. However, due to the limited number and intensity of vortices, the fluid requires a significant distance to achieve complete uniformity. The rectification coefficient of the bent pipe is 4.243, indicating suboptimal rectification performance.

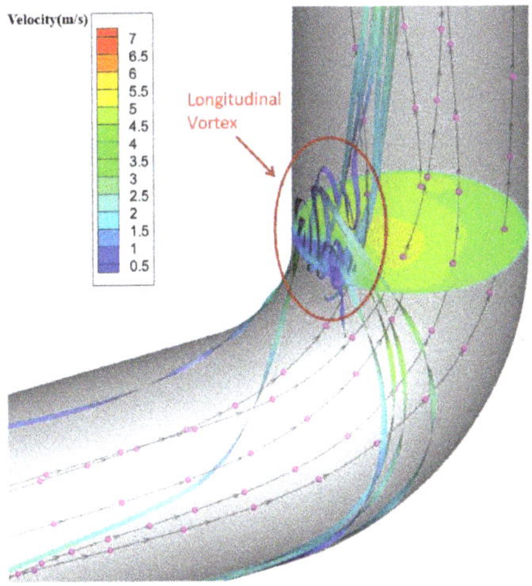


**Figure 15.** 3D streamline plot of the bent pipe.

3.4.2. Flow Characteristics of Oil–Water Two-Phase Flow Inside the Blind Tee

Taking a blind tee with a blind end of 1D as an example, the flow characteristics of oil–water two-phase flow inside were investigated. The inlet velocity was set to 2 m/s with an oil content of 0.2, and other conditions were kept consistent with the bent pipe. Figure 16 shows the velocity and phase distribution contour plots of the oil–water two-phase flow at the z = 0.05 m cross section. At the pipe bend, the flow velocity is similar to that in the bent pipe, with the high-velocity region extending along the upper wall of the bend to the outer wall of the vertical pipe. In this region, the oil phase volume fraction is also higher than in other areas. The velocity at the blind end is slightly lower than in the straight pipe section, with low-velocity regions appearing at the upper right and lower right corners. This is because the high-velocity fluid impacts the blind end, causing a change in direction and a significant decrease in velocity.

Figures 17 and 18 present velocity and phase distribution contours at the outlet cross sections of the blind tee at various y-positions. Post-bend, a low-velocity vortex forms along the left wall, extending to the pipe's center. As the flow progresses, the low-velocity fluid mixes with high-velocity fluid near the right wall, leading to a more uniform cross-sectional velocity distribution. The blind tee induces a more chaotic velocity distribution in the lower vertical section, but as the flow ascends, the velocity distributions of both the blind tee and bent pipe converge, indicating more effective velocity integration in the blind tee. Phase distribution patterns are consistent between the blind tee and bent pipe, though the blind tee exhibits smaller regions of maximum and minimum values at the y = 0.32 m cross section. By y = 0.4 m, the blind tee's low-velocity region diminishes, while the bent pipe still displays a significant low-velocity zone. The blind tee's rectification coefficient is 2.765, lower than the bent pipe's 4.243, confirming its superior rectification performance. The longitudinal vortex is a rotational flow structure extending along the main flow direction, primarily generated by wall shear effects, fluid inertia, and secondary flow induced by geometric perturbations. In the blind tee, the corner-induced longitudinal vortex enhances momentum exchange between the central region and near-wall areas, reducing velocity gradients between different flow layers, effectively preventing the formation of low-speed stagnant zones and thereby improving overall flow uniformity. Additionally, the rotational motion of the vortex induces fluid particles to rotate along the main flow direction, increasing turbulent kinetic energy and enhancing flow disturbances, which further improves internal fluid mixing uniformity. In contrast, the lateral vortex at the blind end is a rotational structure perpendicular to the main flow direction, generated by fluid impact on the structural wall. This vortex induces rolling motion in the fluid, enhancing molecular mixing, promoting interactions between different flow layers, and reducing concentration gradients, ultimately resulting in a more uniform phase distribution. As shown in Figure 19, the combined effect of the longitudinal and lateral vortices enables the blind tee to achieve superior mixing efficiency and rectification performance. Figure 20 compares the internal flow fields of both structures; despite the smaller longitudinal vortex in the blind tee, it extends to the pipe's center, affecting a larger fluid area. The blind end also generates significant vortices, enhancing internal mixing. Numerical simulations align with experimental results, underscoring the simulations' high reference value.

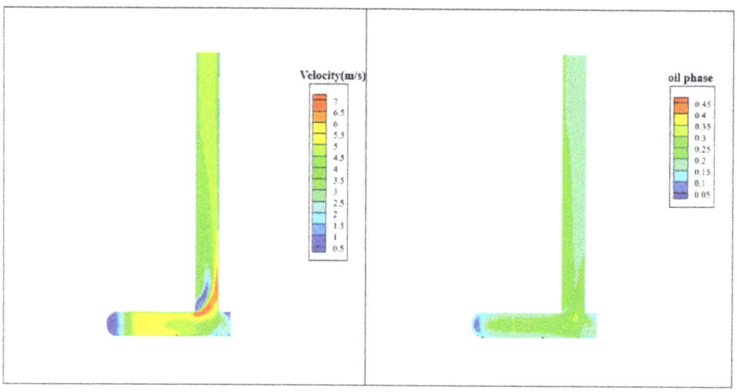


**Figure 16.** Velocity distribution contour plot (**left**) and phase distribution plot (**right**) of oil–water two-phase flow at cross section z = 0.05 m in the blind tee.

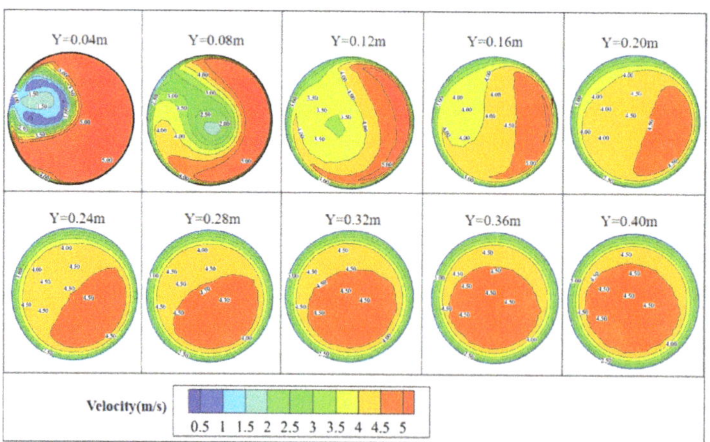


**Figure 17.** Velocity contour plots at cross sections of the outlet sections at different positions in the y-direction of the blind tee.

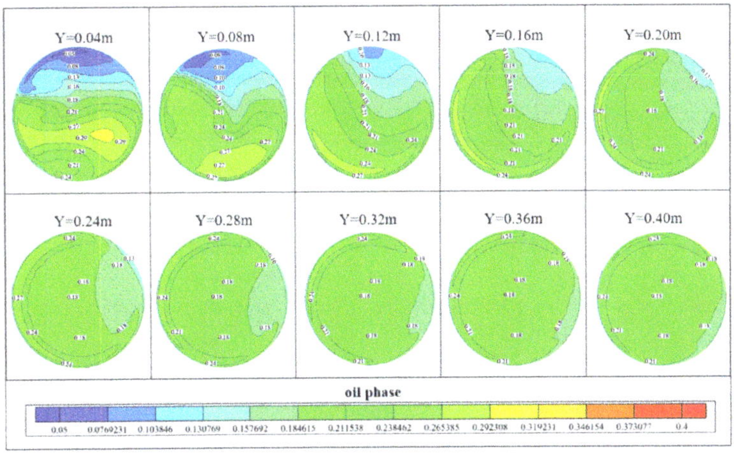


**Figure 18.** Phase distribution contour plots at cross sections of the outlet sections at different positions in the y-direction of the blind tee.

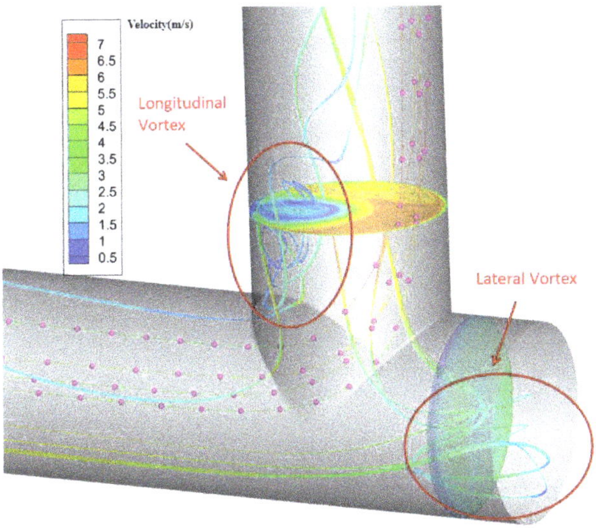


**Figure 19.** 3D streamline plot of the blind tee.

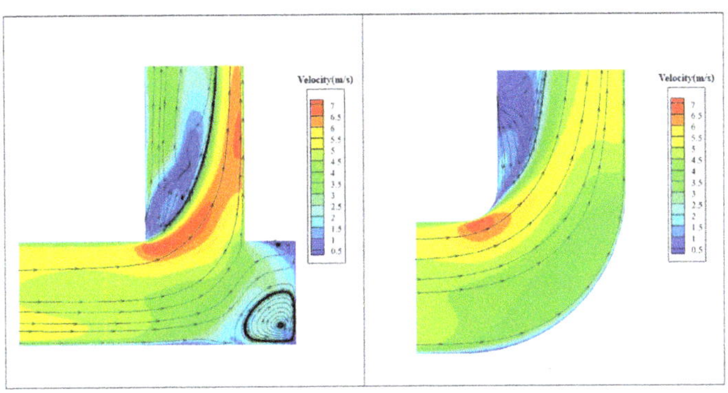


**Figure 20.** Comparative flow field plot of the blind tee and bent pipe.

To compare the differences in vortices between the two structures, vortex identification was performed on both structures. Existing vortex identification methods include the vorticity method, $Q$-criterion, $\lambda_2$ method, and $\Delta$ method. Among these, the $Q$-criterion is commonly used due to its lower computational cost and relatively consistent results. The $Q$-criterion is based on the analysis of the velocity gradient tensor. The formula for the $Q$-criterion is as follows:

$$Q = 0.5 \times \left( \|B\|_F^2 - \|A\|_F^2 \right) \tag{11}$$

In the formula, $\|B\|_F^2$ is the squared norm of matrix $B$, which is equal to the sum of the squares of all elements of matrix $B$. Matrices $A$ and $B$ are the symmetric and antisymmetric parts of the velocity gradient tensor, respectively, defined as follows:

$$A = 0.5 \times \left( \Delta V + \Delta V^T \right) \tag{12}$$

$$B = 0.5 \times \left( \Delta V - \Delta V^T \right) \tag{13}$$

$$\Delta V = \begin{pmatrix} U_X & U_Y & U_Z \\ V_X & V_Y & V_Z \\ W_X & W_Y & W_Z \end{pmatrix} \tag{14}$$

Figure 21 shows the vortex flow obtained using the $Q$-criterion for both the blind tee and the bent pipe. It is evident that the vortex in the vertical pipe section of the blind tee is wider and longer than that in the bent pipe, and a larger vortex region exists at the blind end. This explains why the rectification effect of the blind tee is significantly better than that of the bent pipe. To evaluate the applicability of blind tees in industrial settings, it is necessary to compare their pressure drop characteristics with those of elbows. By measuring the average pressure 50 mm upstream of the outlet, the pressure for the blind tee is found to be 596.311 Pa, while that for the elbow is 535.876 Pa. The comparison indicates that under identical inlet conditions, the pressures near the outlet are similar for both structures. This suggests that the pressure drop characteristics of blind tees are highly consistent with those of elbows, validating the feasibility of using blind tees in industrial applications to enhance the rectification performance of oil–water two-phase flow.

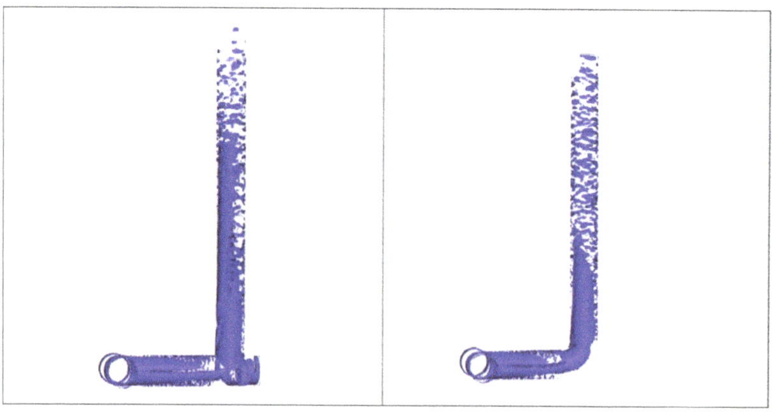

**Figure 21.** The internal vortex patterns of a blind tee (**left**) and an elbow pipe (**right**).

### 3.4.3. Comparison of Experimental and Numerical Simulation Conclusions

By comparing the experimental vector field and the numerical simulation vector field in the blind end region, it can be observed that the numerical simulation results exhibit good agreement with the experimental data in terms of fluid backflow vector direction and rotational motion characteristics (Figure 22). This consistency indicates that the numerical simulation can accurately reproduce the flow characteristics within the blind end region. Furthermore, this agreement demonstrates that numerical simulations not only provide a reasonable prediction of the blind end flow field structure but also offer more detailed flow parameter information, serving as a reliable foundation for further optimization of the rectification performance of the blind tee.

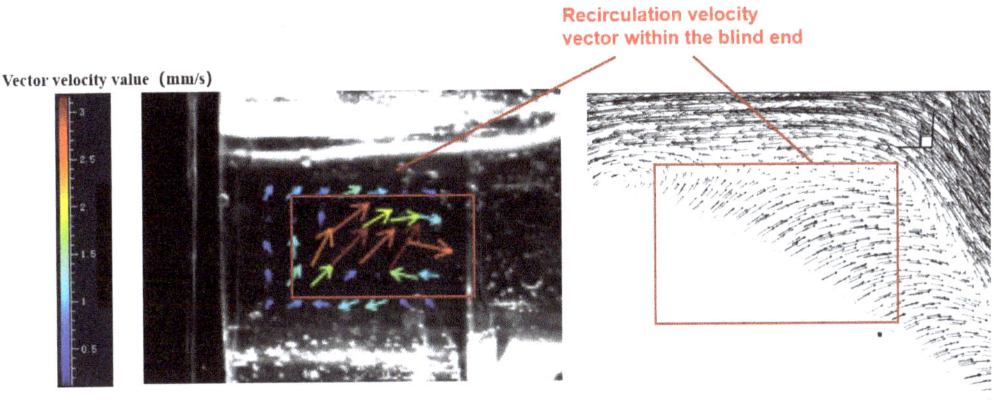


**Figure 22.** The experimental vector field and the numerical simulation vector field in the blind end region.

### 3.5. Effects of Different Conditions on the Oil–Water Rectification Efficiency of Blind Tees

#### 3.5.1. Effect of Blind Tee Length

To investigate the effect of blind tee blind end length (BTL) on oil–water rectification performance, five blind tee models with blind end lengths of D, 1.5D, 2D, 2.5D, and 3D were established and analyzed through numerical simulations. As shown in Figure 23, with increasing blind end length, both the longitudinal vortex at the bend and the lateral vortex at the blind end expand; however, the velocity within the vortex region decreases. Figure 24 presents the phase distribution at y = 0.35 m in the vertical pipe for different blind end lengths, while Figure 25 provides statistical data on rectification coefficients. A comparison reveals that the rectification coefficient S increases with blind end length, indicating that rectification performance deteriorates as the blind end length increases. The optimal blind end length for rectification, corresponding to the lowest S value, was found to be 1.5D. Figure 26 illustrates the cross-sectional average velocity and turbulent dissipation rate at the midpoint of the blind end for each blind tee. By analyzing these parameters, the underlying mechanisms affecting rectification performance can be elucidated. As the blind end length increases, the velocity of the fluid entering the blind end decreases significantly, accompanied by a reduction in turbulent dissipation rate. This suggests that while vortex structures grow larger with increasing blind end length (Figure 23), their intensity weakens considerably, resulting in diminished oil–water mixing and reduced rectification efficiency, albeit with lower energy losses. Conversely, for very short blind ends, both velocity and turbulent dissipation rate within the blind end increase significantly, indicating strong vortex intensity. However, this leads to excessive energy dissipation, causing insufficient residual kinetic energy in the oil–water two-phase flow upon entering the vertical section. Consequently, the remaining energy is inadequate to counteract the combined effects of gravity and pipe wall friction, resulting in a large-phase distribution gradient across the cross section, which adversely affects rectification efficiency. Therefore, in practical engineering applications, an optimal blind end length should be determined by balancing vortex structure size, energy dissipation, and rectification performance to achieve an efficient design.

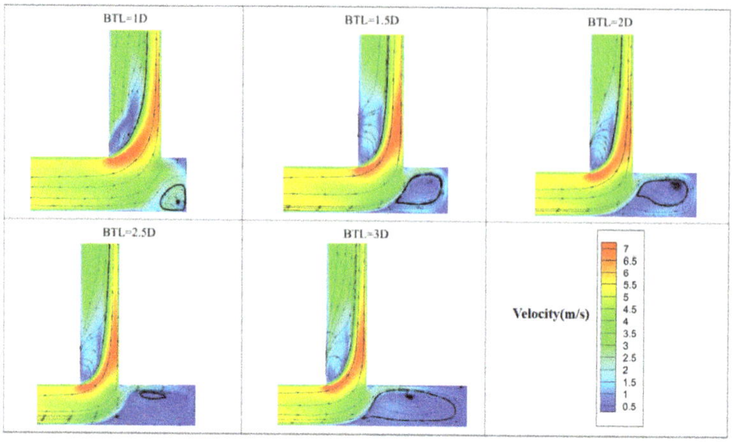


**Figure 23.** Streamline diagrams of blind tees with different blind end lengths.

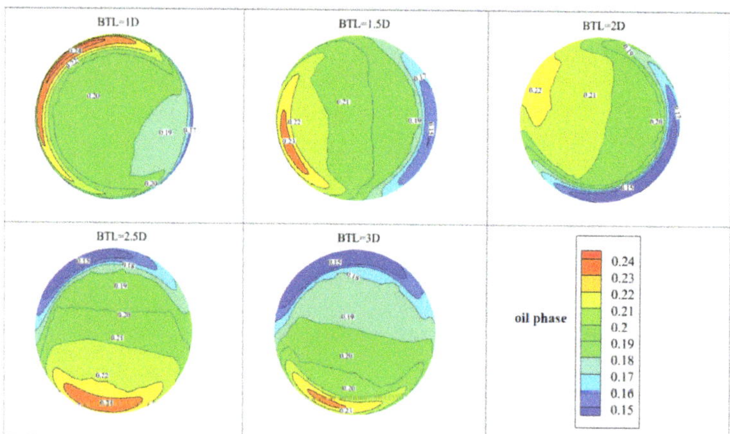


**Figure 24.** Phase distribution contour maps at y = 0.35 m for vertical pipes with different blind end lengths in blind tees.

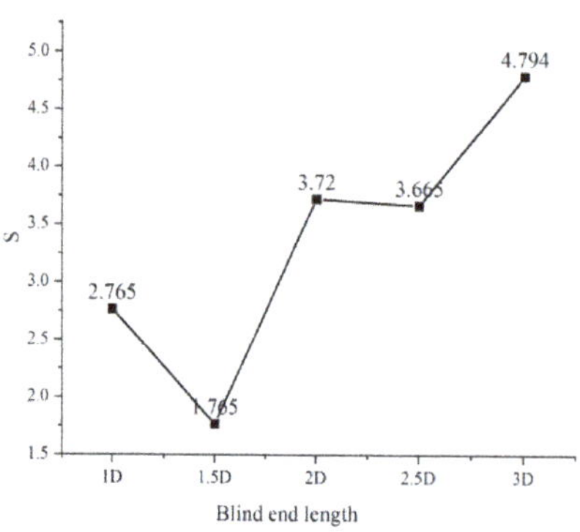


**Figure 25.** Rectification coefficients of blind tees with different blind end lengths.

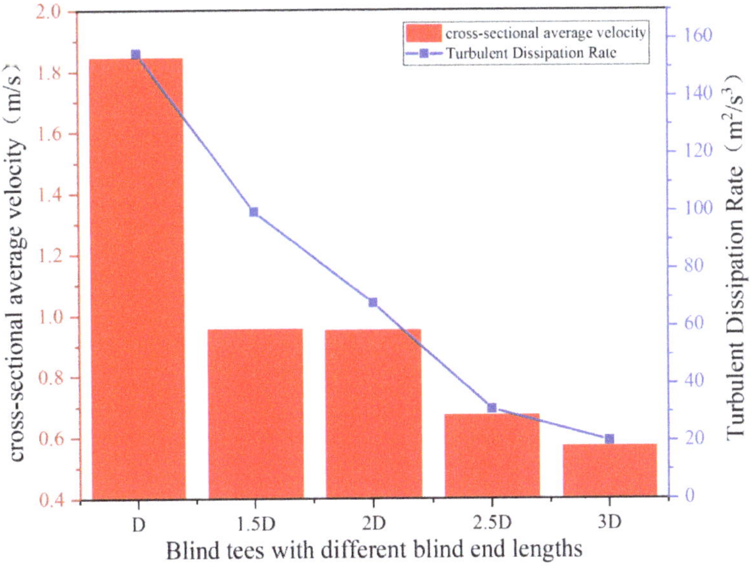


**Figure 26.** Cross-sectional average velocity and turbulent dissipation rate at the blind ends of different types of blind tees.

3.5.2. Effect of Flow Velocity

To investigate the effect of different inlet velocities on the oil–water rectification efficiency of a blind tee with a blind end length of D, oil–water mixtures with inlet velocities of 1 m/s, 2 m/s, 3 m/s, 4 m/s, and 5 m/s were introduced. As shown in Figure 27, when the velocity increases from 1 m/s to 3 m/s, the vortices at the blind end become larger, but further increases in velocity reduce the vortex size because the high-speed fluid entering from the horizontal section suppresses vortex development. From the rectification coefficient at different inlet velocities (Figure 28), it can be observed that the rectification coefficient decreases with increasing velocity from 1 m/s to 5 m/s, but peaks at 3 m/s, indicating the worst rectification efficiency at this velocity. Combining the analysis of Figure 27, Figure 28, Figure 29, it is evident that at 3 m/s, the vortex generated at the blind end has a width along the X-axis approximately equal to the length of the blind end. This results in the blind end being sealed off by the vortex, causing most of the fluid to bypass the blind end and enter the vertical pipe directly, thereby reducing the area affected by the vortex and leading to poor rectification efficiency. This velocity is defined as the blind end sealing velocity. As the velocity increases further, the high-speed fluid from the horizontal section suppresses vortex development, resulting in smaller vortices and an increase in the fluid entering the blind end. Higher velocities enhance the fluid's kinetic energy, reduce the size of the vortices, but increase the vortex intensity and secondary flow velocity, thereby improving the disturbance of non-uniform distribution areas and enhancing oil–water rectification efficiency.

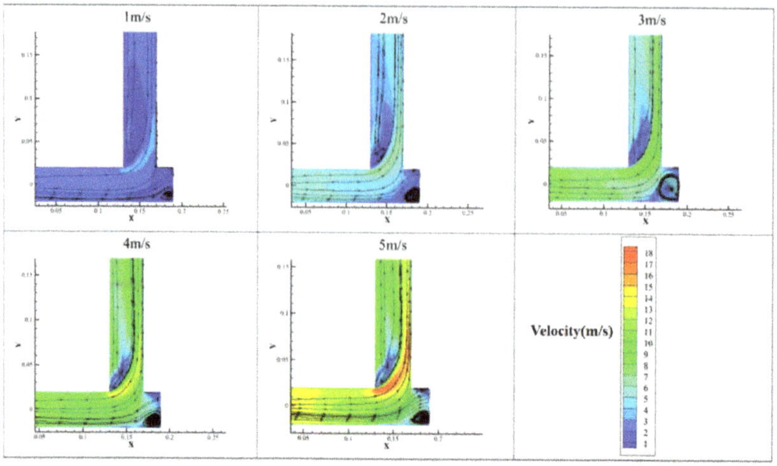


**Figure 27.** Streamline diagrams of blind tees at different inlet velocities.

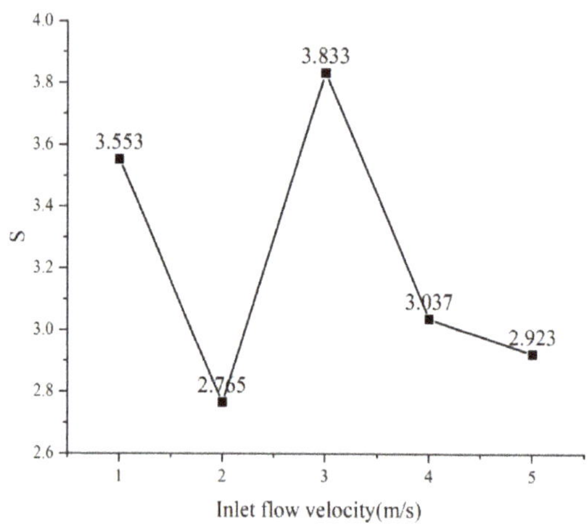


**Figure 28.** Rectification coefficients of blind tees at different inlet velocities.

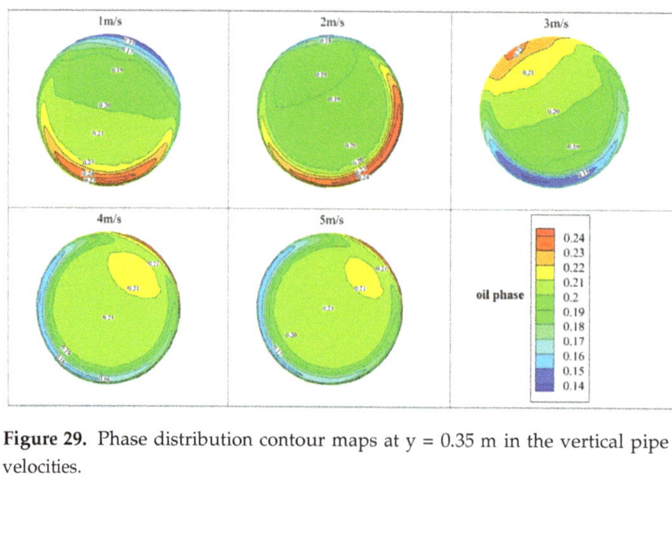


**Figure 29.** Phase distribution contour maps at y = 0.35 m in the vertical pipe for different inlet velocities.

### 3.5.3. Effect of Inner Diameter of the Vertical Pipe

There are still a large number of differential diameter blind tees in offshore oil transportation pipelines. This study investigated the effect of varying the internal diameter of the vertical pipe on oil–water rectification efficiency. Models of blind tees with vertical pipe diameters of 40 mm, 35 mm, 30 mm, 25 mm, and 20 mm were established for internal flow field analysis. As shown in Figure 30, as the vertical pipe diameter increases, the vortices at the blind end gradually become larger, and the flow velocity in the vertical section decreases. Figure 31 demonstrates that with increasing pipe diameter, the phase distribution at the cross section becomes more uneven. When the pipe diameter is 20 mm, the phase distribution achieves a very uniform state, indicating that higher fluid velocities in the vertical pipe lead to earlier uniform phase distribution. Figure 32 compares the rectification coefficients of blind tees with different vertical pipe diameters, revealing that smaller vertical pipe diameters result in lower rectification coefficients and better rectification efficiency. The reason is that a vertical pipe diameter smaller than that of the horizontal pipe acts as a throttle. According to the continuity equation of pipe flow, a smaller diameter results in higher velocity, which in turn increases the longitudinal vortex intensity at the bend, thereby enhancing the effect of promoting more uniform phase distribution.

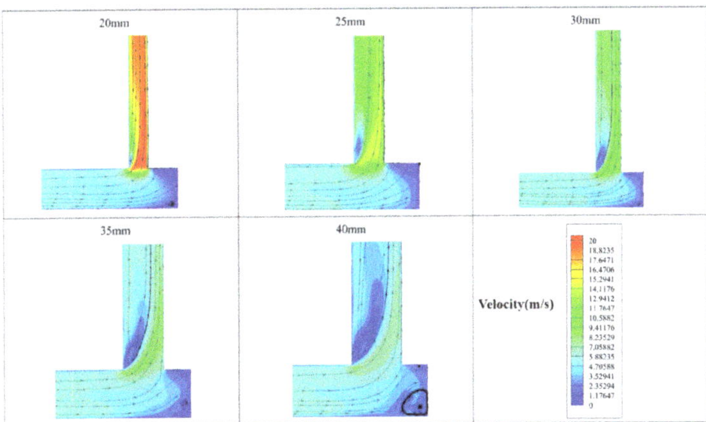


**Figure 30.** Streamline diagrams of blind tees with different vertical pipe diameters.

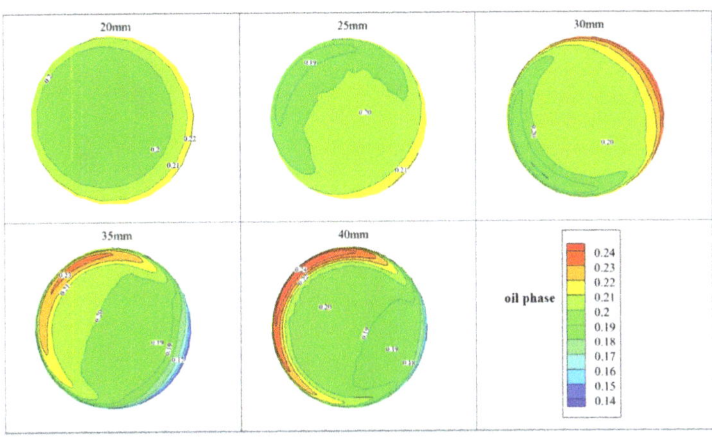


**Figure 31.** Phase distribution contour maps at y = 0.35 m in the vertical pipe for blind tees with different vertical pipe diameters.

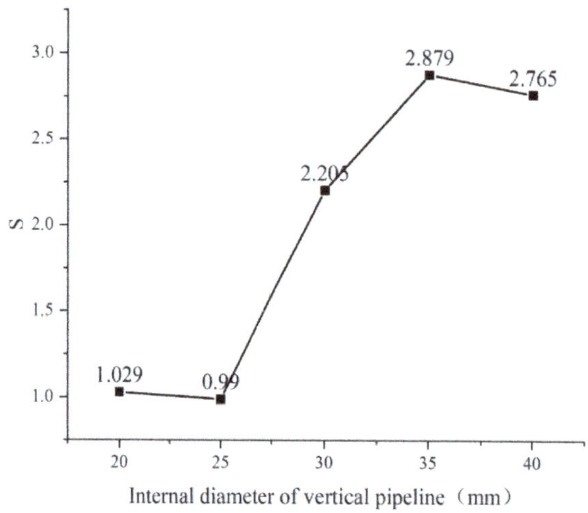


**Figure 32.** Rectification coefficients of blind tees with different vertical pipe diameters.

### 3.5.4. Effect of Oil Content

To investigate the effect of different oil contents on the rectification efficiency of blind tees, the oil content at the blind tee inlet was set to 5%, 10%, 15%, 20%, and 25%. As shown in Figure 33, the size of the vortices formed at the blind end is relatively consistent across different oil contents, but a larger low-velocity region is observed in the case of 25% oil content. Figure 34 clearly shows that higher oil content results in a more uneven phase distribution at the cross section. Figure 35 indicates that higher oil content leads to higher rectification coefficients, implying poorer rectification efficiency. As the oil content increases, the overall fluid density decreases, reducing the kinetic energy of the fluid at the same inlet velocity. This results in lower vortex intensity, which in turn reduces the disturbance to the oil–water two-phase flow, requiring a longer flow region for uniform phase distribution.

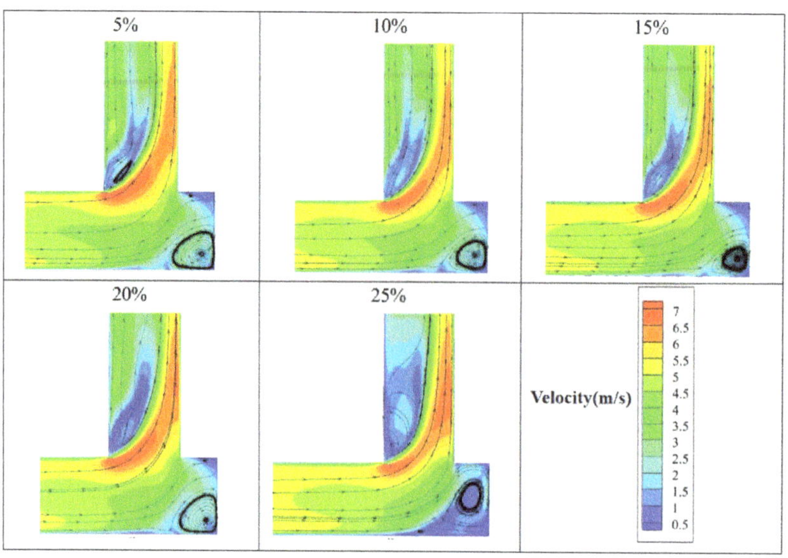


**Figure 33.** Streamline diagrams of blind tees at different oil content levels.

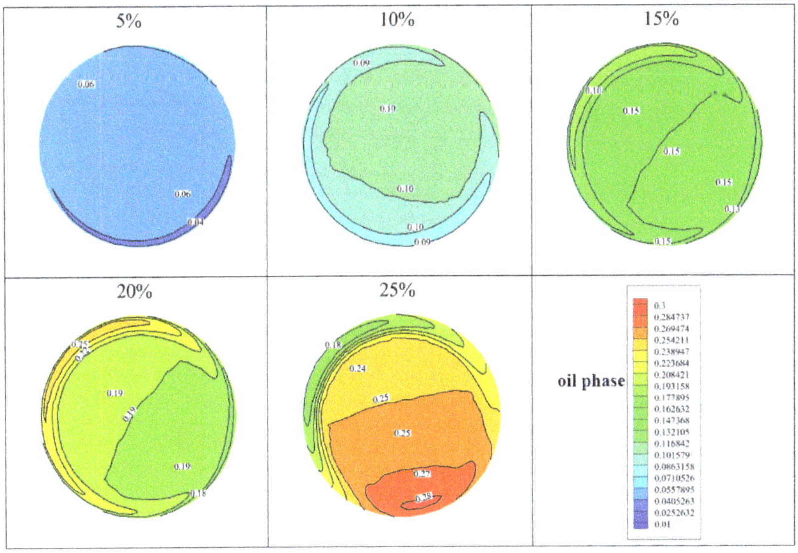


**Figure 34.** Phase distribution contour maps at y = 0.35 m in the vertical pipe for blind tees at different oil content levels.

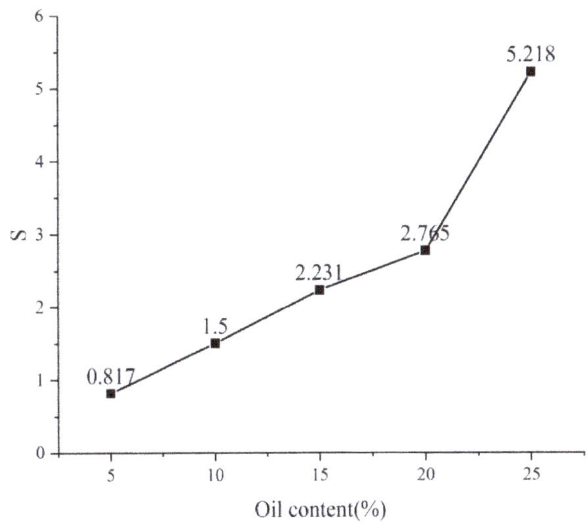


**Figure 35.** Rectification coefficients of blind tees at different oil content levels.

## 4. Conclusions

We defined the rectification coefficient S to assess the rectification performance of the device. Comparing the internal flow characteristics of blind tees and bends in oil–water systems, it was observed that bends generate only one large vortex inside, whereas blind tees generate two to three vortices. Due to the larger number and broader range of vortices inside blind tees, their rectification efficiency is significantly superior to bends. As the length of the blind end increases, the vortices within the blind tee grow larger. However, the elongation of the blind end results in increased wall frictional resistance and greater dissipation of kinetic energy during fluid flow, thereby deteriorating the rectification efficiency. At higher velocities, the fluid's kinetic energy increases. At the velocity typical of blind-tee blind end closures, vortices formed at the blind end obstruct the entry of subsequent fluid, resulting in the poorest rectification efficiency. As velocity continues to increase, the vortices are suppressed, leading to smaller vortices but higher intensity, thereby improving the rectification efficiency of oil–water. Vertical pipes with an inner diameter smaller than horizontal pipeline diameters act as throttling devices. Smaller diameters result in higher velocities, increasing vortex intensity at bends and promoting more uniform phase distribution. Higher oil content correlates with poorer rectification efficiency. Through the research presented in this paper, it was found that replacing the bend with a specific blind tee in the oil–water two-phase flow measurement device under the same operating conditions increases the rectification efficiency by 76.68%, thereby significantly enhancing measurement accuracy.

## 5. Limitations of the Study and Future Research Directions

This study focused only on specific blind tee structures and operating conditions, without covering a broader range of structural forms and flow conditions, which may limit the applicability of the research conclusions. Additionally, due to limitations in the experimental setup, the experimental validation primarily provides qualitative results, lacking high-precision quantitative data support, which may affect the verification of the simulation results.

Future research can be further expanded in the following aspects: 1. Explore the impact of different geometric configurations (e.g., variable cross section blind ends, rounded blind ends) on flow characteristics, and analyze their rectification performance and mixing efficiency under various flow regimes of oil–water two-phase flow; 2. Design adaptive blind tee structures capable of automatically adjusting the opening, closing, or length of the blind end based on real-time flow conditions to meet rectification requirements under different operating scenarios; 3. Investigate the adaptability of blind tee structures under extreme conditions such as high pressure and high temperature.

**Author Contributions:** Conceptualization, C.C.; methodology, C.C.; formal analysis, C.C.; investigation, C.C. and Y.Z.; resources, Q.G.; data curation, Y.S. and Q.G.; writing—original draft preparation, C.C. and Y.C.; writing—review and editing, C.C. and Q.G. All authors have read and agreed to the published version of the manuscript.

**Funding:** This research received no external funding.

**Data Availability Statement:** Data were obtained from the China National Petroleum Corporation (CNPC) and are available from the authors with the permission of the CNPC.

**Conflicts of Interest:** Chen Cheng was employed by the Sinopec Natural Gas Branch. Qingming Gan and Yubin Su were employed by the Changqing Oilfield Branch Company. Yang Cheng was employed by the Jianghan Oilfield Branch Company. The authors declare that the research was conducted in the absence of any commercial or financial relationships that could be construed as potential conflicts of interest.

## Nomenclature

| | |
|---|---|
| $\rho$ | fluid density, kg/m$^3$ |
| $K$ | turbulence kinetic energy, J/kg |
| $t$ | time, s |
| $\overline{u_j}$ | the mean velocity component in the -th direction, m/s |
| $x_j$ | the spatial coordinate in the $j$-direction, m |
| $\mu$ | dynamic viscosity, Pa·s |
| $\mu_t$ | turbulent viscosity, Pa·s |
| $\mathrm{Pr}_K$ | the Prandtl number for turbulence kinetic energy |

| | |
|---|---|
| $P_K$ | the generation rate of turbulence kinetic energy, $W/m^3$ |
| $G_b$ | the turbulence kinetic energy generated by buoyancy, $m^2/s^3$ |
| $\varepsilon$ | the dissipation rate of turbulence kinetic energy, $m^2/s^3$ |
| $Y_M$ | the influence of fluctuating dilation on turbulence dissipation rate, $m^2/s^3$ |
| S | rectification coefficient |
| N | the number of sampling points |
| $\rho_{x_i y_j}$ | the density of sampling points at different locations, $kg/m^3$ |
| $\overline{\rho}$ | the average density of the fluid entering the device, $kg/m^3$ |
| $\alpha_q$ | the volume fraction of phase $q$ |
| $\rho_q$ | the density of phase $q$, $kg/m^3$ |
| $v_q$ | the velocity of phase $q$, m/s |
| $\tau_q$ | the stress tensor of the phase $q$, $N/m^2$ |

## References


1. Yang, Y.; Ha, W.; Zhang, C.; Liu, M.; Zhang, X.; Wang, D. Measurement of high-water-content oil-water two-phase flow by electromagnetic flowmeter and differential pressure based on phase-isolation. *Flow Meas. Instrum.* **2022**, *84*, 102142. [CrossRef]
2. Zhang, J.; Xu, J.Y.; Wu, Y.X.; Li, D.H.; Li, H. Experimental validation of the calculation of phase holdup for an oil–water two-phase vertical flow based on the measurement of pressure drops. *Flow Meas. Instrum.* **2013**, *31*, 96–101. [CrossRef]
3. Brister, C.M. Wet gas performance of a low energy gamma ray multiphase meter. In Proceedings of the Offshore Technology Conference Brasil, Rio de Janeiro, Brazil, 29–31 October 2013; p. OTC-24352-MS.
4. Zheng, G.B.; Jin, N.D.; Jia, X.H.; Lv, P.J.; Liu, X.B. Gas–liquid two phase flow measurement method based on combination instrument of turbine flowmeter and conductance sensor. *Int. J. Multiph. Flow* **2008**, *34*, 1031–1047. [CrossRef]
5. Patankar, S.V.; Pratap, V.S.; Spalding, D.B. Prediction of turbulent flow in curved pipes. In *Numerical Prediction of Flow, Heat Transfer, Turbulence and Combustion*; Pergamon: Oxford, UK, 1983; pp. 147–159.
6. Ito, H. Flow in curved pipes. *JSME Int. J.* **1987**, *30*, 543–552. [CrossRef]
7. Di Liberto, M.; Di Piazza, I.; Ciofalo, M. Turbulence structure and budgets in curved pipes. *Comput. Fluids* **2013**, *88*, 452–472. [CrossRef]
8. Dutta, P.; Nandi, N. Effect of Reynolds number and curvature ratio on single phase turbulent flow in pipe bends. *Mech. Mech. Eng.* **2015**, *19*, 5–16.
9. Noorani, A.; Sardina, G.; Brandt, L.; Schlatter, P. Particle transport in turbulent curved pipe flow. *J. Fluid Mech.* **2016**, *793*, 248–279. [CrossRef]
10. Ayala, M.; Cimbala, J.M. Numerical approach for prediction of turbulent flow resistance coefficient of 90 pipe bends. *Proc. Inst. Mech. Eng. Part E J. Process Mech. Eng.* **2021**, *235*, 351–360. [CrossRef]
11. Jarrahi, M.; Castelain, C.; Peerhossaini, H. Secondary flow patterns and mixing in laminar pulsating flow through a curved pipe. *Exp. Fluids* **2011**, *50*, 1539–1558. [CrossRef]
12. Kim, J.; Yadav, M.; Kim, S. Characteristics of secondary flow induced by 90-degree elbow in turbulent pipe flow. *Eng. Appl. Comput. Fluid Mech.* **2014**, *8*, 229–239. [CrossRef]
13. Pouraria, H.; Seo, J.K.; Paik, J.K. Numerical study of erosion in critical components of subsea pipeline: Tees vs. bends. *Ships Offshore Struct.* **2017**, *12*, 233–243. [CrossRef]
14. Chen, X.; McLaury, B.S.; Shirazi, S.A. Numerical and experimental investigation of the relative erosion severity between plugged tees and elbows in dilute gas/solid two-phase flow. *Wear* **2006**, *261*, 715–729. [CrossRef]
15. Han, F.; Ong, M.C.; Xing, Y.; Li, W. Three-dimensional numerical investigation of laminar flow in blind-tee pipes. *Ocean. Eng.* **2020**, *217*, 107962. [CrossRef]
16. Ma, Y.; Li, C.; Pan, Y.; Hao, Y.; Huang, S.; Cui, Y.; Han, W. A flow rate measurement method for horizontal oil-gas-water three-phase flows based on Venturi meter, blind tee, and gamma-ray attenuation. *Flow Meas. Instrum.* **2021**, *80*, 101965. [CrossRef]
17. Rammohan, A.; Bhakta, A.; Natrajan, V.; Ward, J.; Kumar, M. Flow swirl and flow profile measurement in multiphase flow. In Proceedings of the 31st International North Sea Flow, Tonsberg, Norway, 22–25 October 2013.
18. Millington, D. Evaluation of blind-t mixing effectiveness within two-phase gas-viscous oil flow using electrical capacitance tomography. In Proceedings 10th North American Conference on Multiphase Technology, Banff, AB, Canada, 8–10 June 2016; p. BHR-2016-399.
19. Hjertaker, B.T.; Tjugum, S.A.; Hallanger, A.; Maad, R. Characterization of multiphase flow blind-T mixing using high speed gamma-ray tomometry. *Flow Meas. Instrum.* **2018**, *62*, 205–212. [CrossRef]
20. Valdés, J.P.; Kahouadji, L.; Matar, O.K. Current advances in liquid–liquid mixing in static mixers: A review. *Chem. Eng. Res. Des.* **2022**, *177*, 694–731. [CrossRef]

21. Yan, S.; Lv, F.; Wang, B.; Dong, X.; Yang, X.; Ma, L.; Chen, S.; Han, B.; Bai, Z. Mixing characteristics and intensification mechanism of an improved swirl jet mixer for high phase ratios. *J. Ind. Eng. Chem.* **2024**, *135*, 457–470. [CrossRef]
22. Meng, H.; Meng, T.; Yu, Y.; Wang, Z.; Wu, J. Experimental and numerical investigation of turbulent flow and heat transfer characteristics in the Komax static mixer. *Int. J. Heat Mass Transf.* **2022**, *194*, 123006. [CrossRef]
23. Chakleh, R.; Azizi, F. Performance comparison between novel and commercial static mixers under turbulent conditions. *Chem. Eng. Process. -Process Intensif.* **2023**, *193*, 109559. [CrossRef]

processes

MDPI

*Article*

# A Convolutional Neural Network-Based Method for Distinguishing the Flow Patterns of Gas-Liquid Two-Phase Flow in the Annulus

**Chen Cheng [1,*], Weixia Yang [2], Xiaoya Feng [3], Yarui Zhao [4] and Yubin Su [5]**

1 Sinopec Natural Gas Branch, Beijing 100020, China
2 Jinchang PetroChina Kunlun Gas Co., Ltd., Jinchang 737100, China; 15927888223@163.com
3 Xi'an Changqing Chemical Group Co., Ltd., Xi'an 710018, China; fxya_cq@petrochina.com.cn
4 No. 11 Oil Production Plant, Changqing Oilfield Company, PetroChina, Qinyang 745000, China; zhaoyarui_cq@petrochina.com.cn
5 Oil & Gas Technology Research Institute, Changqing Oilfield Branch Company, PetroChina, Xi'an 710021, China; suyb_cq@petrochina.com.cn
* Correspondence: chengchen_sinopec@163.com

**Abstract:** In order to improve the accuracy and efficiency of flow pattern recognition and to solve the problem of the real-time monitoring of flow patterns, which is difficult to achieve with traditional visual recognition methods, this study introduced a flow pattern recognition method based on a convolutional neural network (CNN), which can recognize the flow pattern under different pressure and flow conditions. Firstly, the complex gas–liquid distribution and its velocity field in the annulus were investigated using a computational fluid dynamics (CFDs) simulation, and the gas–liquid distribution and velocity vectors in the annulus were obtained to clarify the complexity of the flow patterns in the annulus. Subsequently, a sequence model containing three convolutional layers and two fully connected layers was developed, which employed a CNN architecture, and the model was compiled using the Adam optimizer and the sparse classification cross entropy as a loss function. A total of 450 images of different flow patterns were utilized for training, and the trained model recognized slug and annular flows with probabilities of 0.93 and 0.99, respectively, confirming the high accuracy of the model in recognizing annulus flow patterns, and providing an effective method for flow pattern recognition.

**Keywords:** gas–liquid two-phase flow; convolutional neural network; flow pattern identification; CFDs simulation

## 1. Introduction

The gas–liquid two-phase flow pattern significantly influences drilling fluid injection (Figure 1a), gas well drainage and recovery (Figure 1b), and oil and gas gathering and transportation processes, as well as chemical manufacturing. Among them, in the drainage and gas production process of gas wells, the duration of the forward lift and reverse lift processes is alternately used. The flow law in the annulus during the reverse lift process is the theoretical guarantee for the success of construction technology. Understanding the dynamics of flow pattern transitions is essential for mitigating unstable flow phenomena within pipelines. Consequently, the development of flow pattern identification techniques has emerged as a critical component in the technology for managing flow pattern transitions. At present, there are more studies on multiphase flow in the annulus. Caetano [1], Kelessidis [2], Sadatomi [3], and other researchers have carried out some experimental and theoretical studies on the transition of gas–liquid two-phase flow, the prediction of gas content, and the pressure drop in the annulus. Some research progress has been made in the flow pattern recognition for multiphase flow. Shi [4] investigated the fuzzy recognition of flow patterns based on image processing and developed a new recognition method to realize the automatic and accurate recognition of flow patterns.

### *1.1. Identification of Circular Pipe Flow Pattern*

Wang [5] studied churn flow in a 19 mm ID pipe at atmospheric pressure and employed a shadow detection technique to capture high-resolution images of liquid droplets entrained in churn flow. A phase separation method was used to change the inlet flow pattern to a uniformly symmetrical vortex core annular flow; it was possible to isolate the gas–liquid mixture into a clear and smooth interface [6]. Liu et al. [7] investigated the flow recognition of gas–liquid swirling flows using image processing techniques and neural networks, and performed visualization experiments. In 2019, Nnabuife et al. [8] investigated a non-radioactive and non-invasive method to objectively recognize flow patterns using Doppler ultrasonic sensors and machine learning methods. Ligus et al. [9] presented an experimental study to optimize two-phase fluid flow in an airlift pump and developed an orifice friction equivalent length model for predicting the pressure loss of two-phase flow through an orifice. The study used the computational fluid dynamics (CFDs) software Ansys Fluent 19.0 to identify the flow pattern and pressure gradient [10].

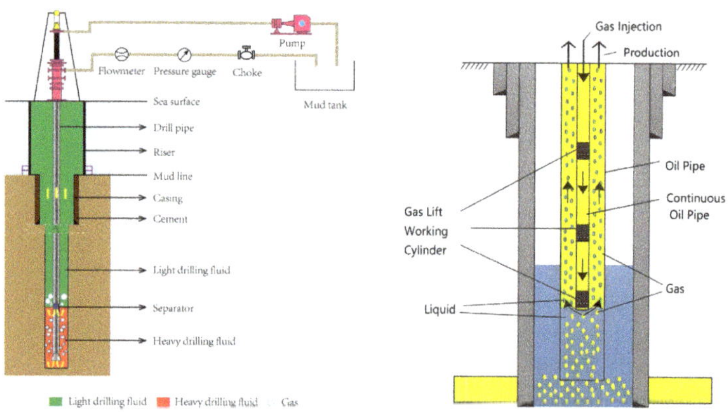


(**a**) (**b**)

**Figure 1.** Typical two-phase flow of annular gas–liquid under operating conditions. (**a**) Wellbore gas–liquid two-phase flow for dual-gradient drilling [11]; (**b**) drainage of liquid–gas wells for gas recovery.

### *1.2. Neural Network Recognition Method*

Veisi et al. [12] developed an artificial neural network (ANN) model for identifying gas–liquid two-phase flow patterns using a multilayer perceptron (MLP) neural network in MATLAB 9.12.0.188 software. Xiao [13] utilized a support vector machine for the feature extraction of conductivity measurements to identify the two-phase flow patterns in a narrow channel, and explored the characteristics of gas–liquid two-phase flow through visualization experiments in rectangular channels. Du et al. [14] proposed a novel interconnected ordinate flow pattern network for studying the spatial coupling behavior of two-phase flow patterns. This network model can help us to deeply understand the characteristics and behaviors of gas–liquid two-phase flow, which is important for the real-time monitoring of gas–liquid pipeline flow in industrial processes. Zhao et al. [15] have proposed an innovative method for flow pattern recognition, which is based on the SSA-BP neural network algorithm. By employing the SSA algorithm (Sparrow Search Algorithm) to optimize the conventional backpropagation algorithm, this technique effectively avoids the issue of premature convergence to local minima that may occur during the training process of BP networks. Consequently, an efficient and accurate flow pattern recognition method based on the SSA-BP neural network has been developed.

To summarize, the research in multiphase flow state recognition in recent years has involved several fields such as image processing, neural networks, ultrasonic sensors, and optical sensors. However, there is a lack of experimental data support under specific working conditions, especially for complex gas–liquid two-phase flow systems, and the algorithms and models need to be further developed and improved.

Based on the CFDs simulation method, this study focused on analyzing the basic characteristics of the slug flow in the annulus and its transition mechanism to the churn flow, and considered the influence of the gas–liquid interface and the multiphase flow characteristics, combining this with deep learning technology to improve the accuracy and stability of the flow pattern identification, which is of great significance to improve the production efficiency and ensure the safe operation of the pipeline system, and it will provide theoretical support for the on-line flow pattern recognition method in the annular region, thereby enriching the theoretical framework of multiphase flow pattern recognition.

## 2. Methodology

### 2.1. Transition Boundaries of Flow Patterns

In the annulus or circular tube, it was first thought that the multiphase flow pattern transformation was caused by bubble merging [16], and the difference was that the gas–liquid distribution showed an asymmetric distribution when the slug flow appeared in the annulus. As shown in Figure 2, with a decrease in the gas volume, the flow pattern in the annulus appeared as annular flow, churn flow, slug flow, and bubble flow. Slug flow is the continuous aggregation of gas bubbles that leads to an increase in the proportion of gas, and the gas changes from small bubbles to Taylor bubbles, thus switching from bubble flow to slug flow. The annular flow is due to the liquid-slugged area of the segment-slugged unit gradually decreasing the liquid holdup and gradually increasing the void friction, resulting in the liquid film area and the liquid-slugged area being interconnected, thus forming an annular flow in the vertical tube.

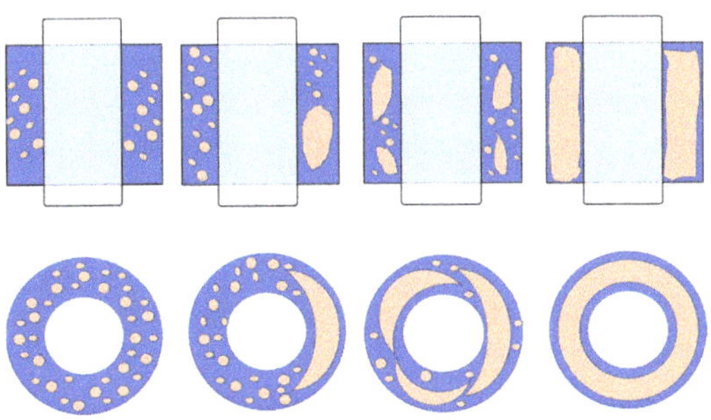

**Figure 2.** Schematic diagram of flow pattern changes in a vertical annulus pipe.

In a concentric annulus pipe, when transitioning from bubbly flow to slug flow, where the void fraction is $\beta = 0.25$, then

$$v_{sw} = 3v_{sg} - \left[\frac{(\rho_w - \rho_g)g\sigma}{\rho_w^2}\right]^{0.25} \quad (1)$$

where $v_{sw}$ is the superficial flow rate of water, m/s; $v_{sg}$ is the superficial flow rate of gas, m/s; $\rho_w$ is the density of water, 998.0 kg/m$^3$; $\rho_g$ is the gas density, kg/m$^3$; and $\sigma$ is the gas–liquid interfacial tension, N/m.

Churn flow occurs when the air slug passes through the liquid slug, and liquid-phase reflux and oscillation occur in the vertical tube. The liquid-phase reflux forms a buildup, which hinders the gas flow and leads to the transient compression of the gas, and finally results in liquid-phase eruption. The relationship between the liquid slug length and the hydraulic equivalent diameter of the annulus under churn flow conditions is

$$\frac{L_s}{D_s} = 96.18v_m + 22.96 \tag{2}$$

$$D_s = D_{oin} - D_{iout} \tag{3}$$

$$v_m = v_{sw} + v_{sg} \tag{4}$$

where $D_s$ is the hydraulic equivalent diameter of the annulus, m; $D_{oin}$ is the internal diameter of the outer pipe, m; $D_{iout}$ is the outer diameter of the inner pipe, m; and $v_m$ is the mixture velocity, m/s.

As the gas flow continues to increase, the proportion of gas in the annulus is much larger than the proportion of the liquid phase, and the churn flow is gradually transformed into annular flow. When annular flow occurs in the annulus, the liquid phase in the tube does not reflux and the liquid film is symmetrically distributed. Scholars believe that the liquid holdup of the slug flow in the middle section of the circular pipe is also applicable to the annulus pipe [2], so then

$$\phi_{churn} = \phi_{slug} = \frac{v_{sg}}{v_{tb}} \tag{5}$$

$$v_{tb} = Cv_m + v_{rc} \tag{6}$$

The void fraction of the annulus can be expressed as

$$\beta_{an} = \frac{A_g}{A_{at}} \tag{7}$$

where $\phi_{churn}$ is the liquid holdup of churn flow, dimensionless; $\phi_{slug}$ is the liquid holdup of slug flow, dimensionless; $v_{tb}$ is the maximum velocity of Taylor bubbles in the slug, m/s; $v_{rc}$ is the rising speed of bubbles in the annulus in the stationary liquid, m/s; $\beta_{an}$ is the void friction of the annular flow, dimensionless; $A_g$ is the area of the gas in the cross section of the annulus, m$^2$; and $A_{at}$ is the total cross section area of the annulus, m$^2$.

When $\phi_{churn} = 1 - \beta_{an}$, there is a transition that happens between the churn flow and the annular flow. Vertically upward two-phase flow in the annulus is important in many industrial fields due to the close relationship between its flow pattern and heat transfer characteristics. However, the asymmetric shape of Taylor bubbles in the annulus leads to the transition between slug flow and churn flow, which may be affected by a variety of factors. Most of the studies have seriously neglected the flow field characteristics of the slug flow and the churn flow, which leads to the fact that the flow pattern criterion of the circular pipe is not fully applicable to the annulus, especially the transition from the slug flow to the churn flow, which is the most different from that in the circular pipe, and should be investigated separately [17].

### 2.2. Flow Field Simulation

The Volume of Fluid (VOF) is used to describe the interfacial behavior and interactions between phases such as liquid and gas [18]. The model is based on the continuous medium assumption and assumes that the sum of the volume fractions of the phases in each space cell is equal to one, i.e., the volume fraction represents the proportion of each phase occupied. In this study, the VOF model was used to track the gas–liquid interface and describe the evolution of the interface position and shape. The VOF is able to take into account physical phenomena such as the interfacial tension and surface tension, and by solving the equations of flow and the governing equations, the motions of the interface, the deformations of the droplets and the bubbles, collisions, mergers, and other phenomena, as in Equation (8):

$$\frac{\partial \beta_i}{\partial t} + \vec{v} \cdot \nabla \beta_i = \frac{S_{\beta_i}}{\rho_i} \tag{8}$$

$$\sum_{n}^{i=1} \beta_i = 1 \tag{9}$$

When the volume ratio of the liquid phase is tracked, then the density in each cell can be calculated using the following, Equation (10):

$$\rho = \phi_w \rho_w + (1 - \phi_w)\rho_g \tag{10}$$

The momentum equation for the gas–liquid phase is

$$\frac{\partial}{\partial t}(\rho \vec{v}) + \nabla \cdot (\rho \vec{v} \vec{v}) = -\nabla p + \nabla \cdot \left[\mu\left(\nabla \vec{v} + \nabla \vec{v}^{T}\right)\right] + \rho \vec{g} + \vec{F} \tag{11}$$

The energy equation is

$$\frac{\partial}{\partial t}(\rho E) + \nabla \cdot (\vec{v}(\rho E + p)) = \nabla \cdot \left(k_{eff} \nabla T\right) + S_h \tag{12}$$

where $\beta_i$ is the volume ratio of phase $i$; $S_{\beta_i}$ is the mass source term, which is zero under default conditions; $\vec{F}$ is the volume force; $\rho$ is the density and effective thermal conductivity; $k_{eff}$ are shared by the phases; $\mu$ is the mixing viscosity; and $\vec{v}$ is the velocity vector.

The turbulent kinetic energy and turbulent dissipation rate in turbulent flows are described using the standard model $k - \varepsilon$, which expresses the turbulent kinetic energy and turbulent dissipation rate of the flow field as $k$ and $\varepsilon$, respectively, and they are simulated by the following two transport equations, which are able to predict the behavior of turbulence better.

Turbulent kinetic energy equation:

$$\frac{\partial}{\partial t}(\rho k) + \frac{\partial}{\partial x_i}(\rho k u_i) = \frac{\partial}{\partial x_j}\left[\left(\mu + \frac{\mu_t}{\sigma_k}\right)\frac{\partial k}{\partial x_j}\right] + G_k + G_b - \rho\varepsilon - Y_M + S_k \tag{13}$$

Diffusion equations:

$$\frac{\partial}{\partial t}(\rho \varepsilon) + \frac{\partial}{\partial x_i}(\rho \varepsilon u_i) = \frac{\partial}{\partial x_j}\left[\left(\mu + \frac{\mu_t}{\sigma_\varepsilon}\right)\frac{\partial \varepsilon}{\partial x_j}\right] + C_{1\varepsilon}\frac{\varepsilon}{k}(G_k + C_{3\varepsilon}G_b) - C_{2\varepsilon}\rho\frac{\varepsilon^2}{k} + S_\varepsilon \tag{14}$$

where $G_k$ denotes the turbulent kinetic energy due to the laminar velocity gradient, $G_b$ is the turbulent kinetic energy due to buoyancy, $Y_M$ denotes the fluctuations due to the diffusion of the transition in compressible turbulence, $C_{1\varepsilon}$, $C_{2\varepsilon}$, and $C_{3\varepsilon}$ are constants, $\sigma_k$ and $\sigma_\varepsilon$ are the turbulent Prandtl numbers for the $k$ and $\varepsilon$ equations, and $S_k$ and $S_\varepsilon$ are user-defined.

In this study, the negative direction of the Y-axis was defined as the direction of gravity; the boundaries were the mass flow inlet and the pressure outlet, and the inlet pressure was atmospheric; the standard wall function and the no-slip wall shear condition were used, and the pressure–velocity coupling was performed using the SIMPLE method.

### 2.3. Flow Pattern Discrimination Method

A convolutional neural network (CNN) is a deep learning model especially designed to process data with a grid-like structure. The CNN is widely used for image classification, target detection, image segmentation, etc. The hierarchical structure and parameter-sharing properties of the CNN make it perform well in processing images and other grid-like data. With a deep stack of convolutional and pooling layers, as well as appropriate fully connected layers, CNNs are able to gradually extract abstract features from the input data and achieve accurate classification, recognition, or prediction.

#### 2.3.1. Modeling

To address the complexity and uniqueness of slug flow, this study employed $3 \times 3$ convolutional kernels, which offer high parameter efficiency, strong feature extraction capabilities, and ease of stacking. Compared to larger kernels, the $3 \times 3$ design possesses fewer parameters, thereby reducing the model complexity and mitigating the risk of overfitting, effectively extracting edge and texture features from images. The architecture comprised three convolutional layers with 32, 64, and 64 kernels, respectively, each tasked with extracting key features from the input feature images through convolutional operations. The model utilized the ReLU activation function, which excels in converting linear inputs into nonlinear outputs, enhancing the model's classification performance. Additionally, ReLU's simplicity contributes to high computational efficiency. The input shape was a $32 \times 32$ RGB image, and the feature image size was reduced by two $2 \times 2$ pooling layers. Before the establishment of the fully connected layer, the output of the convolutional layer was spread out to facilitate the connection, and finally the image processing and classification were carried out by the fully connected layer, which contained 64 neurons.

$$a_j^l = f\left(b_j^l + \sum_{i \in M_j^l} a_i^{l-1} * k_{ij}^l\right) \tag{15}$$

where $a_j^l$ is the output value of the $j$th unit of the convolutional layer $l$, $f$ denotes the activation function, $M_j^l$ denotes the set of input feature maps, $b_j^l$ is the bias, and $k$ denotes the convolutional kernel, which can be represented by the following equation:

$$k = \frac{1}{4}\begin{bmatrix} 1 & 1 \\ 1 & 1 \end{bmatrix} \tag{16}$$

A two-dimensional image has multiple pixels, and the convolution kernel is like a moving scanning lens, which can recognize and learn from the image according to a specific step size. The image used in this case was $32 \times 32$, i.e., $M = 32$, $k = 2$, and the step size was set to 1. According to the formula of the convolutional layer, the size of the image can be obtained as $N$. Generally, the region in the middle of the feature image will be calculated by convolution several times, but the edge position can only be calculated once. In order to ensure the accuracy of the prediction and retain all the information of the image, the filler region was added to the edge of the image. The common filler methods are the Valid and Same fillers.

$$N = \frac{M - k}{s} + 1 = 31 \tag{17}$$

The convolutional layer and pooling layer have a one-to-one correspondence. This study used the Max Pooling function (Max Pooling)' the activation value in pooling layer $l$ is

$$a_j^l = f\left(b_j^l + \beta_j^l MaxPooling(a_j^{l-1}, M^l)\right) \tag{18}$$

$\beta_j^l$ is the multiplicative residual, which is used to improve the representation of the convolutional neural network. In this paper, we used a Softmax classifier with the expression [19]

$$Softmax(a)_i = \frac{e^{a_i}}{\sum\limits_{i \in j} e^{a_j}} \tag{19}$$

where $a_i$ denotes the input to the classifier of $i$, $j$ denotes the slug flow and the annular flow, and $Softmax(a)_i$ denotes the output of the classifier. In this study, the Adam optimizer was used, and the loss function was specified as cross entropy, and since the output values have been Softmax-processed, the cross entropy loss function can be expressed as

$$L = -\sum_{i \in M_j^l} \left( a_i^{l-1} * log\left(a_j^l\right) \right) \tag{20}$$

### 2.3.2. Data Preprocessing and Arithmetic Process

Image preprocessing is crucial when using convolutional neural networks (CNNs) for image recognition. It improves the classification accuracy of the model by removing noise and enhancing the contrast, while unifying the input size to ensure a consistent network structure. Data enhancement techniques (e.g., rotation, scaling, flipping) increase the diversity of training data and improve the generalization ability of the model. The normalization process accelerates the training process, and the problem of vanishing or exploding can be avoided. In addition, color space transformation helps to extract more meaningful features, while filtering processing highlights important features and helps the CNN learn better. In conclusion, proper image preprocessing significantly improves the performance of the CNN in image recognition tasks, ensuring the effectiveness and robustness of the model. In this study, all images were processed as 32 × 32 RGB images, and the pixel values of the images were scaled to between 0 and 1 by using the formula $x = (x - x_{\min})/(x_{\max} - x_{\min})$. The images were then randomly cropped and flipped to further increase the amount and diversity of the training data and to improve the model's generalization ability. Assuming that the original image data used to train the model after data preprocessing have a clear interface, completely consistent RGB values, and ignores color and pixel differences, the specific image preprocessing process is as follows:

a. Data cleaning:

The initial screening of the data was first performed to remove incomplete, anomalous, and duplicate records.

b. Data standardization/normalization:

In order to eliminate the effect of different magnitudes, the numerical-type features were standardized or normalized.

c. Noise reduction processing:

In image data preprocessing, a median filtering noise reduction technique was applied to reduce the image noise and improve the image quality.

d. Feature engineering:

For image data, feature extraction and selection were performed, including features such as the color, texture, shape, etc.

e. Image enhancement:

While selecting the image enhancement techniques, it was ensured that the enhanced image was realistic, no additional noise was introduced, and the enhancement effect significantly improved the model performance. Specific methods such as rotation, scaling, flipping, contrast adjustment, etc., were used to ensure that the model had a better generalization ability to the transformed image.

The processed data were utilized to establish a database, and the following principles were followed: a wide range of flow pattern variations, including the four typical flow patterns, and multifactorial flow influences, mainly including the gas–liquid flow rate and tube inclination angle. Intelligent recognition was performed on the experimental images obtained based on the above principles, and the specific process is shown in Figure 3:

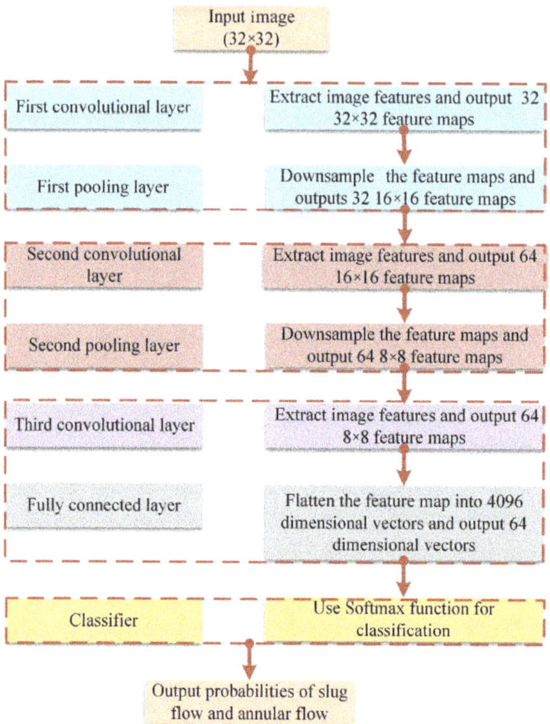


**Figure 3.** Flowchart of model operation.

The model extracts image features through a convolutional layer, a pooling layer for downsampling, a fully connected layer for classification, and has a final output of the probability that the image belongs to a slug flow or an annular flow.

## 3. Results and Discussion

### 3.1. Gas–Liquid Interface

For the slug flow with an inclination angle of 45°, the gas–liquid phase distribution is shown in Figure 4, with a red color for gas and a blue color for liquid. The distribution of gas and liquid phases in the pipe is non-axisymmetric, and the void fractions in different cross sections have great differences. Due to gravity, the liquid phase in different cross sections remains at the bottom of the pipe, while the gas is mostly present in the upper part of the horizontal diameter. In individual cases, the gas in the pipe aggregates and forms large Taylor bubbles, which are distributed in a spiral shape, when the gas is dispersed and the liquid phase is continuous. As shown in the phase distribution cloud diagrams of cross sections F2 and F3, the regions of gas distribution in the pipeline are obviously different, in that most of the gas in cross section F2 is distributed on the right side of the Y-Z plane, while the gas distribution in cross section F3 is just the opposite, which verifies the assumption of the asymmetric distribution of Taylor bubbles.

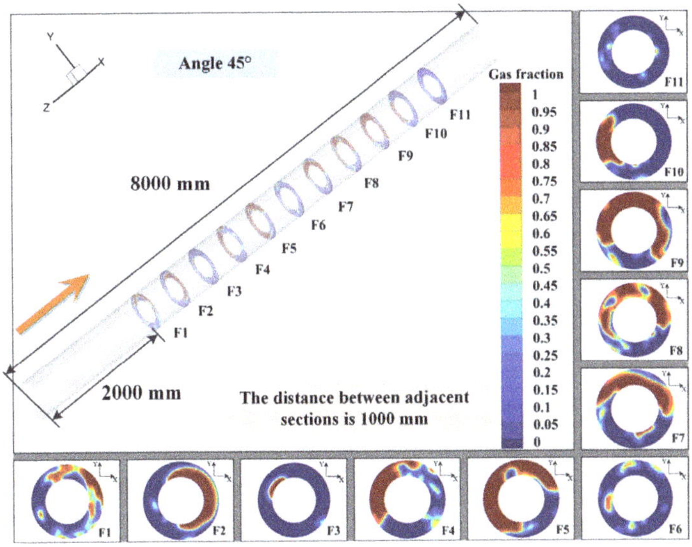


**Figure 4.** Gas–liquid phase distribution in 45° inclined pipe with different cross sections.

Since the two gas–liquid phases appear to have an asymmetric distribution in the annulus, this will lead to a large error in the flow pattern discrimination of the visualization method. Unlike the circular tube, the gas–liquid see–slip phenomenon is more obvious in the annulus, and accurate flow pattern data cannot be obtained through image analysis alone. Figure 5 shows the cross section of phase distribution at different distances from the inlet, from 2200 mm to 2700 mm for the gas–liquid distribution law of a slug unit in the annulus. As can be seen from the figure, the Taylor bubble is close to the outer wall of the tubing, and the liquid phase slips in the gap between the inner wall of the casing and the Taylor bubble. The reason for this is that the inner wall of the casing has a large wetted circumference, the contact area between the liquid phase and the pipe wall is large, and it is not easy for the Taylor bubble to push the liquid slug upward and cause it to flow back. Therefore, from the same-angled camera recording gas–liquid two-phase flow pattern changes, the downward slippage of the liquid film will be misleading for flow pattern discrimination; it is recommended that at least two machine positions should simultaneously capture and record the gas–liquid distribution in the annulus.

### 3.2. Velocity Field

For the multiphase flow in the annular, the flow lines of the two phases are very different from those of the circular pipe. As can be seen in Figure 6, the flow line is spiraling upward, and the movement direction of the fluid is different at different positions. Therefore, the change characteristics of the flow pattern in the inclined pipe are complicated, and if the gas–liquid distribution at a certain moment is used to determine the flow pattern map, it will produce great errors. In this study, we analyzed the gas–liquid distribution and its velocity change at different positions (4300–5700 mm, 3400–4800 mm) and on different planes (X-Z, Y-Z) of the pipe, and found that under the condition of a constant inlet gas–liquid flow rate, the gas–liquid velocity and phase distribution in different pipe sections are closely related. The velocity vector diagrams at the gas–liquid interface show that the liquid phase stagnates, reflows, and rotates here. In flow pattern identification, the real trajectory of the liquid phase can only be found under dynamic conditions, and the flow pattern cannot be defined by the gas–liquid distribution at a certain moment.

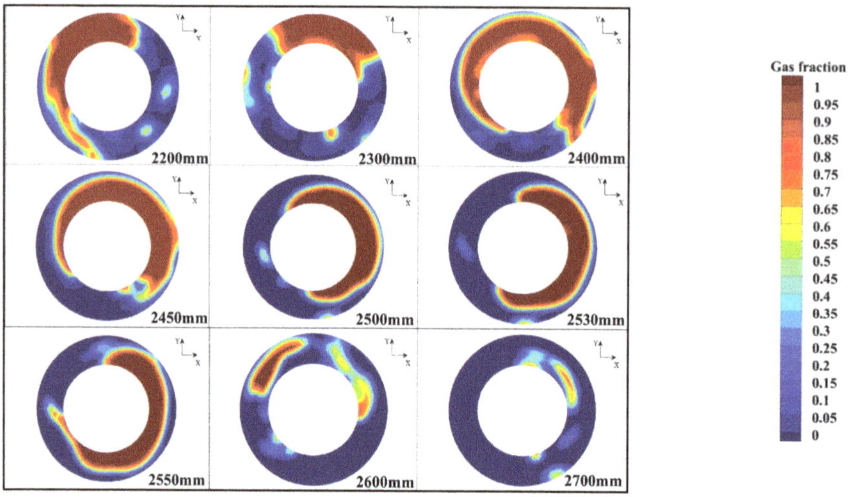


**Figure 5.** Gas–liquid distribution pattern of the slug unit in the annulus.

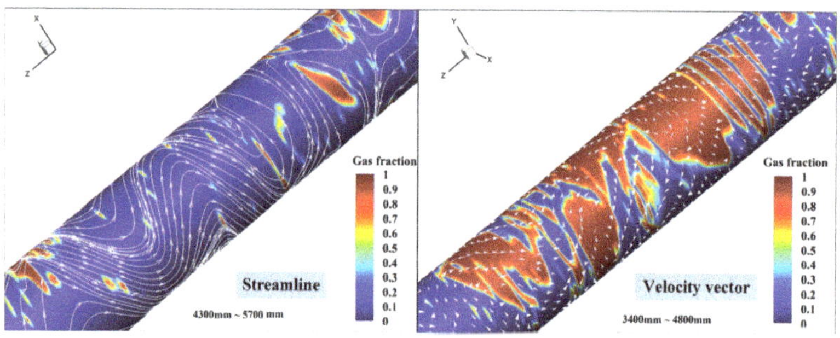


**Figure 6.** Streamlines and velocity vectors at 45° inclination angle.

With the increase in the gas flow rate, the proportion of gas increases, the Taylor bubbles are connected to each other to form a continuous phase, and the liquid phase is squeezed tightly against the pipe wall. The liquid phase exists in the form of an annular liquid film, and the friction between the gas and the pipe wall is converted into friction between the gas and the liquid film, and the pressure drop inside the pipe decreases gradually. If the gas flow rate continues to increase, the liquid phase is accelerated, the repositioning pressure drop decreases while the friction pressure drop increases, and the total pressure drop gradually decreases [17].

Through the above analysis, it was found that flow pattern recognition based on an image method cannot be completed only at a certain moment or angle, which will result in inaccurate flow pattern discrimination and affect the selection of pressure drop calculation methods. In summary, this study obtained experimental flow patterns from multiple angles by summarizing previous research results. Through the deep learning techniques of convolutional neural networks, the flow patterns were identified to test the recognition accuracy of our model.

### 3.3. Uncertainty Analysis

Uncertainty is mainly divided into cognitive uncertainty and accidental uncertainty. For this study, accidental uncertainty is the key factor affecting the prediction accuracy of CNN models. Due to the noise or unpredictability of the data themselves, it cannot be eliminated by increasing the amount of data. Given the complex gas–liquid distribution of annular slug flow and turbulent flow, the prerequisite for using this method for flow pattern recognition is to obtain clear flow pattern photos of the phase interface from experiments or practical applications on site. The uncertainty analysis process is as shown in Figure 7.

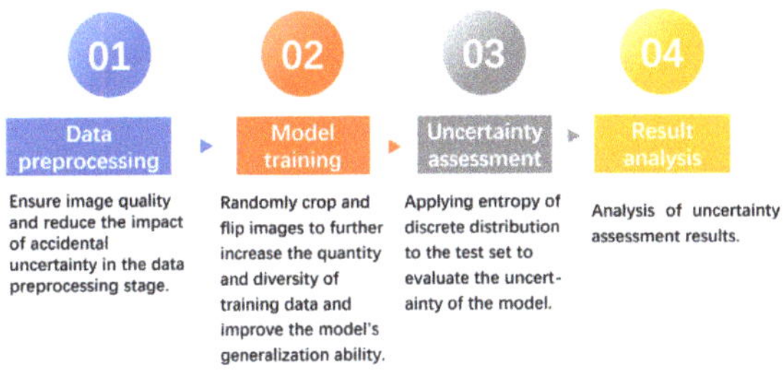


**Figure 7.** Uncertainty analysis process.

According to information theory [20], the entropy of discrete distributions is one of the methods for measuring uncertainty. The actual obtained flow pattern data are $P_s = (x, y)$ and generate a finite dataset after data preprocessing $C_s$, expressed as

$$C_s = \left\{x^{(i)}, y^{(i)}\right\}_{i=1}^{N} \tag{21}$$

$$\left\{x^{(i)}, y^{(i)}\right\} \sim P_s(x, y) \tag{22}$$

The uncertainty of the input random flow pattern image is the entropy of its true posterior distribution $P_s(y|x)$, expressed as

$$H[P_s(y \mid x)] = -\sum_{c=1}^{K} P_s(y = w_c \mid x) \ln P_s(y = w_c \mid x) \tag{23}$$

Therefore, the accidental uncertainty generated by the entire distribution can be represented by its expected conditional entropy, that is

$$E_{P_{s(x)}}[H[P_s(y \mid x)]] \tag{24}$$

Through uncertainty analysis, it was found that the flow pattern identification error involved in this study is within the allowable range.

### 3.4. CNN Prediction Accuracy

According to Cazarez-Candia [21], Zhao [22], and Yang [23], it was found that the slug flow has different motion states due to different pressure conditions. Under the same gas–liquid conditions, when the pipe pressure is higher, the liquid film region of the slug is compressed, the liquid slug gradually becomes longer, and the frequency of slug generation becomes larger. In order to ensure the accuracy of the model prediction, the slug flow under the same pressure conditions was used for training first, and then the trained model was used to predict the slug flow under different pressure conditions. Through literature research to obtain flow type photos under different working conditions, to use slug identification as an example, we first used the churn flow, bubble flow, and slug flow for model training, used the churn flow and bubble flow to interfere with the model through a number of convolution and pooling operations, and finally used Softmax classification to achieve the identification of the slug results, as shown in Table 1.

**Table 1.** Recognition accuracy of CNN for slug flow.

| Flow Pattern | Sample Size | Class Probability of Softmax Output | Prediction of Flow Patterns | Number of Hits |
|---|---|---|---|---|
| Slug flow | 125 | 0.896 | Slug flow | 112 |
| Churn flow | 180 | 0.04 | Non-churn flow | 0 |
| Bubble flow | 145 | 0.0014 | Non-bubble flow | 0 |

The model obtained through training performs well and is able to accurately predict slug flow and annular flow. Since the annular flow is less affected by pressure and the annular flow pattern under different pressure conditions does not change much, the annular flow under atmospheric pressure conditions was used for model training, and the new model recognized the annular flow under other pressure conditions. Sixty photographs of slug flow, churn flow, and annular flow obtained through literature research are shown in Figure 8, in which the experimental temperature is 11–14 °C, the tube operating pressure is 0.9 MPa, and the superficial gas flow rate is 0.4–56.4 m/s, and the superficial liquid flow rate is 0.008–0.075 m/s. In Figure 8a, the gas–liquid interface within the slug flow in the central segment is distinctly delineated, with a pronounced reflux of the liquid phase observable at the base of the pipeline. Notably, only a partial liquid film is present on the upper wall of the conduit. As the gas flow rate intensifies, the liquid film progressively migrates towards the upper wall of the pipeline, culminating in the formation of a fully developed annular flow, as depicted in Figure 8c. The liquid holdup in churn flow is intermediate between that of slug flow and annular flow, indicating an absence of liquid-phase reflux and the absence of a complete annular liquid film. Consequently, churn flow imposes more stringent demands on CNN methodologies, which significantly diminishes the precision of flow pattern recognition. Fifty pictures containing the different flow patterns were used for training, and the rest of them were used as the model for detection, and the overall recognition results are shown in Table 2.

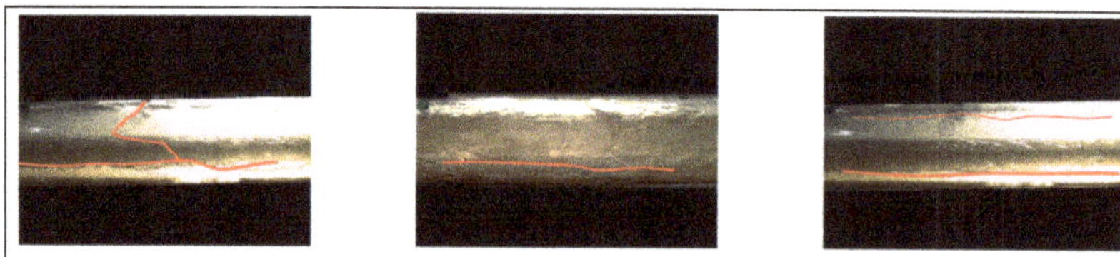

(**a**) Slug flow (**b**) Churn flow (**c**) Annular flow

**Figure 8.** Photograph of typical flow pattern in the annulus (the red line is the gas–liquid interface).

As can be seen from Table 2, the prediction probability for annular flow is basically above 0.98, and the average prediction probability is 0.99. The new model obtained after training has better performance, which confirms that the model has better recognition accuracy. Due to the special characteristics of the slug flow, the recognition accuracy for the slug flow under different pressure conditions is basically kept above 0.92, with an average prediction probability of 0.93, and the individual prediction probability is less than 0.9, which is caused by the fact that the slug flow generated under different experimental conditions, with different characteristics, produces a large deviation in the recognition process.

**Table 2.** Results of stream type identification.

| Sample Size | Flow Pattern | Class Probability of Softmax Output | | |
|---|---|---|---|---|
| | | Slug Flow | Annular Flow | Churn Flow |
| 136 | Slug flow | 0.9786 | 0.0115 | 0.0100 |
| | | 0.9001 | 0.0127 | 0.0872 |
| | | 0.9317 | 0.0113 | 0.0570 |
| | | 0.8897 | 0.0108 | 0.0995 |
| | | 0.9324 | 0.0161 | 0.0515 |
| 129 | Annular flow | 0.0002 | 0.9878 | 0.0120 |
| | | 0 | 0.9804 | 0.0196 |
| | | 0.0001 | 0.9868 | 0.0131 |
| | | 0 | 0.9941 | 0.0059 |
| | | 0.0015 | 0.9873 | 0.0112 |

This study reveals that slug flow is primarily caused by unstable gas–liquid flow rates, leading to blurred gas–liquid interfaces in flow pattern images. Even after initial image processing, it remains challenging to eliminate the instability of the gas–liquid interface caused by flow fluctuations. This blurriness is a major impediment to the recognition accuracy of convolutional neural network (CNN) models. In contrast, annular flow exhibits a clear gas–liquid interface, where the liquid and gas phases can flow stably within the pipe and maintain a fixed flow pattern, resulting in a high recognition accuracy for annular flow.

By comparing our findings with those of Zhao et al. [15], we note that despite both studies employing neural network models to predict gas–liquid two-phase flow patterns, there is a significant difference in prediction accuracy. Zhao's research indicates that the flow pattern recognition accuracy of the traditional BP neural network is 87.5%, while the SSA-optimized BP neural network improves the recognition accuracy to 91.66%. In our study, the trained CNN model achieved prediction accuracies of 0.9265 for slug flow and 0.9873 for annular flow, respectively. This demonstrates that the model employed in our study has a clear advantage in terms of prediction accuracy.

## 4. Conclusions

In this paper, the gas–liquid two-phase flow characteristics in the annulus were investigated using CFDs simulation, focusing on the gas–liquid distribution law and its velocity field change characteristics. The shortcomings of the flow pattern recognition technology based on the image method were analyzed, and this study used a convolutional neural network to recognize the flow patterns obtained under different conditions, and it was found that the use of the CNN method to recognize the flow patterns has a better performance. Its good local sensing ability can effectively capture local features such as the liquid film, liquid flow, and gas–liquid interface in the flow pattern, and it can accurately discriminate the slug flow and annular flow, and the Max Pooling function can accurately extract the high-level features of the image or flow pattern data, which improves the classification accuracy of the Softmax classifier.

In this study, the recognition probability of the CNN model for slug flow and annular flow was 0.93 and 0.99, respectively. In light of the intricate gas–liquid distribution characteristics and pronounced transitional behavior inherent in churn flows, there exists a discrepancy between the flow patterns predicted by the convolutional neural network approach and the actual observations. Moving forward in our research endeavors, we intend to employ intelligent algorithms for the identification of flow patterns. Additionally, auxiliary judgment will be rendered by incorporating data such as differential pressure signals and the liquid holdup. The synthesis of various methodologies to analyze the flow parameters associated with flow patterns is anticipated to enhance the predictive accuracy of potential flow configurations within the annulus. It is important to acknowledge the limitations of the model, particularly with respect to the computational expense and hardware requirements, which may render it less suitable for direct applications in practical industrial settings.

**Author Contributions:** Conceptualization, C.C.; methodology, C.C.; formal analysis, C.C.; investigation, C.C.; resources, Y.Z.; data curation, Y.S., W.Y. and X.F.; writing—original draft preparation, C.C. and W.Y.; writing—review and editing, C.C. and W.Y. All authors have read and agreed to the published version of the manuscript.

**Funding:** This research received no external funding.

**Data Availability Statement:** Restrictions apply to the availability of these data. Data were obtained from the CNPC and are available from the authors with the permission of the CNPC.

**Acknowledgments:** The authors are grateful to Zhihui Wang for her help with the preparation of figures in this paper, and we thank the editors and the reviewers for their useful feedback, which improved this paper.

**Conflicts of Interest:** Chen Cheng was employed by Sinopec Natural Gas Branch. Weixia Yang was employed by Jinchang PetroChina Kunlun Gas Co., Ltd. Xiaoya Feng was employed by Xi'an Changqing Chemical Group Co., Ltd. Yarui Zhao was employed by Changqing Oilfield Company. Yubin Su was employed by Changqing Oilfield Branch Company. The authors declare that the research was conducted in the absence of any commercial or financial re-lationships that could be construed as a potential conflict of interest.

## Nomenclature

| | |
|---|---|
| $v_{sw}$ | Superficial flow rate of water, m/s |
| $\rho_w$ | Density of water, kg/m$^3$ |
| $\sigma$ | The gas–liquid interfacial tension, N/m |
| $D_{iout}$ | Outer diameter of the inner pipe, m |
| $v_m$ | Mixture velocity, m/s |
| $\phi_{slug}$ | Liquid holdup of slug flow, dimensionless |
| $v_{rc}$ | Rising speed of bubbles in the annulus in the stationary liquid, m/s |
| $A_g$ | Area of the gas in the cross section of the annulus, |
| $\beta_i$ | Volume ratio of phase i |
| $\vec{F}$ | Volume force, N |
| $k_{eff}$ | Effective thermal conductivity, dimensionless |
| $\vec{v}$ | Velocity vector, m/s |
| $G_b$ | Turbulent kinetic energy due to buoyancy |
| $\sigma_k$ | Turbulent Prandtl numbers for the $k$ equations |
| $a_j^l$ | Output value of the jth unit of convolutional layer l |
| $M_j^l$ | Set of input feature maps |
| $k$ | Convolutional kernel |
| $C_s$ | Generate a finite dataset after data preprocessing |
| $v_{sg}$ | Superficial flow rate of water, m/s |
| $\rho_g$ | Gas density, kg/m$^3$ |
| $D_s$ | Hydraulic equivalent diameter of the annulus, m |
| $D_{oin}$ | Internal diameter of the outer pipe, m |
| $\phi_{churn}$ | Liquid holdup of churn flow, dimensionless |
| $v_{tb}$ | Maximum velocity of Taylor bubble in slug, m/s |
| $\beta_{an}$ | Void friction of the annular flow, dimensionless |
| $A_{at}$ | Total cross section area of the annulus, m$^2$ |
| $S_{\beta_i}$ | Mass source term, 0 |
| $\rho$ | Density, kg/m$^3$ |
| $\mu$ | Mixing viscosity, Pa·s |

| | |
|---|---|
| $G_k$ | Turbulent kinetic energy due to the laminar velocity gradient |
| $Y_M$ | Fluctuations due to the diffusion of the transition |
| $\sigma_\varepsilon$ | Turbulent Prandtl numbers for the *e* equations |
| $f$ | Activation function |
| $b_j^l$ | Bias |
| $\beta_j^l$ | Multiplicative residual |
| $a_i$ | Input to the classifier of i |

## References


1. Caetano, E.F.; Shoham, O.; Brill, J.P. Upward Vertical Two-Phase Flow Through an Annulus-Part I: Single-Phase Friction Factor, Taylor Bubble Rise Velocity, and Flow Pattern ediction. *J. Energy Resour. Technol.* **1992**, *114*, 1–13. [CrossRef]
2. Kelessidis, V.C.; Dukler, A.E. Modeling flow pattern transitions for upward gas-liquid flow in vertical concentric and eccentric annuli. *Int. J. Multiph. Flow* **1989**, *15*, 173–191. [CrossRef]
3. Saruwatari, M.S.S. Two-phase flow in vertical noncircular channels. *Int. J. Multiph. Flow* **1982**, *8*, 641–655. [CrossRef]
4. Shi, L. Fuzzy Recognition for Gas-liquid Two-phase Flow Pattern Based on Image Processing. In Proceedings of the IEEE International Conference on Control & Automation, Guangzhou, China, 30 May–1 June 2007; IEEE: Piscataway, NJ, USA, 2007.
5. Wang, K.; Ye, J.; Bai, B. Entrained droplets in two-phase churn flow. *Chem. Eng. Sci.* **2017**, *164*, 270–278. [CrossRef]
6. Yang, Y.; Wang, D.; Niu, P.; Liu, M.; Wang, S. Gas-liquid two-phase flow measurements by the lectromagnetic flowmeter combined with a phase-isolation method. *Flow Meas. Instrum.* **2018**, *60*, 78–87. [CrossRef]
7. Liu, L.; Bai, B. Flow regime identification of swirling gas-liquid flow with image processing technique and neural networks. *Chem. Eng. Sci.* **2019**, *199*, 588–601. [CrossRef]
8. Nnabuife, S.G.; Pilario, K.E.S.; Lao, L.; Cao, Y.; Shafiee, M. Identification of gas-liquid flow regimes using a non-intrusive Doppler ultrasonic sensor and virtual flow regime maps. *Flow Meas. Instrum.* **2019**, *68*, 101568. [CrossRef]
9. Ligus, G.; Zając, D.; Masiukiewicz, M.; Anweiler, S. A New Method of Selecting the Airlift Pump Optimum Efficiency at Low Submergence Ratios with the Use of Image Analysis. *Energies* **2019**, *12*, 735. [CrossRef]
10. Sukamta, S. Simulation of High Viscosity Gas-Liquid Two-Phase Flow in a Horizontal Mini Pipe. *Semesta Tek.* **2019**, *22*, 95–102. [CrossRef]
11. Wang, J.; Li, J.; Liu, G.; Li, M.; Luo, K.; Zhang, G. Development and application of transient gas-liquid two-phase flow model considering sudden density change. *Energy Sci Eng.* **2020**, *8*, 1209–1219. [CrossRef]
12. Veisi, A.; Shahsavari, M.H.; Roshani, G.H.; Eftekhari-Zadeh, E.; Nazemi, E. Experimental Study of Void Fraction Measurement Using a Capacitance-Based Sensor and ANN in Two-Phase Annular Regimes for Different Fluids. *Axioms* **2023**, *12*, 66. [CrossRef]
13. Xiao, Q. Identifying Flow Patterns in a Narrow Channel via Feature Extraction of Conductivity Measurements with a Support Vector Machine. *Sensors* **2023**, *23*, 1907. [CrossRef] [PubMed]
14. Du, M.; Wei, J.; Li, M.Y.; Gao, Z.K.; Kurths, J. Interconnected ordinal pattern complex network for characterizing the spatial coupling behavior of gas-liquid two-phase flow. *Chaos Interdiscip. J. Nonlinear Sci.* **2023**, *33*, 063108. [CrossRef] [PubMed]
15. Zhao, X.; Song, H.; Wang, M. Research on Flow Pattern Identification Method for Horizontal Well Gas Water Two Phase Production Logging Based on Artificial Intelligence Algorithm. *Contemp. Chem. Res.* **2023**, *8*, 173–175. [CrossRef]
16. Zhao, J.-F. Influence of Bubble Initial Size on Bubble-to-Slug Transition. *J. Eng. Thermophys.* **2005**, *26*, 3. [CrossRef]
17. Yin, B.; Li, X.; Sun, B.; Zhang, H. Hydraulic model of steady-state multiphase flow in wellbore annuli. *Pet. Explor. Dev.* **2014**, *41*, 8. [CrossRef]
18. Wang, Z.; Luo, W.; Liao, R.; Xie, X.; Han, F.; Wang, H. Slug Flow Characteristics in Inclined and Vertical Channels. *Fluid Dyn. Mater. Process.* **2019**, *15*, 583–595. [CrossRef]
19. Bengio, Y. Learning deep architectures for AI. *Found. Trends Mach. Learn.* **2009**, *2*, 1–12. [CrossRef]
20. Thomas, M.T.; Joy, A.T. *Elements of Information Theory*; John Wiley &Sons: Hoboken, NJ, USA, 2006.
21. Cazarez-Candia, O.; Benítez-Centeno, O. Comprehensive experimental study of liquid-slug length and Taylor-bubble velocity in slug flow. *Flow Meas. Instrum.* **2020**, *72*, 101697. [CrossRef]
22. Zhao, P. Investigation of Recognition and Active Control of the Two-Phase Flow Pattern. Ph.D. Thesis, China Institute of Metrology, Beijing, China, 2014.
23. Yang, W.; Liang, R.; Fang, Z.; Ding, Q.; Liao, R. Research on Flow Pattern Transformation of Gas-liquid Two-phase-flow in Inclined Pipe. *Sci. Technol. Eng.* **2021**, *21*, 9353–9359.

processes

MDPI

Article

# Investigation into the Variation Law of Network Fracture Conductivity in Unconventional Oil and Gas Reservoirs

Jianfa Ci [1], Lei Wang [2,3,*], Feiyang Xiong [2,3] and Yanxin Zhao [2,3]

1 Petroleum Engineering Technology Research Institute, Sinopec Southwest Oil and Gas Company, Deyang 618000, China
2 School of Petroleum Engineering, Yangtze University, Wuhan 430100, China
3 Key Laboratory of Drilling and Production Engineering for Oil and Gas, Wuhan 430100, China
* Correspondence: wang-lei@yangtzeu.edu.cn; Tel.: +86-18971168785

**Abstract:** The network fracturing technique is a key technology for increasing effective reservoir volume and enhancing production in shale oil and gas. The fracture network's conductivity is one of the crucial factors affecting the efficient development of shale gas. To evaluate the variation patterns and influencing factors of the conductivity of network fractures, this study employed a proppant conductivity evaluation system and an equivalent theory testing method. It investigated the conductivity of propped and self-propped fractures under different angles and numbers of fractures. Experimental results showed that fractures with proppant support had higher conductivity than unsupported fractures. Smaller angles between secondary and main fractures resulted in greater conductivity. The conductivity of multi-fracture structures increased with the number of fractures. Under low-closure pressure conditions, self-propped fractures exhibited significantly higher conductivity than propped fractures, but this trend reversed under high closure pressure. This experimental research provides guidance for constructing network fractures with high conductivity in unconventional oil and gas reservoir fracturing.

**Keywords:** network fracturing; propped and self-propped fractures; number and angle of fractures; conductivity variation patterns

## 1. Introduction

China possesses enormous shale gas resource potential, yet exploration and development are still in their infancy, with hydraulic fracturing technology continuously evolving and improving [1]. Currently, the prevalent method involves multi-stage hydraulic fracturing in horizontal wells, adjusting fluid and proppant volumes to enhance artificial fracture complexity and improve reservoir fluid flow dynamics [2]. The degree of fracture development is crucial to shale gas reservoir quality, as fractures provide storage space and serve as primary pathways for gas migration within shale formations.

Different types and scales of fractures play varying roles in shale development [3]. Previous studies by Cui Mingyue [4] and others have characterized reservoir modification through complex fracture networks, analyzing factors influencing the formation of these networks. Research indicates that the hydraulic complexity index of reservoirs is influenced by both geological and engineering factors. Han Huifen [5] and Sun Chen [6], among others, simulated supported and self-propping artificial core fractures with varying fracture porosity. They utilized API (American Petroleum Institute) shale permeability testing to reveal degradation patterns in hydraulic conductivity under factors such as fluid type, fracture slip and number of fractures. Fredd [7], Zhang J [8], Cipolla [9] and others characterized the effects of fracture conductivity through measurements of electrical conductivity across fractures, and conducted tests on varying fracture spacings and proppant distributions, discovering that certain natural fracture spacings can lead to enhanced electrical conductivity, thus demonstrating improved flow capabilities. Researchers including

Gao Shuaiqiang [10], Liu Jiajie [11], Ning Wenxiang [12], Cao Haitao [13], Zhu Haiyan [14], Wen Qingzhi [15] and others undertook conductivity testing on single fracture types under different effective stress conditions, comparing and analyzing variations in conductivity between supported and self-propping fractures. Zhang Han [16], applying principles of hydrodynamic similarity, categorized post-fracturing complex fracture networks into supported and self-propping types. They proposed the concept of equivalent fractures, establishing an optimization model for conductivity based on fracture morphology and elucidating the relationships between the conductivity of these two fracture types. Notably, these studies did not consider variations in the conductivity of self-propping and proppant-supported fractures under different closure pressures.

This study employs a proppant long-term conductivity system evaluation device, utilizing an API standard flow chamber and experimentally fabricated shale samples with varying angles and numbers of fissures. By comparing the conductivity differences between propped fractures and self-propped fractures at different closure pressures, the impact of fracture network morphology on conductivity is analyzed. This research aims to provide guidance and reference for optimizing parameters in fracture network fracturing processes.

## 2. Experimental Installation and Experimental Procedures

### 2.1. Experimental Installation

In this research, a device for long-term proppant conductivity capacity system evaluation was used to conduct experiments on testing fracture conductivity. This device consists of a model testing system, a pressure compensation system, a fluid injection system, a data collection system (computer) and other auxiliary devices. The maximum closing pressure in the diversion chamber is 100 MPa, the maximum working temperature is 150 °C and the maximum liquid flow rate is 0.2 L/min. Laboratory equipment is shown in Figure 1.

**Figure 1.** Equipment photos.

### 2.2. Cutting Plan for the Experimental Rock Plate

In this paper, the effects of the closure pressure, the angle of branching joints and the number of branching joints on the fracture conductivity were comprehensively considered, indoor tests on the conductivity of fractures were carried out to compare the conductivity difference between the two types of fractures (proppant-supported fracture and sliding self-supported fracture) and 12 sets of experiments were carried out. The specific experiment scheme is shown in Table 1.

**Table 1.** Experimental plan for different fracture network structures.

| Number | Main Fracture and Secondary Fracture Angles/Degrees | Number of Secondary Fractures | Proppant |
|---|---|---|---|
| 1 | 0 | 0 | Mixing 70/140 and 40/70 ceramic beads in a 3:2 ratio, with a main fracture sand concentration of 5 kg/m$^2$ and secondary fracture concentration of 3 kg/m$^2$ |
| 2 | 30 | 1 | |
| 3 | 30 | 2 | |
| 4 | 30 | 3 | |
| 5 | 30 | 4 | |
| 6 | 60 | 2 | |
| 7 | 0 | 0 | Self-supporting fractures do not add proppants |
| 8 | 30 | 1 | |
| 9 | 30 | 2 | |
| 10 | 30 | 3 | |
| 11 | 30 | 4 | |
| 12 | 60 | 2 | |

The self-supported and proppant-supported experimental slabs were processed based on the experimental scheme. The cutting schemes for the main joint and branching joint of the two types of crack test rock slabs are the same, and both of them were processed by the line-cutting method. The splitting method was adopted for the main joint of the self-supported crack to simulate cracking. The angle between the branching joint and the main joint (hereinafter abbreviated as the branching joint angle) and the amount of branching joints are consistent with the proppant crack test scheme. Taking a proppant crack as an example, the rock slab was cut into two sections of 176 mm in length, 37 mm in width and 25 mm in height. Sand was spread evenly in the middle of the rock slabs, and this group (without branched joint) was set as the basic control group (Figure 2aA).

The position 19 mm to the right of the inlet end of the test fluid (the position without radian) shall be deemed the starting point of measurement. The setting method of the branch seam should be the main seam, which is the bottom symmetrical cut on the datum plane. One branch seam is set at the position 69 mm away from the starting point, and the branch seam angle should be 30°. Two branch seams shall be set and, on the basis of this seam, a branch seam shall be added with an angle of 30° at the starting point and branch seam. Three branch seams shall be set at locations of 0 mm, 46.5 mm and 93 mm away from the starting point, and the branch seam angle shall be 30°. Four branch seams shall be set at locations of 0 mm, 34 mm, 68 mm and 102 mm away from the starting point, and the branch seam angles should all be 30°. One branch seam shall be set at the same location as that in Figure 2aC, and the branch seam angle shall be 60°.

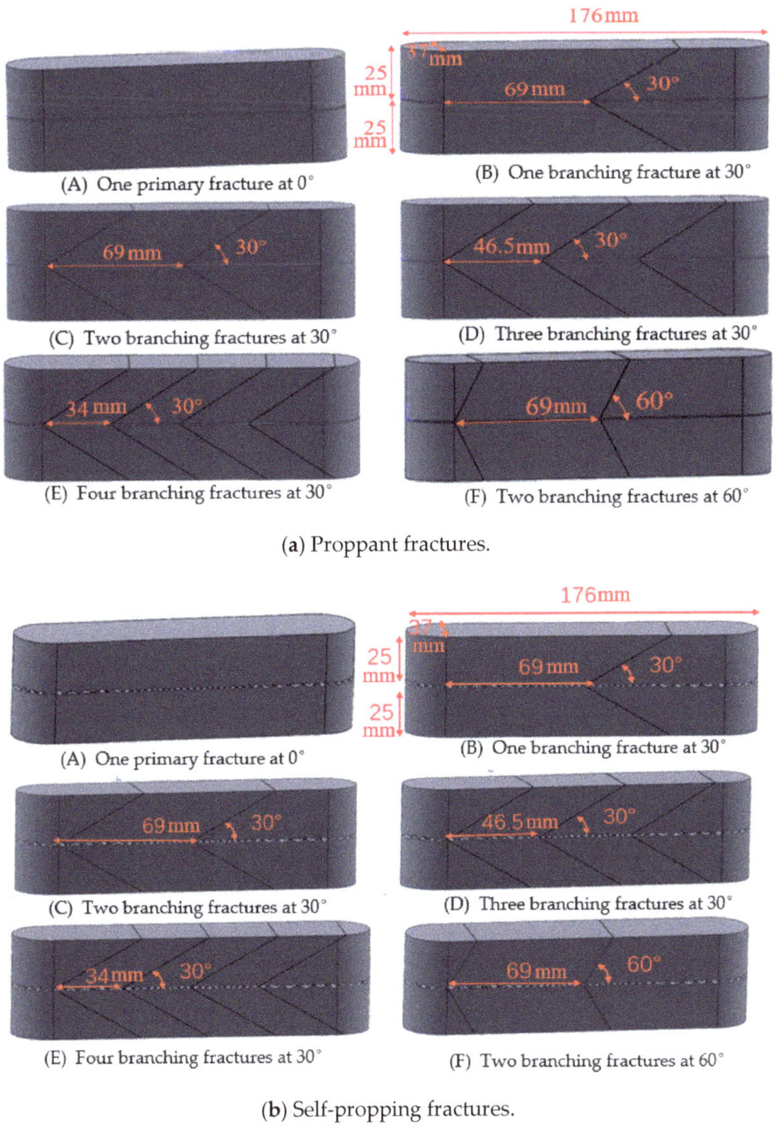


**Figure 2.** Cutting sketch of rock plate for testing conductivity capacity.

### 2.3. Experimental Procedures

The systematic evaluation device for proppant long-term conductivity was used to evaluate the conductivity of netted fractures. The main steps are as follows: (1) Install the lower cover plate and the lower rock plate in the diversion chamber in turn, spread the proppant evenly on the lower rock plate of the diversion chamber and install the upper rock plate and the upper cover plate of the diversion chamber. (2) Install the flow guide chamber on the hydraulic press, and adjust the closing pressure to 14 MPa. (3) Put fresh water into the pumping liquid container, set a liquid flow rate of 5 mL/min and start pumping liquid. (4) After pumping the liquid for 15 min, wait for the pressure difference to stabilize and record the upstream and downstream pressure difference. Adjust the closing pressure to 27

MPa. Record the pressure difference again when the pressure difference stabilizes after 15 min. (5) Keep changing the closing pressure until the slab displacements and upstream and downstream pressure difference under all target closing pressures have been recorded. (6) Shut down the pumps and release the sealing pressure. The rock slab was taken out and the proppant on the surface was brushed away; photos were taken to record the changes in the rock surface.

The proppant used in the experiment was configured by proportioned mixing; 70/140 mesh and 40/70 mesh ceramisite were mixed at 3:2 (Figure 3 right). The concentration of sanding in the main joint was 5 kg/m$^2$, and the concentration of sanding in the support joint was 3 kg/m$^2$.

**Figure 3.** Images show 70/140 mesh ceramic beads (**left**), 40/70 mesh ceramic beads (**center**) and 3:2 mixture (**right**).

## 3. Analysis of Conductivity Experiments

### 3.1. Experimental Principle

The propping cracks were filled with proppant in the flow diversion chamber to form a proppant filling layer flow diversion channel. The self-supporting cracks relied on the rough protrusion of the cracks caused by the shear sliding of the split surface to support the cracks and establish the flow diversion channel. Using the equivalence theory, the flow diversion ability of the cracks of different seam network structures could be equivalent to the flow diversion ability of the main seam. Under different closing pressures, the test fluid passed through the flow conducting channel, the flow rate and pressure difference at both ends of the test rock plate were recorded and the fluid viscosity was found; the principle of flow conducting capacity testing can be expressed as Darcy's law:

$$k = \frac{Q\mu L}{A\Delta P} \tag{1}$$

In the equation, ($k$) [17] represents the permeability of supported or self-supporting fractures, μm$^2$; ($Q$) stands for the flow rate within the fracture, cm$^3$/s; ($\mu$) denotes fluid viscosity, mPa·s; ($L$) is the distance between pressure measurement points (test ports), cm; ($A$) represents the cross-sectional area of the fracture, cm$^2$; and ($\Delta P$) indicates the pressure differential between the two ends of the test section, kPa.

The proppant long-term conductivity system evaluation device utilizes an API standard diversion chamber. The equivalent fracture network conductivity can be expressed as

$$k = \frac{5.555Q\mu}{\Delta P w_f} \tag{2}$$

In the equation, $W_f$ represents the width of the fracture, cm.

The equivalent conductivity of the fracture can be expressed as

$$kW_f = \frac{5.555Q\mu}{\Delta P} \tag{3}$$

In the experiments, the conductivity of the fracture can be determined based on the measured pressure differential and flow rate using the equation above.

*3.2. The Experiment on Proppant Fracture Conductivity*

To investigate the variation in fracture conductivity of proppants with different fracture network structures, six pairs of propped fracture experimental rock plates were prepared. The fracture network structures consisted of one primary fracture at 60° (no branching fractures), one primary fracture at 30° with branching fractures, two branching fractures at 30°, three branching fractures at 30°, four branching fractures at 30° and two branching fractures at 60°. Proppants used were a 3:2 mixture of 70/140 and 40/70 ceramics, with a sand concentration of 5 $kg/m^2$ for the primary fracture and 3 $kg/m^2$ for each angled branching fracture, uniformly applied. The rock plate with one primary fracture at 0° (no branching fractures) was selected as the experimental control group. Water was used as the testing fluid, and the fracture conductivity was tested under five different closure pressures: 14 MPa, 27 MPa, 41 MPa, 55 MPa and 69 MPa. Experimental results are shown in Figure 4.

(**a**) One primary fracture at 0°.

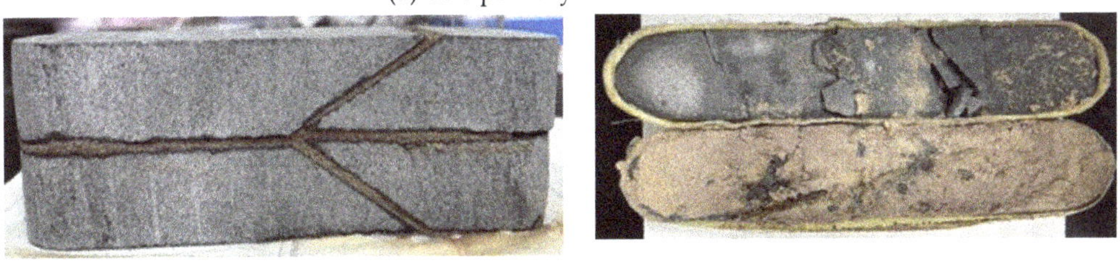

(**b**) One branching fracture at 30°.

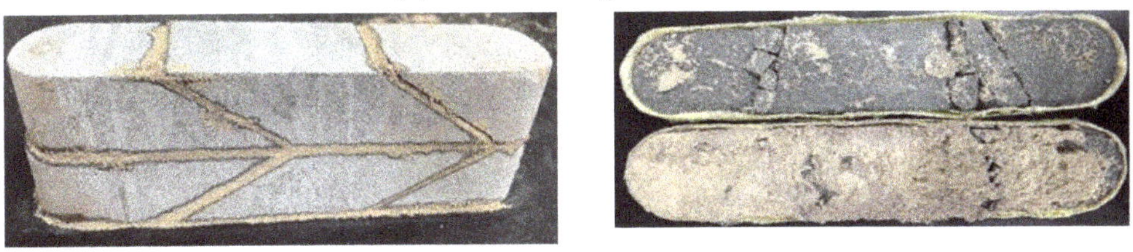

(**c**) Two branching fractures at 30°.

(**d**) Three branching fractures at 30°.

**Figure 4.** *Cont.*

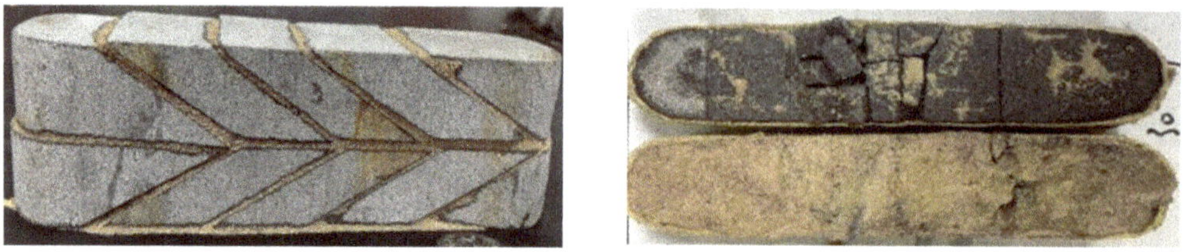

(**e**) Four branching fractures at 30°.

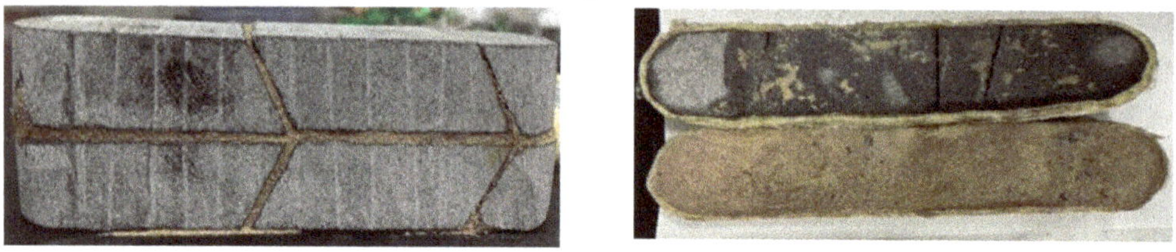

(**f**) Two branching fractures at 60°.

**Figure 4.** Comparison image of proppant fracture conductivity before (**left**) and after (**right**) experimental testing.

Based on experimental data, plots were created to illustrate the variations in proppant fracture conductivity with different numbers and angles of branching fractures. From the conductivity trends in proppant fractures in Figure 5, the following observations can be made: (1) The trend in fracture conductivity of the supported joint structure and unsupported joint structure with increasing closure pressure was the same, and the fracture conductivity of both structures decreased gradually with increasing closure pressure. (2) The fracture conductivity capacity decreased significantly at low closure pressure (less than 41 MPa) and decreased slightly at high closure pressure (more than 41 MPa). (3) Regardless of the number of supported joints and the angle between the main supported joints, the conductivity of fractures with supported joints was always higher than that of fractures without supported joints. (4) The higher the number of branched fractures, the greater the conductivity, and the conductivity of the fracture network with two branched fractures was significantly higher than that of the fracture with only one branched fracture. However, when the number of branch joints was further increased, that is, when the number of branch joints was more than two, the change in conductivity was not obvious. (5) If there were also two branching joints, the measured fracture conductivity in support of the 30° jointed structure was clearly higher than that of the 60° jointed structure.

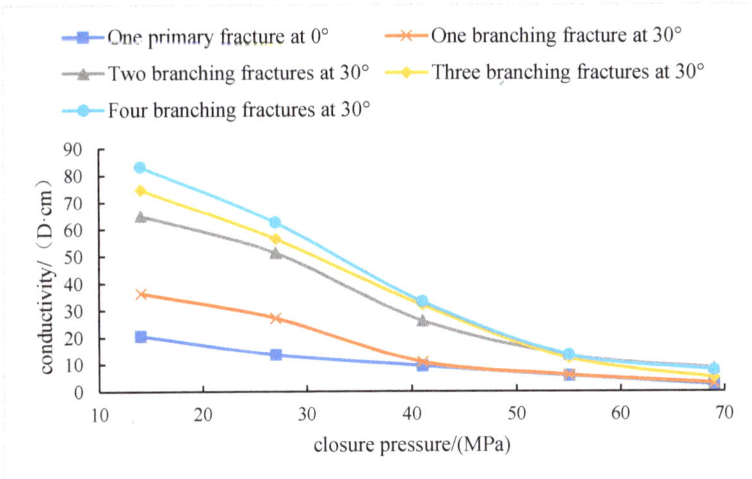


(**a**) Number of different branching fractures.

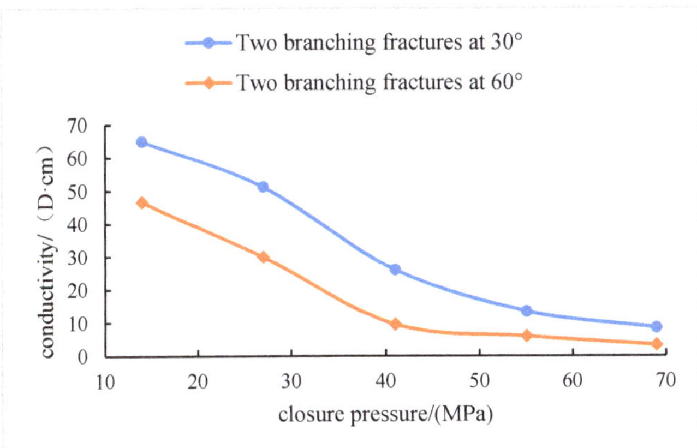


(**b**) Different angles of branching fractures.

**Figure 5.** Variation curve of proppant fracture conductivity with closure pressure.

### *3.3. Experimental Study on the Conductivity of Self-Propping Fractures*

To further investigate the variations in conductivity of self-propping fractures with different numbers of branching fractures, six pairs of self-propping fractures were set with the upper and lower rock plates moved relative to each other by 1 mm. These fractures relied on rough fractures with shear sliding along the fracture surface to protrude and support them. The 0° primary fracture panel was used as the basic control group. The fracture network configurations included one primary fracture (no branching fractures), one branching fracture at 30°, two branching fractures at 30°, three branching fractures at 30°, four branching fractures at 30° and two branching fractures at 60°. Water was used as the testing fluid to measure the conductivity of self-propping fractures under five different closure pressures: 14 MPa, 27 MPa, 41 MPa, 55 MPa and 69 MPa. The experimental results are shown in Figure 6, and conductivity variation curves (Figure 7) were plotted based on the experimental data.

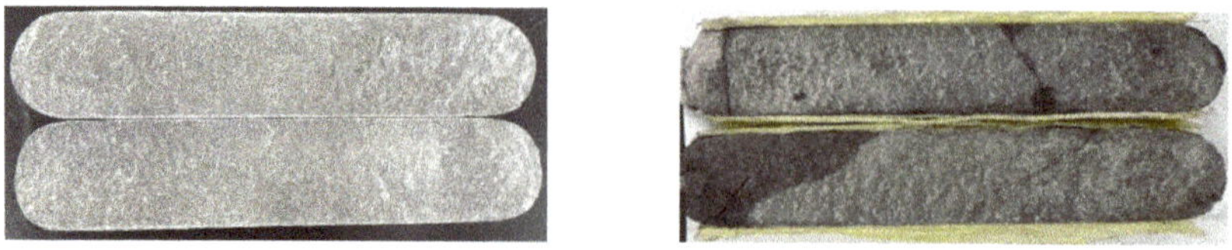

(**a**) One primary fracture at 0°.

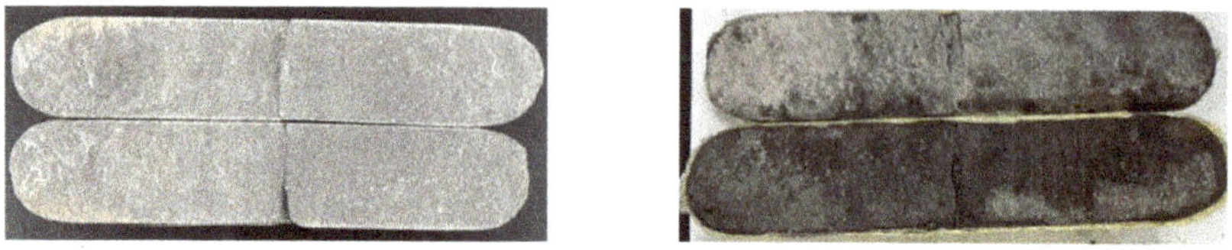

(**b**) One branching fracture at 30°.

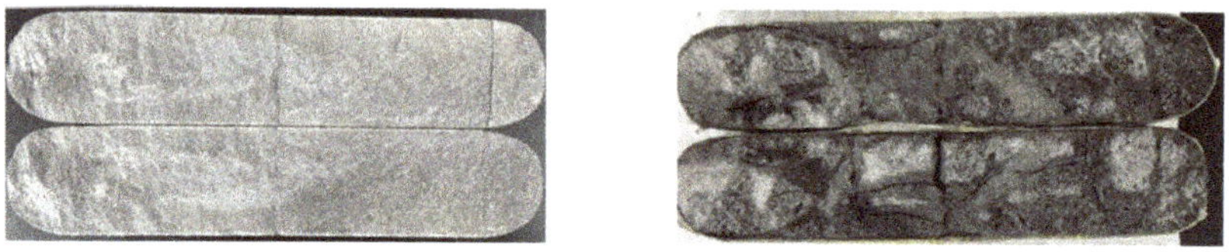

(**c**) Two branching fractures at 30°.

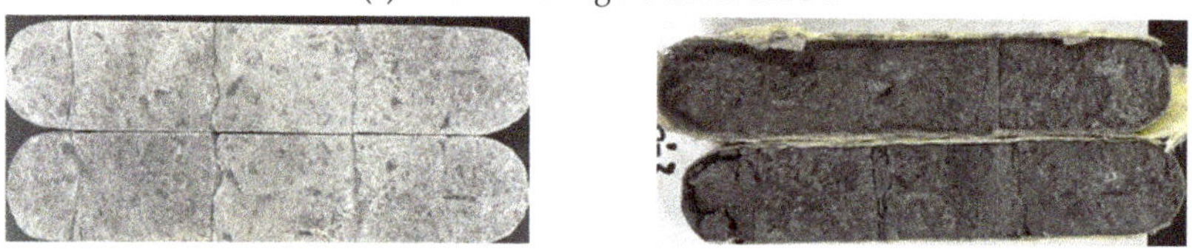

(**d**) Three branching fractures at 30°.

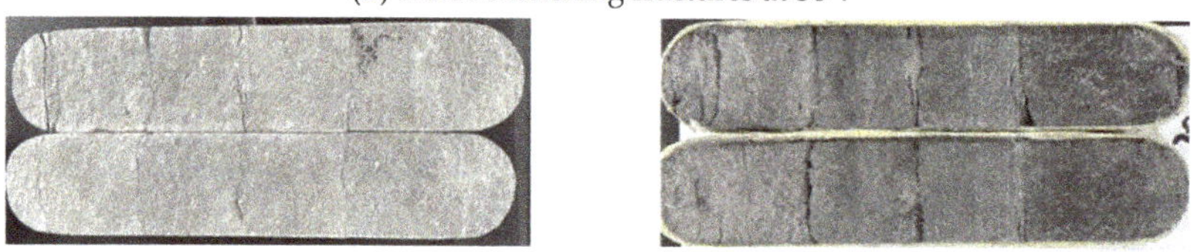

(**e**) Four branching fractures at 30°.

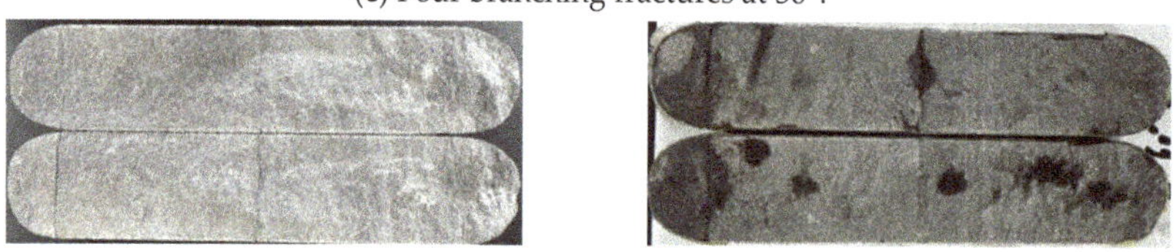

(**f**) Two branching fractures at 60°.

**Figure 6.** Comparison images of the conductivity of self-propping fractures before (**left**) and after (**right**) experimental testing.

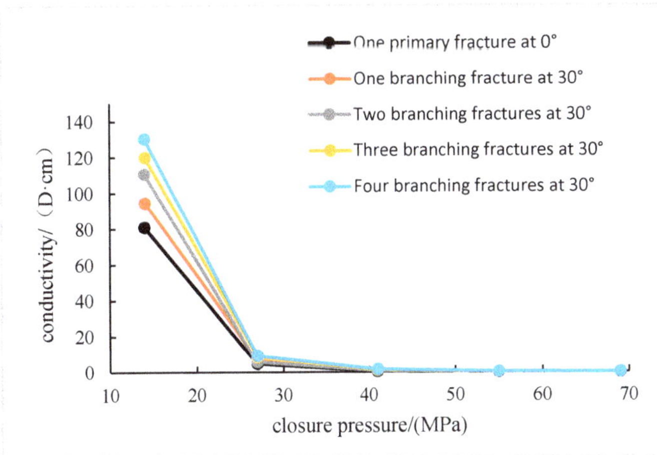


(**a**) Number of different branching fractures.

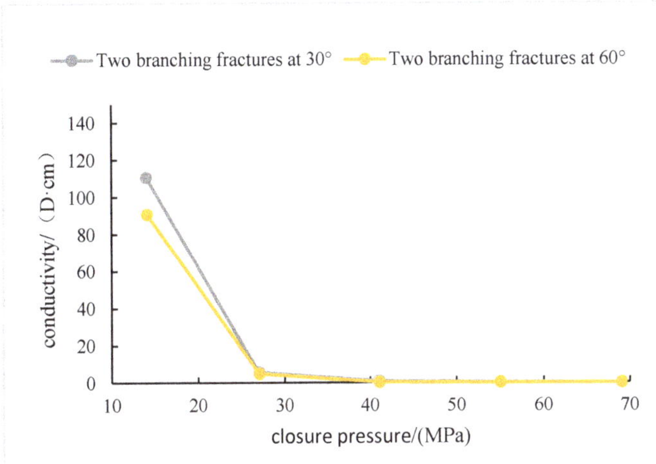


(**b**) Different angles of branching fractures.

**Figure 7.** The conductivity curve of self-propping fractures as a function of closure pressure.

According to Figure 7, the following were found: (1) The flow conductivity of the self-supporting crack with the supported joint was similar to that of the self-supporting crack with the unsupported joint, but the flow conductivity of the self-supporting crack with the supported joint was significantly higher than that of the self-supporting crack with the unsupported joint. (2) Under different closing pressures, the conductivity of multi-supported joints increases with the increase in the number of supported joints. Under medium and low closing pressures (less than 55 MPa), the support joints have a significant effect on increasing the conductivity, about 2–5 times, while under high closing pressures, the effect of supporting joints on increasing the conductivity is less; about 1–3 times. (3) Under different closure pressures, the conductivity of self-supporting fractures with different buttressed angles has different characteristics. The conductivity of two 30° buttressed fractures is a little higher than that of two 60° buttressed fractures. (4) As the closure pressure increases from 14 MPa to 27 MPa, there is a significant decrease in the conductivity of self-propping fractures, with a decrease of approximately 1–2 orders of magnitude at each pressure point.

### *3.4. Comparison and Discussion of Proppant and Self-Propping Fracture Conductivity*

As shown in Figure 8, the experimental results of the conductivity of proppant fractures and those of self-proppant fractures were compared and analyzed in detail. The results show that the conductivity of proppant fractures decreases linearly with the increase in the closure pressure. However, the conductivity of self-proppant fractures decreases sharply with the increase in the closure pressure. When the value of self-proppant fractures reaches 41 MPa, the conductivity of self-proppant fractures decreases to less than 1D · cm. It is noteworthy that the conductivity of self-proppant fractures is superior under the condition of low closure pressure, even surpassing that of proppant fractures. However, the fracture is extremely sensitive to the change in pressure, and the conductivity of self-proppant fractures is negligible when the closure pressure exceeds 41 MPa. This indicates that self-proppant fractures cannot provide sufficient conductivity under the condition of high closure pressure.

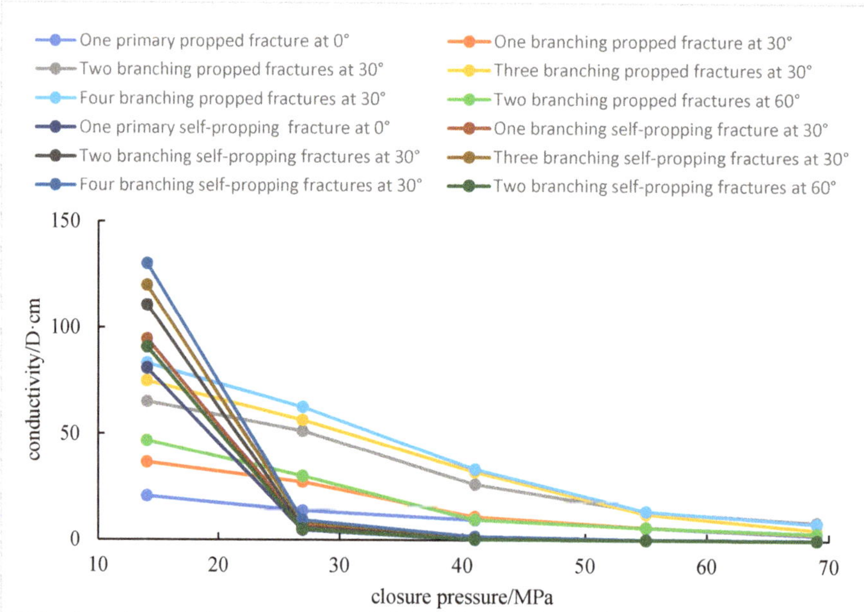


**Figure 8.** Comparison of conductivity between propped fractures and self-propping fractures in different fracture network structures.

In the multiple branch seam structure, with the increase in the number of joints, the overall conductivity is also enhanced, because the fluid can flow through the joints and enhance the conductivity. When the angle between the main and secondary joints decreases, the flow resistance decreases and the conductivity increases as the angle between the main and secondary joints decreases.

## 4. Conclusions and Recommendations

Previous research endeavors on fracture conductivity, particularly pertaining to varied fracture network architectures, have predominantly concentrated on individual factors, for instance, the angle or the count of branch fractures. In contrast, this study undertakes an experimental investigation into the fracture conductivity of two distinct fracture types:

proppant-supported fractures and slip self-supported fractures. Furthermore, it delves into the variation in fracture conductivity across varying levels of closure pressure. The influence of fracture network shape on fracture conductivity is analyzed, so as to provide reference for the optimization of fracture network fracturing parameters.

(1) As closure stress escalates, the conductivity of self-supported fractures significantly diminishes, whereas the conductivity of proppant fractures experiences a more gradual reduction. At low closure pressures, self-supported fractures demonstrate superior conductivity compared to proppant fractures; however, at medium to high closure pressures, proppant fractures exhibit greater conductivity than their self-supported counterparts.
(2) The flow-conducting capacity of cracks with multiple branch seams increases with the number of branch seams, and the branch seams have a significant effect on enhancing the flow-conducting capacity at low and medium closure pressures, while the effect of branch seams on enhancing the flow-conducting capacity is small at high closure pressures.
(3) When the closure pressure and the quantity of branch seams are identical, the inflow capacity is influenced by the angle of the branch seams. Specifically, an increase in the angle of the branch seams correlates with a decrease in the inflow capacity of the fractures. This is attributable to the fact that a larger angle between the secondary seams and the primary seams results in greater resistance to fluid flow, thereby reducing the inflow capacity.
(4) Under the condition of low closure pressure, increasing the number of self-supporting fractures can be considered by increasing the amount of slippery water and optimizing the sand-adding parameters. Under the condition of high closure pressure, attention should be paid to optimizing the proppant placement by means of full-scale sand-spreading, so that all the fractures can develop effective fracture conductivity.
(5) In order to increase the number of branch joints, more complex joints should be adopted in the process of reconstruction. At the same time, the angle between branch joints and main joints can be reduced by means of temporary blocking and turning, so as to improve the overall flow capacity.

**Author Contributions:** Investigation, J.C.; methodology, L.W.; project administration, L.W.; writing—original draft, F.X.; revising—original draft, Y.Z. All authors have read and agreed to the published version of the manuscript.

**Funding:** This study was funded by the SINOPEC Science and Technology Research Project 'Key Technology of Gas Accumulation Fracturing in Lake-phase Shale', Grant number P19025-1.

**Data Availability Statement:** The original contributions presented in the study are included in the article; further inquiries can be directed to the corresponding author.

**Conflicts of Interest:** Author Jianfa Ci is employed by the Sinopec company. The remaining authors declare that the research was conducted in the absence of any commercial or financial relationships that could be construed as potential conflicts of interest.

## References

1. Zheng, J.W.; Sun, D.Q.; Li, X.Y.; Zhang, J. Advances in Exploration and Exploitation Technologies of Shale Gas. *Nat. Gas Geosci.* **2011**, *22*, 511–517.
2. Liang, S.; Wan-Yi, Z.; Xiao-Ling, L. The Key Reservoir Fracturing Transformation Technology and Development Trend about the Shale Gas. *Geol. Sci. Technol. Inf.* **2013**, *32*, 156–162.
3. Nie, H.; Zhang, J. Types and characteristics of shale gas reservoir: A case study of Lower Paleozoic in and around Sichuan Basin. *Pet. Geol. Exp.* **2011**, *33*, 219–250.
4. Mingyue, C.U.; Yuzhang, L.I.; Nailing, X.I.; Yunhong, D.I.; Xuemei, Y.A.; Jingjing, D.O.; Haifeng, F.U. Analysis of factors affecting the formation of effective stimulated reservoir volume (ESRV). *Oil Drill. Prod. Technol.* **2014**, *36*, 82–87.
5. Huifen, H.; Junliang, P.; Jian, W. Test method for long-term flow conductivity of propped shale fractures and modifications made on test equipment. *Drill. Prod. Technol.* **2018**, *41*, 55–58.

6. Sun, C. Study on the Decline Characteristics and Influencing Factors of Hydraulic Fractures Conductivity in Shale Reservoirs. Master's Thesis, Xi'an Shiyou University, Xi'an, China, 2023.
7. Fredd, C.N.; McConnell, S.B.; Boney, C.L.; England, K.W. Experimental study of fracture conductivity for water-fracturing and conventional fracturing applications. *SPE J.* **2001**, *6*, 288–298. [CrossRef]
8. Zhang, J.; Kamenov, A.; Hill, A.D.; Zhu, D. Laboratory Measurement of Hydraulic-Fracture Conductivities in the Barnett Shale. *SPE Prod. Oper.* **2014**, *29*, 216–227. [CrossRef]
9. Cipolla, C.L.; Lolon, E.P.; Mayerhofer, M.J.; Warpinski, N.R. Fracture Design Considerations in Horizontal Wells Drilled in Unconventional Gas Reservoirs. In Proceedings of the SPE Hydraulic Fracturing Technology Conference, The Woodlands, TX, USA, 19–21 January 2009.
10. Gao, S. Study on Conductivity of Different Artificial Fracture Types in Shale. Master's Thesis, University of Petroleum, Beijing, China, 2023.
11. Liu, J.; Ning, W.; Wang, P. Influence of the Fracture Intersection Angle on Conductivity of Fracture Network: A Case Study of "Bu" Type Structure. *Sci. Technol. Eng.* **2023**, *23*, 15465–15472.
12. Ning, W. Research on Conductivity of Shale "Y"-Shaped cross Fractures. Master's Thesis, Sichuan University, Chengdu, China, 2021.
13. Cao, H.T.; Zhan, G.W.; Zhao, Y.; Yu, X.Q. Comparative of conductivity between support fracture and self-support fracture of deep shale reservoir in southern Sichuan basin. *Sci. Technol. Eng.* **2019**, *19*, 164–169.
14. Zhu, H.; Shen, J.; Zhou, H. Numerical simulation on propped fracture conductivity. *Acta Pet. Sin.* **2018**, *39*, 1410.
15. Qingzhi, W.; Shuting, W.; Jinjian, G.; Xiaofei, D.; Feng, W.; Xinjia, L.; Liu, Y. Research on flow conductivity experiment in complex fracture network. *Pet. Geol. Recovery Effic.* **2016**, *23*, 116–121.
16. Zhang, H. Conductivity optimization research of complex network fractures in shale reservoirs of Long maxi Formation in Sichuan Basin. *Nat. Gas Geosci.* **2019**, *30*, 955–962.
17. Yan, G.; Shi, H.; Ma, Y.; Scheuermann, A.; Li, L. Intrinsic permeabilities of transparent soil under various aqueous environmental conditions. *Géotechnique Lett.* **2022**, *12*, 225–231. [CrossRef]

processes

MDPI

Article

# Study on Optimization of Stimulation Technology of Heterogeneous Porous Carbonate Reservoir

**Kangjia Zhao [1,2,3], Hualei Xu [1,2,3,*], Jie Wang [1,2,3,*], Houshun Jiang [1,2,3] and Liangjun Zhang [1,2,3]**

1 Cooperative Innovation Center of Unconventional Oil and Gas, Yangtze University, Wuhan 430100, China; 2022710305@yangtzeu.edu.cn (K.Z.); jhs_hust@yangtzeu.edu.cn (H.J.); zhangliangjun0320@163.com (L.Z.)
2 Hubei Key Laboratory of Drilling and Production Engineering for Oil and Gas, Yangtze University, Wuhan 430100, China
3 State Key Laboratory of Petroleum Resources and Prospecting, China University of Petroleum, Beijing 102206, China
* Correspondence: hualeixu@yangtzeu.edu.cn (H.X.); wangjie@yangtzeu.edu.cn (J.W.); Tel.: +86-136-4716-0273 (H.X.); +86-136-4546-9808 (J.W.)

**Abstract:** Mishrif (M) reservoir of Faihaa (F) oilfield in Iraq is a heterogeneous porous carbonate reservoir. The reservoir properties of each reservoir unit differ greatly, and the distribution of porosity and permeability is non-uniform. Some reservoir units have the problem that the expected production cannot be achieved or the production decline rate is too fast after matrix acidification. This work optimized and compared the process of acid fracturing and hydraulic fracturing techniques. The Mishrif B (MB) and Mishrif C (MC) layers are selected as the target units for fracturing and the perforation intervals are optimized. The acid fracturing process adopted the acid fracturing technology of guar gum pad fluid and gelled acid multi-stage injection. According to the wellhead pressure limit and fracture propagation geometry, the pumping rate is optimized. The recommended maximum pumping rate of acid fracturing is 5.0 $m^3$/min, and the optimized acid volume is 256.4 $m^3$. The pressure changes during hydraulic fracturing and acid fracturing are different. It is recommended that the maximum hydraulic fracturing pump rate is 4.5 $m^3$/min for MB and MC layers, and the amount of proppant in MB and MC layers is 37.5 $m^3$ and 43.7 $m^3$, respectively. The production prediction of two optimized processes is carried out. The results showed that the effect of acid fracturing in MB and MC layers is better than hydraulic fracturing, and it is recommended to adopt acid fracturing technology to stimulate MB and MC layers. Acid fracturing operation is carried out in the X-13 well, and better application results are achieved. The results of this study provide optimized reference ideas for reservoir stimulation in heterogeneous porous reservoirs.

**Keywords:** porous carbonate reservoir; acid fracturing; multi-stage injection; hydraulic fracturing; process optimization

## 1. Introduction

It is estimated that more than 60% of the world's oil reserves are contained in carbonate reservoirs [1]. Taking the Middle East as an example, the region is dominated by carbonate reservoirs, with about 70% of oil and 90% of nsatural gas distributed in these reservoirs [2–4]. However, the non-uniform porosity and permeability distribution and strong heterogeneity of the reservoir are the main factors affecting the stimulation effect of carbonate reservoirs [5]. Matrix acidification has always been the main stimulation technology for eliminating near-well damage and restoring natural productivity in carbonate reservoirs [6–8]. The carbonate reservoir is affected by sedimentation and diagenesis during the formation process, which shows strong heterogeneity. General acidification is often unable to effectively stimulate low-permeability areas, resulting in the improvement of single well production when matrix acidification is not obvious [9]. The diverting agent is added to the acid fluid system to block the high-permeability layer and make the acid fluid flow to the low-permeability area [10–12]. Due to the small scale of matrix acidification and the limited flow range of acid fluid, it cannot be further improved in the reservoir, and only the matrix near the wellbore has a good improvement effect. At present, some reservoir units of the M reservoir in the F oilfield are unable to achieve the expected effect of increasing production after acidification. On the basis of fully understanding the reservoir characteristics and stimulation potential, the field stimulation process optimization is carried out to explore the effective method of low-permeability reservoir stimulation in the Middle East M reservoir.

Carbonate reservoirs in the Middle East have high calcite content, so acid treatment is generally used to improve the porosity and permeability distribution [13]. The common treatment method for low-permeability carbonate reservoirs is acid fracturing. The evaluation indexes of acid fracturing effect mainly include acid etched fracture length and acid etched fracture conductivity [14]. The M carbonate reservoir has strong heterogeneity and mineral homogeneity. Reducing acid fluid filtration and acid rock reaction rate can effectively increase fracture length and conductivity. Most of the existing studies have evaluated the adaptability and process optimization of acid fracturing operations from two aspects: acid fluid combination [15] and acid fluid systems [16]. By adding a variety of additives to the acid fluid, the rheology of the acid fluid system is changed, and the liquid filtration and acid–rock reaction rate are reduced. Liu et al. [17] experimentally tested the reaction rates of three different acid fluid systems: conventional acid, diverting acid, and gelled acid. The experimental results show that the gelled acid with better reaction rate delay effect and lower friction can produce longer acid etched fractures and conductivity. Wang et al. [18] prepared a low-viscosity microlactic acid suitable for the field based on the characteristics of carbonate reservoirs in the Middle East. Compared with hydrochloric acid, emulsified acid, and surface-active acid by rotating disk and core displacement experiments, it is proved that the acid has the ability to etch the surface of the fracture to form high-conductivity fractures and has good field applicability. The above research focuses on the optimization of acid performance, and there are relatively few studies on acid fracturing treatment parameter optimization. In the process of acid fracturing, the influence of treatment parameters on the length and conductivity of acid etched fractures is very important.

In the process of carbonate reservoir stimulation, Iraq's Halfaya oilfield has successfully adopted hydraulic fracturing operation and achieved good stimulation effect [19,20]. This provides a new idea for optimizing the stimulation process of carbonate reservoirs in the Middle East. The hydraulic fracturing process of carbonate reservoirs is affected by many factors, which makes the fracture propagation mechanism complex [21,22], and it is difficult to control fracture propagation. Liu et al. [23] modeled the distribution of natural fractures and cavities in carbonate reservoirs using the element partition method to study the interaction law of hydraulic fractures and cavities under different injection conditions. In order to clarify the mechanism of hydraulic fracture extension in carbonate reservoirs, some scholars have used the extended finite element method to establish a fully coupled model to study the effects of natural fractures and cavities on fracture extension in carbonate reservoirs [24,25].

At present, most of the studies on carbonate reservoir stimulation are related to fracture extension. Fewer scholars optimize the specific parameters in the operation process. The aim of this paper is to establish a set of stimulation technology optimization processes for heterogeneous porous carbonate reservoirs. We use numerical simulation to optimize the operation parameters of acid fracturing and hydraulic fracturing. In the optimization of acid fracturing process parameters, we will take the rate of change in dimensionless conductivity as an evaluation index. The final choice of acid fracturing or hydraulic fracturing technology will be decided based on the production change after stimulation. The field application further verifies the accuracy of the optimization results.

## 2. Reservoir Characteristics

The M reservoir is the most important bearing carbonate reservoir in the Mesopotamian basin in southeastern Iraq [26,27]. The oil reserves of the reservoir account for 30% in Iraq's total oil reserves. The M reservoir depth in the F oilfield is about 2800 m, and the continuity of reservoir thickness is good, with an overall thickness of about 300 m. The reservoir is a porous carbonate reservoir, which is characterized by medium porosity ($\Phi$ = 4~10%) and low permeability (Kg = 0.01~25 mD), with wide distribution of porosity and permeability. Figure 1 shows the diversity of pore types, mainly composed of moldic pores, interparticle pores, and dissolution pores, with a small amount of microfractures. Although the M reservoir contains many pore types in the vertical direction, the pores are filled with microcrystalline calcite cement, resulting in poor connectivity between pores and low reservoir permeability. According to the analysis of reservoir minerals, the M reservoir is mainly composed of calcite, as the content of calcite is more than 95%, containing a small amount of dolomite, quartz, etc., and the content of clay minerals is less than 5%. Due to the high content of calcite, additives such as corrosion inhibitors, diverting agents, and gelling agents are often added during the acid treatment process to slow down the acid–rock reaction rate, increasing the effective distance of the live acid and improving the acid etched fracture conductivity [28].

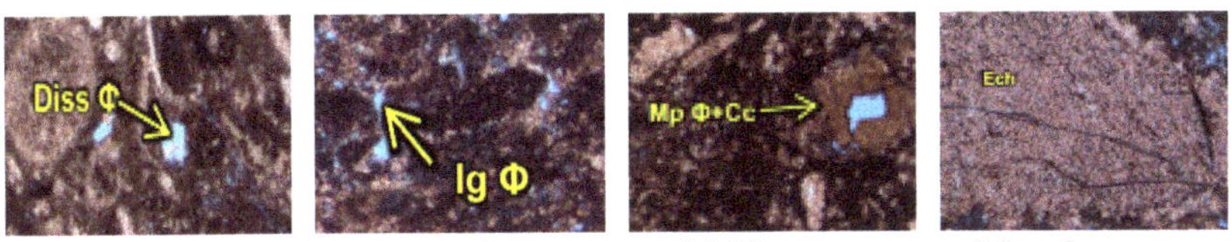


Dissolution pore    Interparticle pore    Moldic pore    Microfractures

**Figure 1.** Pore types of cast thin sections in M reservoir of F oilfield.

Based on GR/Sonic/DEN logging response, the M reservoir in the Faihaa (F) oilfield can be divided into four reservoir units: Mishrif B (MB), Mishrif C (MC), Mishrif D (MD), and Mishrif E (ME), as shown in Table 1. The thickness of the MB layer is the largest and the MC layer is the smallest. The reservoir reserves of the MB and MC layers account for about 70% of the total reserves of the M reservoir, the MD layer accounts for 23%, and the ME layer accounts for 6%. The physical properties of the four reservoir units are quite different. In the MB layer, there is only a thin stable layer in the shallower section, and the pore permeability distribution is poor in the deeper position. The results of a triaxial stress compression test show that the overall performance of the M reservoir includes a low Young's modulus and medium Poisson's ratio. The fracture initiation and propagation mechanism in the reservoir is complex, and it easily forms short fractures.

**Table 1.** Core rock mechanics test results of different reservoir units.

| Layer | Depth/m | Compressive Strength/MPa | Young's Modulus/MPa | Poisson's Ratio | Vertical Stress/MPa | Horizontal Stress/MPa | |
|---|---|---|---|---|---|---|---|
| | | | | | | Maximum | Minimum |
| MB | 2860.55 | 185.61 | $18.81 \times 10^3$ | 0.26 | 75.76 | 71.63 | 66.69 |
| MC | 2915.21 | 117.63 | $22.38 \times 10^3$ | 0.24 | 77.27 | 72.15 | 67.21 |
| MD | 2958.98 | 223.51 | $16.22 \times 10^3$ | 0.23 | 78.46 | 72.81 | 67.02 |
| ME | 3024.50 | 94.20 | $16.33 \times 10^3$ | 0.23 | 80.00 | 75.04 | 69.87 |

## 3. Selection of Target Layers

Reasonable selection and evaluation of the treatment interval are critical steps in the success of fracturing. From the perspective of the overall reservoir physical properties of the M reservoir, the overall gas permeability of MB, MC, MD, and ME layers is about 1.95, 4.31, 10.31, and 15.6 mD, respectively. On the whole, the permeability of MB and

MC layers is lower than that of MD and ME layers. Based on stimulation history analysis, the MD and ME layers of the X-5 well are acidified after drilling and completion. The production of MD and ME layers reached a good level after acidification, and the stable yield effect is better. In the X-10 well, the production of MB and MC layers cannot reach the expected level after acidification, and the production decreases rapidly. It is believed that the low-permeability layer cannot be effectively utilized. Therefore, it is recommended to use fracturing stimulation for MB and MC layers in low-permeability areas. For MD and ME layers with good permeability distribution, acidification is still used to stimulate.

In addition, in the edge area of the reservoir structure, edge and bottom water are developed. For the M reservoir, there are four reservoir units from top to bottom: MB, MC, MD, and ME layers (Figure 2). The crude oil is marked as green and the water is marked as blue. From the vertical distribution position of the reservoir unit, the ME layer is located at the bottom and close to the edge and bottom water. When the ME layer is stimulated by fracturing, there is a risk of fractures communicating with edge and bottom water, resulting in wellbore flooding. Therefore, acid fracturing or hydraulic fracturing is not recommended for the ME layer. In summary, MB and MC are the target layers for reservoir stimulation. The above is the process of selecting the target layers for the Mishrif reservoir. Figure 3 shows the detailed flow chart of the target layers selection, which contains three parties.

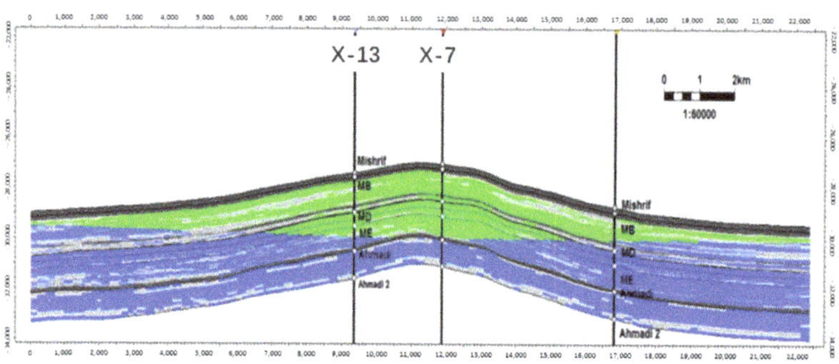


**Figure 2.** Distribution of M reservoir layers (green—crude oil, blue—water).

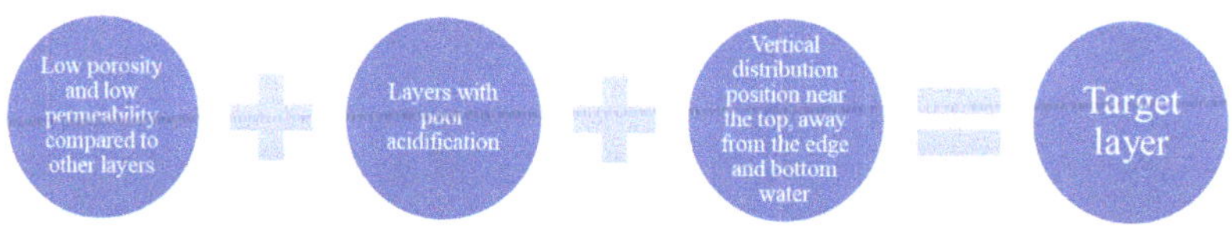


**Figure 3.** Target layer selection process.

## 4. Optimization of Stimulation Process Parameters

Well X-13 is selected as a test well for acid fracturing and hydraulic fracturing. The well is a new drilling well, and the target layers of MB and MC are not formally put into production. The reservoir thickness of the MB layer is 102 m, the average porosity is 15%, and the permeability is 0.9 mD. The minimum stress is 44.5~46.5 MPa, and the API weight of crude oil is about 21.5. The reservoir thickness of the MC layer is 39 m, the average porosity is 17%, and the permeability is 3.52 mD. The minimum principal stress is 44~46 MPa, and the API weight of crude oil is about 20. In the acid fracturing operation of carbonate reservoirs in the Middle East, viscous pad fluid acid fracturing technology is often used to effectively improve the conductivity of acid etched fractures. The effect of the pad fluid is to generate sufficient scale fracture geometry, reduce the temperature of the fracture wall surface, reduce the acid filtration, and form a longer action distance. Subsequently, acid fluid is injected into the formation. Because the viscosity of the pad fluid is higher than that of the acid fluid, the acid fluid will produce "fingering" during the displacement of the fracturing fluid, which helps the acid fluid to carry out non-uniform etching on the fracture wall [29,30]. Alternate injection of pad fluid and acid fluid will reduce the filtration rate of subsequent injection of acid fluid. At the same time, multiple "fingering" is formed in the displacement stage, which increases the action distance of acid fluid and is more conducive to the formation of longer and higher-conductivity acid etched fractures. Figure 4 shows the mechanism of etching the fracture surface by retarded acid (gelled acid) during acid fracturing. This can effectively control the acid filtration, so that the acid fluid fully etches the fracture surface to form a non-uniform etching cake to prevent fracture closure. This acid fracturing operation adopts a multi-stage injection acid fracturing process.

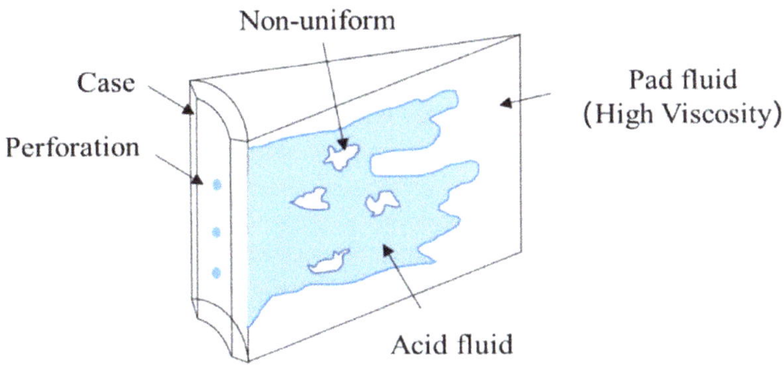


**Figure 4.** Acid fracturing process of viscous pad fluid [30].

In carbonate oil and gas reservoirs, the main minerals are calcium carbonate and calcium magnesium carbonate. Carbonate oil and gas reservoirs are usually acidified with hydrochloric acid, and their acid–rock reaction equations are:

$$2HCl + CaCO_3 = CaCl_2 + CO_2\uparrow + H_2O \tag{1}$$

$$4HCl + MgCa(CO_3)_2 = CaCl_2 + MgCl_2 + 2CO_2\uparrow + 2H_2O \tag{2}$$

The rate of acid–rock reaction can be expressed by the reduction of acid concentration per unit of time, according to the law of mass action: when the temperature and pressure are constant, the rate of chemical reaction is proportional to the product of the reactant concentration to the power of n. Since the acid–rock reaction is a complex-phase reaction, the rock reactant concentration can be regarded as a constant value. Therefore, the acid–rock reaction rate can be expressed as:

$$J = KC_w{}^n \tag{3}$$

where $J$ is the reaction rate, mol/s·cm$^2$; $K$ is the reaction rate constant (mol·L)$^{-n}$·mol/(cm$^2$·s); $C_w$ is the acid concentration, mol·L$^{-1}$; $n$ is the reaction order.

The effects of temperature on the proppant reaction rate constant $K$ are described by the Arrhenius equation:

$$K = K_0 \exp[E_a - (RT)] \tag{4}$$

where $K_0$ is the pre-exponential factor, which is the reaction rate constant at an infinitely high temperature, (mol/L)$^{-n}$·L/(cm$^2$·s); $E_a$ is the activation energy, J/mol; $R$ is the gas constant; J/mol·K; $T$ is the absolute temperature, K.

The fracture conductivity principles of acid fracturing and hydraulic fracturing are different. During the acid fracturing process, the acid fluid etches the surface of the fracture to form non-uniform corrosion, which can support the fracture to prevent closure, form an oil flow channel, and achieve the effect of increasing the fracture conductivity. Hydraulic sand fracturing prevents the fracture from closing by injecting proppant fillers into the fracture. Figures 5 and 6 show the two different processes.

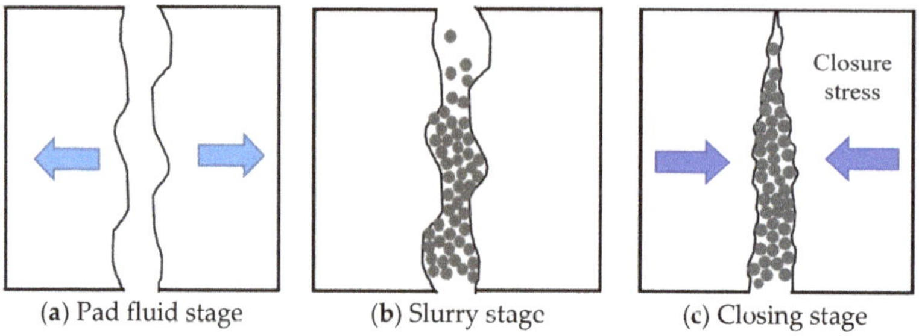


**Figure 5.** Hydraulic fracturing fracture generation process.

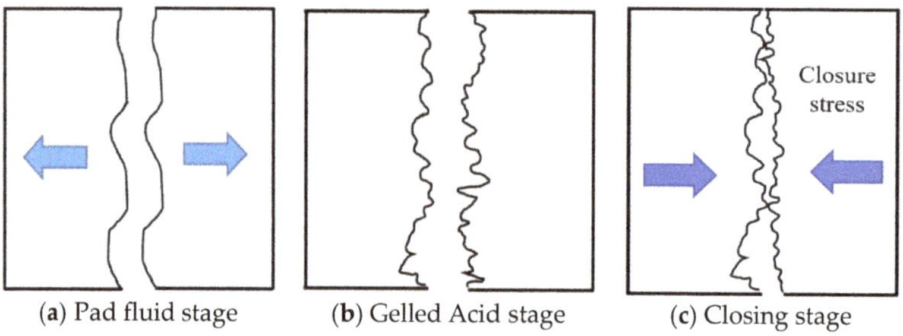


**Figure 6.** Acid fracturing fracture generation process.

The proppant transport modeling during hydraulic fracturing is based on computational fluid dynamics methods. The fluid/proppant flow model simulates the flow of fluid and proppant in a two-dimensional vertical fracture. The properties of the fluid and proppant are allowed to vary along the fracture length and in its height. The width can vary spatially and temporally.

The mass conservation equation for the fracturing fluid is:

$$\frac{\partial m_l}{\partial t} + \frac{m_l}{V}\frac{\partial V}{\partial t} + \frac{1}{W}\left(\frac{\partial}{\partial x}m_l u_l W + \frac{\partial}{\partial z}m_l w_l W\right) = \rho_l(q_{lI} - q_{lL}) \quad (5)$$

and for each proppant it is:

$$\frac{\partial m_p}{\partial t} + \frac{m_p}{V}\frac{\partial V}{\partial t} + \frac{1}{W}\left(\frac{\partial}{\partial x}m_p u_p W + \frac{\partial}{\partial z}m_p w_p W\right) = \rho_p q_{pI} \quad (6)$$

where $m_l$ is the local liquid mass per unit fracture volume, kg; $m_p$ is the proppant mass per unit volume, kg; $V$ is the fracture volume, m$^3$; $W$ is the fracture width, m; $u_l$ and $w_l$ are the liquid velocities in the x- and z-directions, m/s; $u_p$ and $w_p$ are the proppant velocities in the x- and z-directions, m/s; $\rho_l$ and $\rho_p$ are the material densities of the liquid and proppant, respectively, kg/m$^3$; $q_{lI}$ is the volume injection rate of clear liquid, m$^3$/min; $q_{lL}$ is the volume loss rate of liquid due to leakoff, m$^3$/min; $q_{pI}$ is the volume injection rate of proppant, m$^3$/min.

The momentum conservation of the liquid in the x-direction is given by:

$$m_l\frac{\partial u_l}{\partial t} = -\alpha_l\frac{\partial p}{\partial x} - \alpha_l\frac{\partial \tau}{\partial x} - \sum_i C_{d_i}(u_l - u_{p_i}) - m_l g_x \quad (7)$$

The momentum equation for the proppant is:

$$m_{pi}\frac{\partial u_{pi}}{\partial t} = -\alpha_{pi}\frac{\partial p}{\partial x} - \alpha_{pi}\frac{\partial \tau}{\partial x} - \sum_i C_{di}(u_{pi} - u_l) - m_{pi}g_x \tag{8}$$

where $p$ is the pressure, MPa; $\tau$ represents the wall friction stress, MPa; $C_d$ is the drag coefficient between the liquid and proppant, $g_x$ is the acceleration due to gravity in the x-direction, m/s$^2$.

During fracturing, the proppant particles are carried by the fluid and the force of the fluid acting on the particles is expressed as:

$$F = \frac{1}{2}C_d\rho_l u_{rel}^2 A \tag{9}$$

where $u_{rel}$ is the relative velocity between the fluid and particle, m/s; $A$ is the frontal area of the particle, m$^2$.

Friction on the walls of the fracture can lead to a decrease in pressure within the fracture. The pressure drop arising from the shear stress at the wall can be related to the friction factor:

$$\frac{\Delta p}{\Delta x} = \frac{1}{2}f\rho_l u^2\frac{1}{D} \tag{10}$$

where $f$ is friction factor.

The resultant friction factor $f_p$ is modeled as:

$$f_p = f \times CF \tag{11}$$

The proppant multiplier factor is:

$$CF = \mu_r^{0.55}\rho_r^{0.45} \tag{12}$$

where $\mu_r$ is dimensionless relative slurry viscosity, Pa·s; $\rho_r$ is relative slurry density, kg/m$^3$.

### 4.1. Perforation Intervals

In acid fracturing, the selection of perforation position has a great influence on the scale of the fracture. The following aspects need to be considered in the selection process of perforating well section: ① in the selection process of perforation section, the interval with lower in situ stress should be selected, which is conducive to fracture initiation. ② Considering the distribution of crude oil in the reservoir, the position with high oil abundance should be selected as far as possible for perforation to ensure the maximum economic benefit of production increase. ③ The perforation length is controlled between 10 and 15 m. If the perforation length is too long, the stress distribution of the reservoir will be destroyed, which will easily cause the fracture height to be out of control and communicate with the upper and lower adjacent layers.

Based on the above criteria, the perforation position is adjusted and optimized by the fracturing simulation software, and the perforation interval of the MB layer is finally determined to be 2868.0~2883.0 m, and the perforation interval of the MC layer is 2913.0~2923.0 m, as shown in Figure 7. Perforation at the upper perforation position is conducive to fracture initiation and can effectively control the fracture height to achieve a better fracture scale.

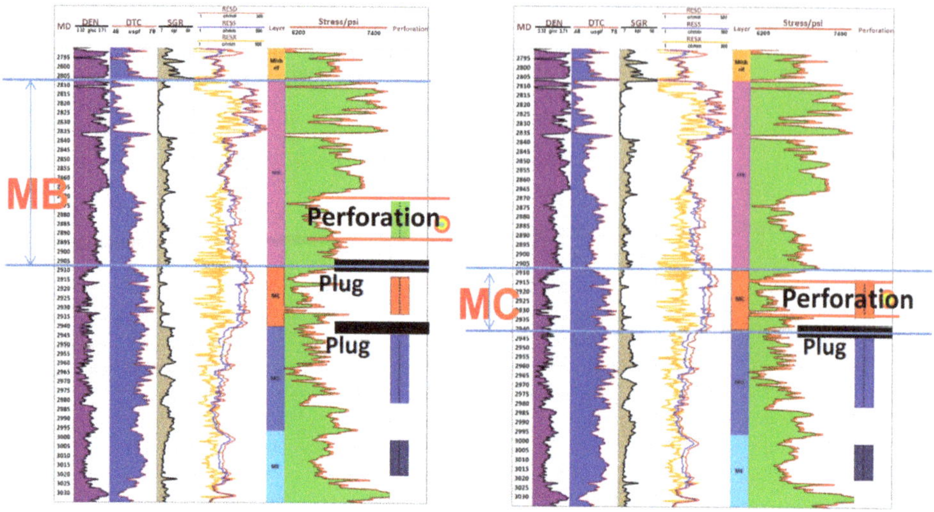


**Figure 7.** Perforation interval optimization.

### 4.2. *Optimization of Acid Fracturing Process Parameters*

The acid fluid system and fracturing fluid system should be determined before the optimization. In this optimization, gelled acid is used as the main acid fluid [31,32], and guar gum fracturing fluid is used as the pad fluid. The volume ratio of fracturing fluid and gelled acid is 1:2. After determining the fluid system and fluid ratio for treatment, the optimization of acid fracturing process parameters mainly considers the optimization of treatment pumping rate and acid volume. Before the fracturing design, it is often necessary to combine the well structure and wellhead device to determine the reasonable treatment pressure range. The reasonable pumping rate range is determined by injection pressure prediction. Next, Fracpro PT (CARBO Ceramics Inc., Houston, TX, USA. Version: 10.10.13) is used for optimization simulation of treatment parameters.

#### 4.2.1. Pumping Rate

The wellhead device pressure is limited to 55.16 MPa during field treatment. The pressure prediction in Table 2 shows that the surface pressure will be close to 55.16 MPa when the pumping rate is 5 $m^3$/min. Therefore, it is recommended that the reasonable treatment pumping rate is 5 $m^3$/min. However, the fracturing pumping rate of the same reservoir in adjacent blocks in the Middle East is up to 6 $m^3$/min. In order to ensure the comprehensive optimization of treatment parameters, the pumping rate optimization range is set to 3~7 $m^3$/min. The change trend of fracture geometry and fracture conductivity in the simulation process is shown in Figures 8 and 9.

**Table 2.** Surface pressure and horsepower forecast.

| Breakdown Pressure Gradient (MPa/m) | 2900 m Downhole Pressure Prediction (MPa) | Wellhead Pressure with Different Pumping Rate (MPa) | | |
|---|---|---|---|---|
| | | 3 $m^3$/min | 4 $m^3$/min | 5 $m^3$/min |
| 0.017 | 49.39 | 31.99 | 40.69 | 46.49 |
| 0.018 | 52.19 | 34.79 | 43.49 | 49.29 |
| 0.019 | 55.19 | 37.79 | 46.49 | 52.28 |
| 1000 m ID2.75″ tubing friction gradient estimation, MPa | | 4.00 | 7.00 | 9 |
| 2900 m total tubing friction, MPa | | 11.60 | 20.30 | 26.10 |
| 2900 m hydrostatic pressure, MPa | | 29.00 | 29.00 | 29.00 |
| Wellhead device pressure limitation, MPa | | | 55.16 | |

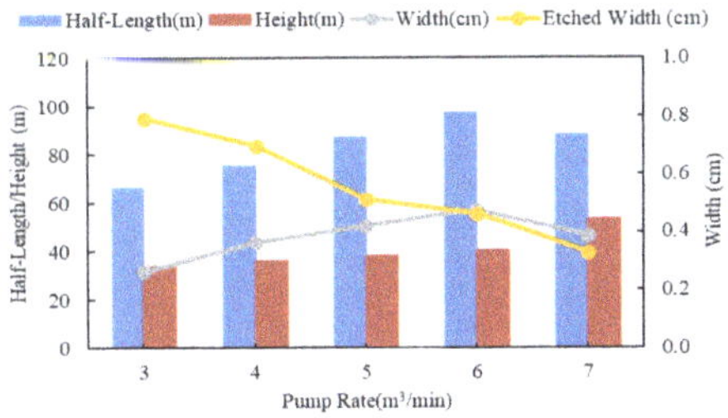


**Figure 8.** Relationship between acid etched fracture geometry and pump rate (MB layer).

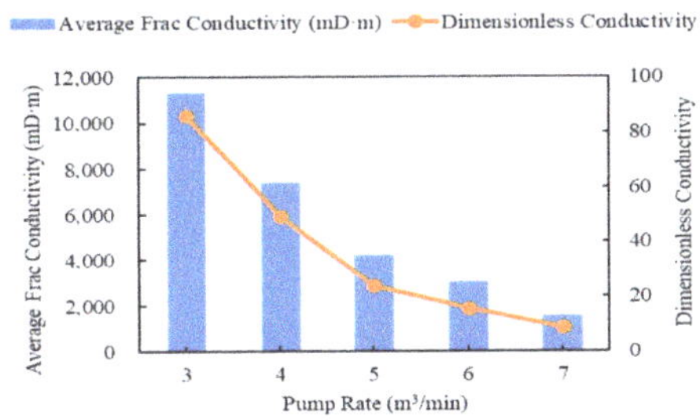


**Figure 9.** Relationship between acid etched fracture conductivity and pump rate (MB layer).

The fracture width in Figure 8 refers to the fracture width of the pad fluid stage. The acid etched fracture width is the fracture width after the closure. MB layer simulation results show that when the pumping rate is less than 6 $m^3/min$, as the pumping rate increases, the fractures mainly propagate along the fracture length direction. Hindered by the stress interlayer, the fracture height development is not obvious. When the pumping rate reaches 7 $m^3/min$, the fracture breaks through the stress barrier, the fracture height breaks through the limit, and the fracture half-length becomes shorter. The change in fracture width is consistent with the change in fracture half-length. Therefore, it is considered that the pumping rate should not exceed the critical value of 6 $m^3/min$. When the pumping rate exceeds the critical value, the risk of interlayer channels increases sharply. In the process of fracturing in low-permeability reservoirs, increasing fracture length is more beneficial to increase production than increasing fracture conductivity. When the pumping rate is 6 $m^3/min$, the maximum surface pressure will reach 62.17 MPa, which is much higher than the pressure limiting condition of the wellhead device. During the treatment process, the pumping rate needs to be appropriately reduced. Therefore, the recommended pumping rate is 5 $m^3/min$, at which time the acid corrosion fracture half-length has reached the expected requirements. The optimization process of pumping rate in the MC layer is basically the same as that in the MB layer. The fracture height of the MC layer is well controlled in the process of changing the pumping rate. The fracture half-length and fracture width of the MC layer increase with the increase in pumping rate, and the fracture height is within the controllable range. The recommended pumping rate of the MC layer is also 5 $m^3/min$. When the pumping rate is 5 $m^3/min$, the surface pressure reaches 51.89 MPa, which is close to the maximum pressure limit of the wellhead device.

The acid etched fracture conductivity is another important index to evaluate the effect of acid fracturing. With the gradual increase in pumping rate, the acid etched fracture conductivity and dimensionless conductivity decrease. In the process of optimization simulation, the total amount of liquid and the amount of acid fluid are unchanged. With the increase in acid pump rate, the fracture length and fracture height gradually increase, and the fracture surface area increases. The contact area between the acid fluid and the fracture surface increases, resulting in a decrease in the etching degree of the acid fluid on the fracture wall. After the pump is stopped, some of the etching cakes on the surface of the fracture cannot withstand the closure stress, resulting in a decrease in the width of the acid etched fracture, which further leads to a decrease in the fracture conductivity. Figure 10 shows the principle that the conductivity of acid etched fracture decreases with the increase in pumping rate.

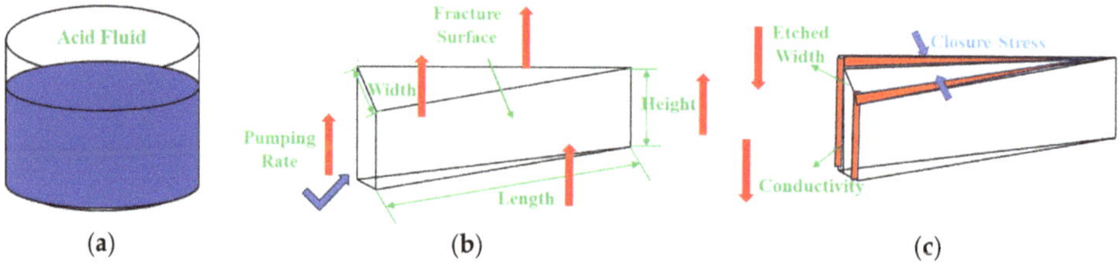


**Figure 10.** Variation of acid etched fracture width and fracture conductivity. (**a**) The amount of acid is constant; (**b**) The fracture size increases with the increase of acid pumping rate; (**c**). The etched width decreases and the fracture conductivity decreases.

4.2.2. Acid Volume

After determining the pumping rate, the reasonable volume of acid fluid is optimized. In the process of acid fracturing, the acid fluid mainly plays the role of etching the fracture surface. The acid fluid reacts with the fracture surface to form a non-uniform etching surface, so that the fracture surface can withstand greater closure stress and increase the fracture conductivity. Figures 11 and 12 show the changes in fracture scale and fracture conductivity in the process of acid fracturing optimization.

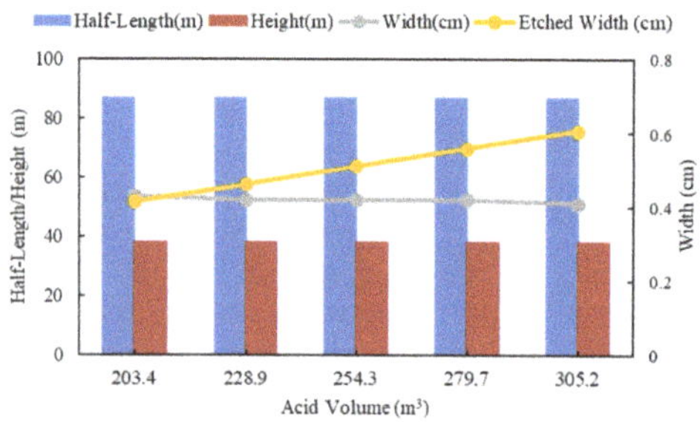


**Figure 11.** Relationship between acid etching fracture geometry and acid volume (MB layer).

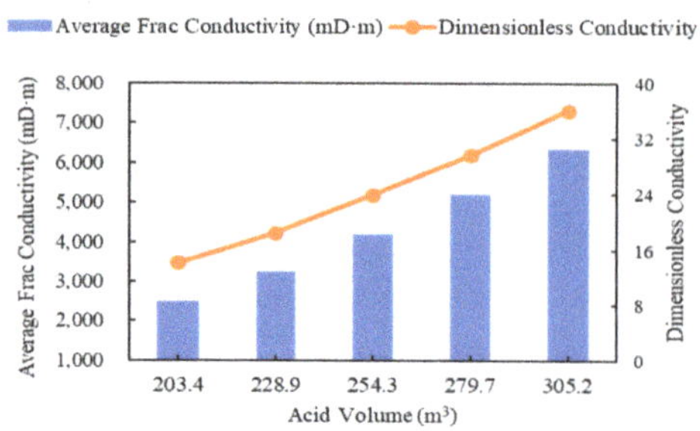


**Figure 12.** Relationship between acid etching fracture conductivity and acid volume (MB layer).

The simulation results (Figure 11) show that under the condition of keeping the amount of pad fluid unchanged and only changing the volume of acid volume, the acid etching fracture half-length, fracture height, and average fracture width mostly do not change. It is considered that the amount of acid fluid has little effect on the fracture scale. The amount of pad fluid mainly affects the fracture scale. After increasing the amount of acid volume, the acid fluid contact per unit fracture area increases. The amount of rock dissolved by acid increases, resulting in a large number of non-uniform convex bodies on the fracture surface, preventing the fracture from closing, increasing the etching fracture width, and improving the conductivity after the fracture is closed. In order to accurately describe the influence of changing the amount of acid fluid on fracture conductivity, a dimensionless conductivity change rate is established to evaluate the contribution of acid to fracture conductivity as shown in Figure 13.

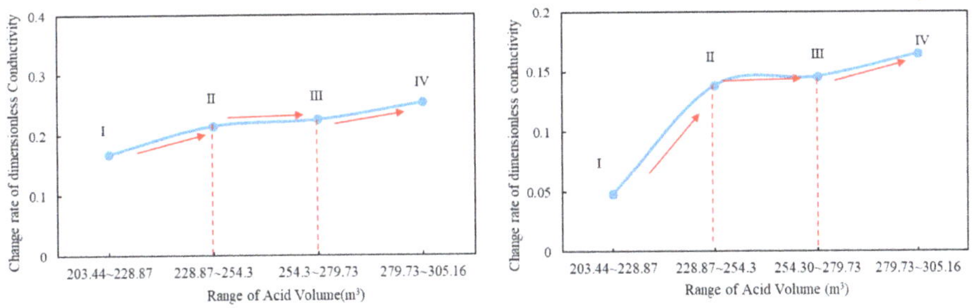


(**a**) (**b**)

**Figure 13.** Dimensionless conductivity change rate curve: (**a**) MB layer, (**b**) MC layer.

The change rate of fracture dimensionless conductivity is defined as:

$$r_{Fcd} = \frac{Fcd_{x_1} - Fcd_{x_2}}{V_{x_1} - V_{x_2}} \tag{13}$$

where $Fcd_{x_1}$ is the dimensionless conductivity when the volume of acid is $x_1$, $Fcd_{x_2}$ is the dimensionless conductivity when the volume of acid is $x_2$, $V_{x_1}$ and $V_{x_2}$ are acid volumes $x_1$ and $x_2$, respectively.

### *4.3. Optimization of Hydraulic Fracturing Process Parameters*

#### 4.3.1. Pumping Rate

During hydraulic fracturing, the proppant carried by the fracturing fluid continuously enters the formation and accumulates, resulting in different pressure changes during hydraulic fracturing and acid fracturing. There are few cases of hydraulic fracturing in carbonate reservoirs. With reference to the treatment parameters of hydraulic fracturing successfully implemented in Tarim oilfield in China and Halfaya oilfield in the Middle East under similar conditions, the maximum pumping rate is no higher than 5.5 $m^3/min$. Combined with the previous prediction of pumping pressure (Table 2), the optimal range of hydraulic fracturing pumping rate is selected as 3.5–5.5 $m^3/min$. The changes in fracture geometry and fracture conductivity under different pumping rates are shown in Figures 14 and 15.

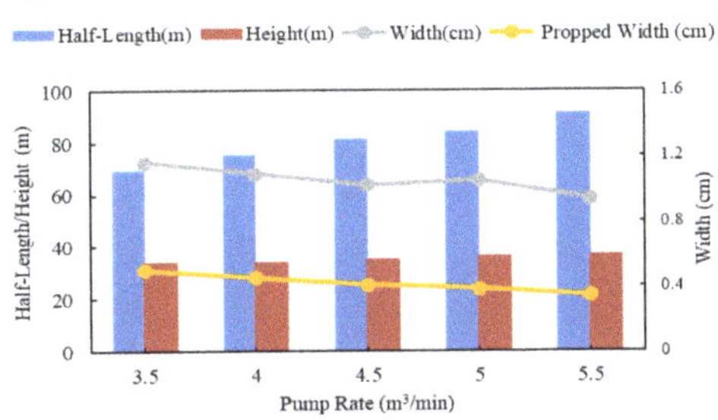


**Figure 14.** Relationship between propped fracture geometry and pump rate (MB layer).

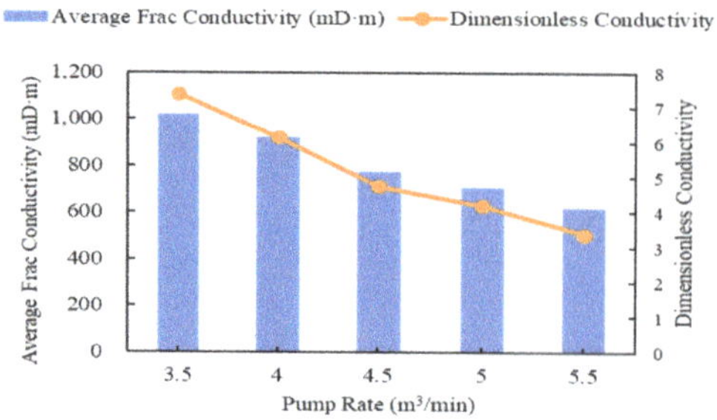


**Figure 15.** Relationship between propped fracture conductivity and pump rate (MB layer).

The simulation results in Figure 14 show that fracture length and fracture height increase as pumping rate increases. The width of the propped fracture tends to decrease under the influence of the fracture height and fracture length. In Figure 15, the fracture conductivity decreases as the pumping rate increases, keeping the total proppant volume constant while the pumping rate is optimized. With the increase in pumping rate, fracture half-length, fracture height, fracture surface area, and fracture volume increase. As the total amount of proppant is constant, the effective proppant placement area increases, which results in a decrease in propped fracture width. The change in fracture conductivity is affected by the width of the propped fracture, so the fracture conductivity also shows a decreasing trend.

Pumping rate also needs to be further combined with treatment pressure changes to optimize. Considering the limited pressure of 55.16 MPa on the wellhead device, when the pumping rate is 4.5 $m^3$/min, the maximum treatment surface pressure on the MB layer is 51.30 MPa, the treatment surface pressure on the MC layer is 53.23 MPa, and the surface pressure is close to 55.16 MPa. Continuing to increase the pumping rate will exceed the wellhead pressure limit and is not recommended. Therefore, it is recommended to have a reasonable pumping rate of 4.5 $m^3$/min for hydraulic fracturing. The optimization process of MC layer pumping rate is similar to that of MB, and it is recommended that the hydraulic fracturing pumping rate of the MC layer is also 4.5 $m^3$/min.

### 4.3.2. Proppant Dosage

During the fracturing process, proppant mainly plays a role in supporting fractures to prevent closure, expanding oil flow channels, and increasing fracture conductivity. In the process of fracturing scale optimization, it is necessary to adjust and optimize the amount of fracturing fluid and proppant. There is often a correlation between fracturing fluid and proppant usage. The two parameters generally need to be adjusted at the same time and cannot be quantitatively analyzed. Therefore, the amount of fracturing fluid and proppant can be adjusted simultaneously by setting the proppant concentration in the fracture during the simulation. Figures 16 and 17 show the change in propped fracture geometry and fracture conductivity as a function of proppant concentration in the fracture.

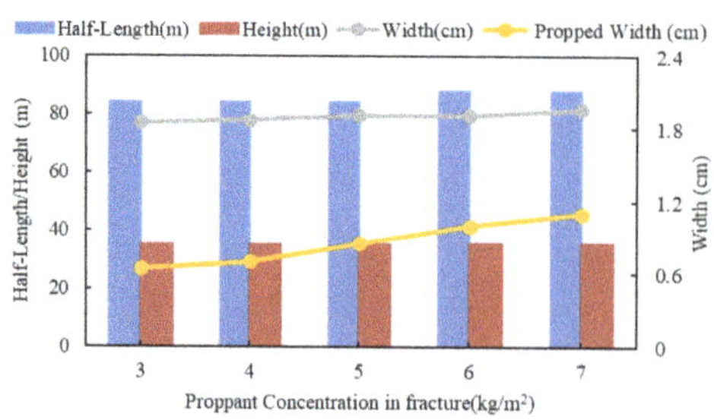


**Figure 16.** Relationship between propped fracture geometry and proppant concentration (MB layer).

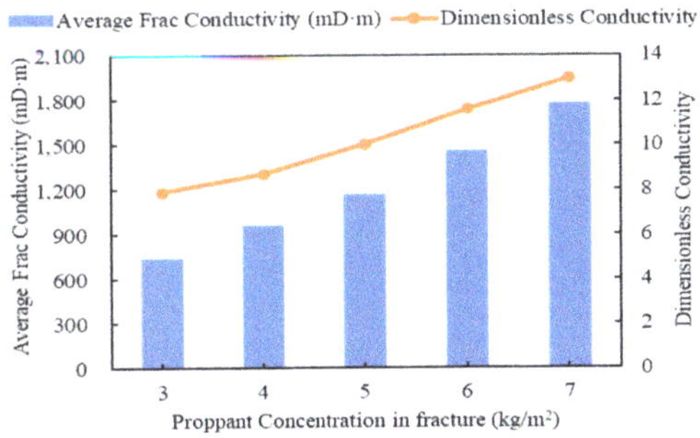


**Figure 17.** Relationship between propped fracture conductivity and proppant concentration (MB layer).

According to the simulation results (Figure 16), the change in proppant dosage has little effect on the fracture half-length and fracture height, while the proppant dosage mainly affects the propped fracture width, that is, the width of the fracture after closure. Increasing the proppant concentration within a certain range means increasing the amount of proppant in the fracture. The proppant fills the fracture sufficiently to support the fracture and prevent it from closing. When the proppant concentration in MB is greater than 4 $kg/m^2$ and the proppant concentration in MC is greater than 5 $kg/m^2$, sand plugging occurs during fracturing. The surface pressure rises sharply after sand plugging. Therefore, the recommended optimal proppant dosages for the MB layer and MC layer are 37.5 $m^3$ and 43.7 $m^3$, respectively.

### *4.4. Comparison of Different Stimulation Measures*

The optimized treatment parameters are simulated for fracture extension, and the simulation results are imported into the reservoir simulator to predict the production changes after different stimulation techniques. Figure 18 shows the comparison of stimulation effects after different stimulation processes in MB and MC layers. The results show that the accumulated oil production of the MB layer after acid fracturing is 18.9~19.7% higher than that of hydraulic fracturing, and that of the MC layer after acid fracturing is 10.9~15.9% higher than that of hydraulic fracturing. After comprehensive analysis, it is concluded that the stimulation effect of acid fracturing is better than that of hydraulic fracturing. The M carbonate reservoir of the F oilfield in Iraq is more suitable for acid fracturing, which is conducive to acid–rock reaction to form acid etched fractures and improve low-permeability conditions. Although hydraulic fracturing can have a certain improvement effect, considering that the porous carbonate reservoir has high filtration loss, sand plugging easily occurs, and the fracture height is not easy to control, which affects the stimulation effect after fracturing. Therefore, hydraulic fracturing is not the preferred stimulation method.

The optimal treatment parameters were imported into the fracture analysis module of Fracpro PT to simulate fracture propagation. For the MB layer, the reservoir pressure is 31.80 MPa, porosity is 15%, water saturation is 22%, the total compressibility is $1.15 \times 10^{-3}$ 1/MPa, and the fluid viscosity is 1.64 mPa·s. For the MC layer, the reservoir pressure is 33.2 MPa, porosity is 20%, water saturation is 23%, the total compressibility is $2.25 \times 10^{-3}$ 1/MPa, and fluid viscosity is 1.51 mPa·s. The rock mechanical parameters of the reservoir are imported into the well logging curve. Figures 19 and 20 show fracture propagation profiles designed for fracturing parameters. Fracture scale meets the requirements and fracture conductivity distribution is good.

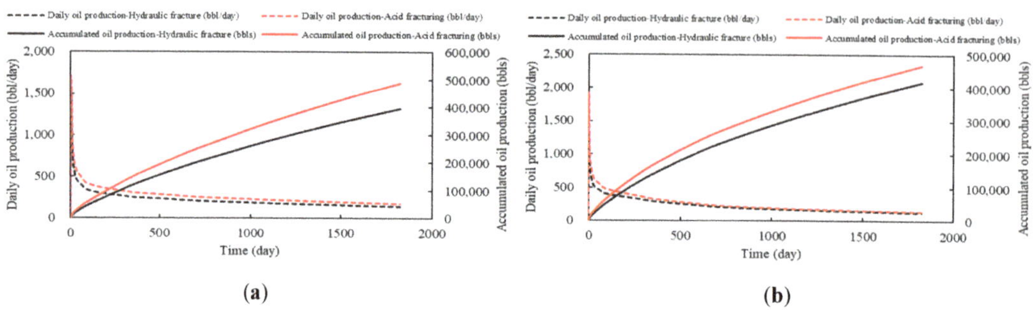


**Figure 18.** Comparison of stimulation effects after different stimulation processes: (**a**) MB layer, (**b**) MC layer.

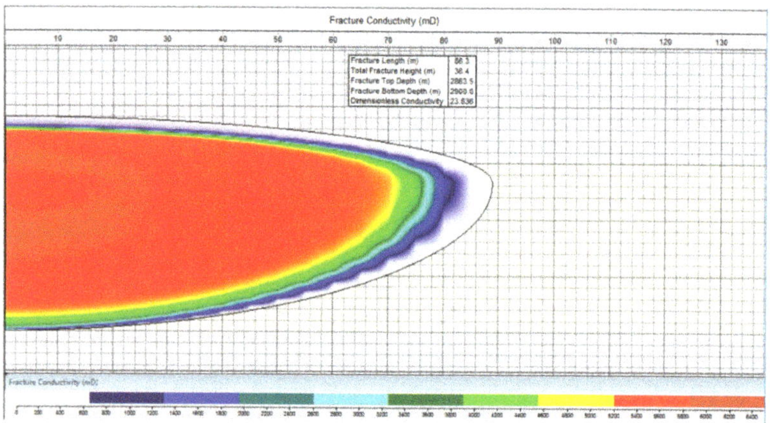


**Figure 19.** Acid etched fracture extension profile (MB).

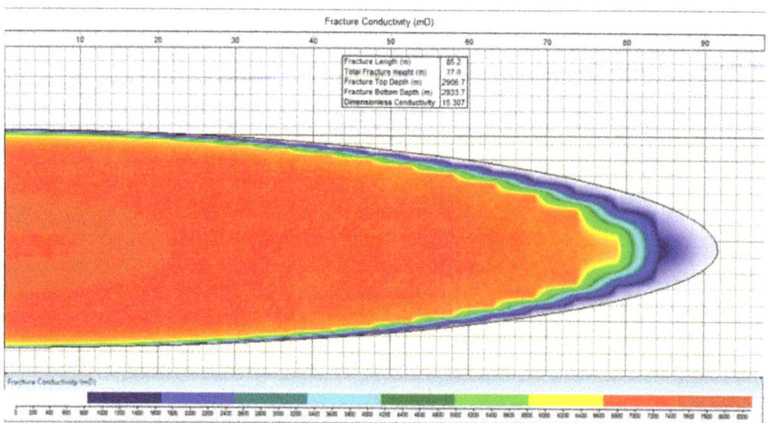


**Figure 20.** Acid etched fracture extension profile (MC).

## 5. Field Application

The field stimulation well is the X-13 well, which is not perforated after completion. MB and MC are undeveloped layers. The depth range of the MB layer is 2807.2~2906.9 m, and the depth range of the MC layer is 2906.9~2949.1 m. The optimized perforation parameters are used for the perforation interval, with a perforation interval of 2868.0~2883.0 m for the MB layer and 2913.0~2923.0 m for the MC layer. The field treatment adopts segmented acid fracturing, which involves performing acid fracturing on the MC layer, followed by inserting a packer, and then perforation and acid fracturing on the MB layer. Before the main acid fracturing operation, a Mini Frac test and a diagnostic fracture injection test were conducted, as shown in Table 3. The test results provide optimization guidance for the main acid fracturing treatment.

The MB and MC layers adopt the pad–acid fluid multi-stage injection acid fracturing process. The target reservoir temperature is 96 °C. The pad fluid uses guar gum fracturing fluid, which has good viscoelasticity and fluid loss reduction properties. Basic formula of pad fluid: 0.27% SRG-1 (guar gum) + 0.3% ADN-1 (claim stabilizer) + 0.1% ADB-1 (non-emulsifier) + 0.01% ADS-1 (highly concentrated fungicide) + 0.1% ADPJ-1B (cross-linking regulator) + 0.3% ADPJ-1A (cross-linking agent). As the main acid system, gelled acid can effectively reduce acid–rock reaction rate and increase the effective action distance of acid fluid. Acid fluid system formula: 20% HCl + 0.6% ADJ-1 (gelling agent) + 1.0~1.5% ADH-1 (corrosion inhibitor) + 1.5% ADT-1 (multi-function). Indoor experiments were conducted to test the viscosity, temperature resistance, shear resistance, and acid fluid etching rate of the pad and acid fluid systems. The experimental results indicate that the pad and gelled acid systems meet the performance requirements for acid fracturing fluids.

**Table 3.** DFIT and Min Frac result.

| Parameter | MB | | MC | |
|---|---|---|---|---|
| | DFIT | Mini Frac | DFIT | Mini Frac |
| Frac Gradient | 0.025 MPa/m | 0.012 MPa/m | 0.013 MPa/m | 0.016 MPa/m |
| Closure Gradient | 0.018 MPa/m | 0.013 MPa/m | 0.012 MPa/m | 0.014 MPa/m |
| Fluid Efficiency | 51.45% | 69.77% | 34.72% | 36.83% |
| Reservoir Pressure | 33.07 MPa | 33.32 MPa | 32.37 MPa | 33.63 MPa |
| Permeability | 0.65 mD | - | 3.75 mD | - |

The main acid fracturing operation adopts three-stage alternating injection. The total liquid consumption is 382.2 $m^3$, of which the amount of gelling acid is 254.3 $m^3$, and the maximum pumping rate is 5 $m^3$/min. During the first stage of alternate injection, the pumping rate is first raised to 4 $m^3$/min, at which time the maximum surface pressure did not exceed the wellhead pressure limit. When the wellbore is fully filled with fracturing fluid, the pressure begins to stabilize and geometric fracture shapes begin to be induced at net pressure. After acid injection, the pressure gradually decreases, mainly due to differences in hydrostatic pressure and friction between fracturing fluid and acid, as well as the reaction between acid and carbonate rocks in the reservoir. In the second stage of alternate injection, the fracturing fluid is still injected at the pumping rate of 4 $m^3$/min, and the pumping rate is increased to 5 $m^3$/min when the acid is injected. Increasing the pumping rate can effectively extend the acid action distance and deeply improve reservoirs. In the third stage of alternate injection, the fracturing fluid is initially injected at the same pumping rate of 5 $m^3$/min, and the pumping rate is reduced to 4.5 $m^3$/min as the surface pressure gradually approaches the maximum pressure limit of the wellhead device. Subsequent acid injection continued at a displacement of 4.5 $m^3$/min. The subsequent acid fluid will continue to be injected at a pumping rate of 4.5 $m^3$/min. After the acid fracturing is completed, the pump is stopped for a period of time. After the fracture is closed, 15% HCl solution is used for closed fracture acidizing, which can effectively improve the conductivity of acid eroded fractures after closure.

As shown in Figures 21 and 22, the wellhead pressure is maintained above 2.42 MPa during the production process after the acid fracturing test, and the initial production after fracturing of the MB layer is 271.1 $m^3$/d in 32/64 "choke conditions". The MC layer produced 306.8 $m^3/d$ at 38/64 "choke". The daily production is much higher than that of other acid wells in the same region. After continuous production for four months, the wellhead pressure is stable, the production decline is small, and the effect of reservoir stimulation is significant.

In addition, according to the company's production requirements, further reservoir stimulation measures are required when the daily production is below the critical value of 40 $m^3/d$. The results of the program simulations indicate that further operations will be considered after 4 years of continuous production after stimulation. Such production is maintained for a longer period of time.

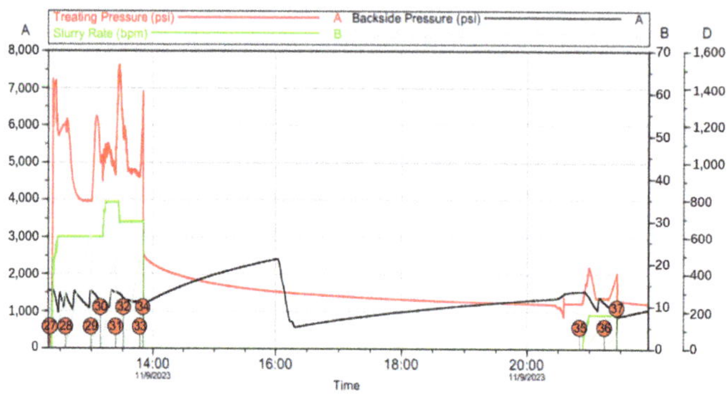


**Figure 21.** Pressure variation curve of MB layer acid fracturing operation.

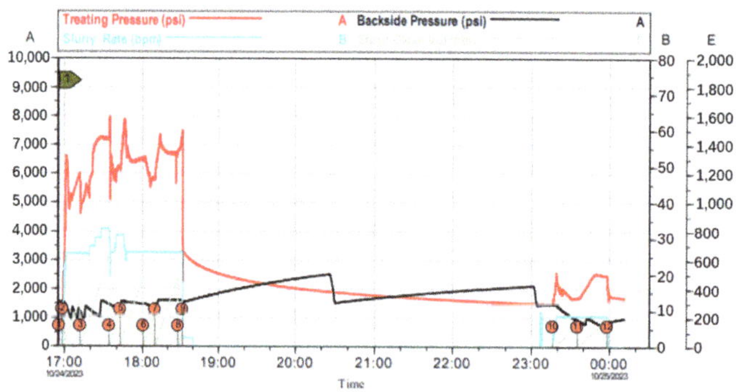


**Figure 22.** Pressure variation curve of MC layer acid fracturing operation.

## 6. Conclusions

In order to solve the problem of poor stimulation effect after acidification in some reservoir units, this paper takes the M reservoir of the F oilfield in Iraq as a study object, the feasibility is analyzed, and the treatment parameters of different stimulation processes in low-permeability reservoirs are optimized. In addition, they have been successfully applied to the X-13 well in the field. The main conclusions are as follows:

(1) Heterogeneous porous carbonate reservoirs have poor general acidification effect. Reservoir porosity and permeability distribution are non-uniform. After a comprehensive study of the M reservoir, it is considered that MB and MC layers with poor porosity and permeability distribution and physical property conditions are target layers for fracturing. The ME layer is close to the edge and bottom water, and MD and ME layers have good porosity and permeability distribution, so it is recommended to continue matrix acidizing operations.

(2) The high filtration loss during hydraulic fracturing and the sensitivity of carbonate rock to sand concentration can easily cause sand plugging. The variation of hydraulic fracturing operation pressure is higher than that of acid fracturing (taking the MB layer as an example, when the pump rate is 5 $m^3/min$, the operation pressure of acid fracturing is 52.38 MPa, and the hydraulic fracturing operation pressure is 55.93 MPa). Generally, the hydraulic fracturing treatment pumping rate (4.5 $m^3/min$) is slightly lower than the acid fracturing pumping rate (5 $m^3/min$). Combined with the analysis of the changes in production after different stimulation measures, the maximum daily oil production after acid fracturing reaches 557.90 $m^3$, and the daily oil production after hydraulic fracturing is 441.97 $m^3$, so acid fracturing is more recommended for heterogeneous porous carbonate reservoirs.

(3) The "guar gum fracturing fluid + gelled acid multi-stage alternating injection + closed fracture acidizing" process is tested in the field. The results show that this technology can effectively increase the distance of acid action, which is conducive to generating high conductivity fractures.

**Author Contributions:** Writing—original draft preparation, K.Z.; writing—review and editing, H.X.; conceptualization, K.Z. and J.W.; methodology, K.Z. and J.W.; software, K.Z.; validation, J.W. and H.X.; formal analysis, K.Z.; investigation, H.J.; data curation, K.Z. and L.Z.; visualization, H.X. and H.J.; supervision, J.W.; project administration, H.X.; funding acquisition, H.J. All authors have read and agreed to the published version of the manuscript.

**Funding:** This research was funded by the National Natural Science Foundation of China (No: 52274028) and State Key Laboratory of Petroleum Resources and Prospecting, China University of Petroleum (No: PRP/open-2205).

**Data Availability Statement:** The original contributions presented in the study are included in the article, further inquiries can be directed to the corresponding authors.

**Acknowledgments:** The authors thank China University of Petroleum (Beijing) for the support with the project experimental equipment.

**Conflicts of Interest:** The authors declare no conflict of interest.

## Nomenclature

| | |
|---|---|
| M reservoir | Mishrif reservoir |
| F oilfield | Faihaa oilfield |
| MB | Mishrif B |
| MC | Mishrif C |
| MD | Mishrif D |
| ME | Mishrif E |
| DFIT | Diagnostic fracture injection test |

## References

1. Al-Shargabi, M.; Davoodi, S.; Wood, D.A.; Ali, M.; Rukavishnikov, V.S.; Minaev, K.M. A critical review of self-diverting acid treatments applied to carbonate oil and gas reservoirs. *Pet. Sci.* **2023**, *20*, 922–950. [CrossRef]
2. Mazeel, M.A. Hydrocarbon reservoir potential estimated for Iraq bid round blocks. *Oil Gas J.* **2012**, *110*, 42–54.
3. Cantrell, D.; Shah, R.; Ou, J.; Xu, C.; Phillips, C.; Li, X.; Hu, T. Depositional and diagenetic controls on reservoir quality: Example from the upper cretaceous mishrif formation of Iraq. *Mar. Pet. Geol.* **2020**, *118*, 104415. [CrossRef]
4. Alarji, H.; Clark, S.; Regenauer-Lieb, K. Wormholes effect in carbonate acid enhanced oil recovery methods. *Adv. Geo-Energy Res.* **2022**, *6*, 492–501. [CrossRef]
5. Li, R.; Chen, Q.; Deng, H.; Fu, M.; Hu, L.; Xie, X.; Zhang, L.; Guo, X.; Fan, H.; Xiang, Z. Quantitative evaluation of the carbonate reservoir heterogeneity based on production dynamic data: A case study from cretaceous mishrif formation in halfaya oilfield, Iraq. *J. Pet. Sci. Eng.* **2021**, *206*, 109007. [CrossRef]
6. Kiani, S.; Jafari, S.; Apourvari, S.N.; Mehrjoo, H. Simulation study of wormhole formation and propagation during matrix acidizing of carbonate reservoirs using a novel in-situ generated hydrochloric acid. *Adv. Geo-Energy Res.* **2021**, *5*, 64–74. [CrossRef]
7. Alameedy, U.; Fatah, A.; Abbas, A.K.; Al-Yaseri, A. Matrix acidizing in carbonate rocks and the impact on geomechanical properties: A review. *Fuel* **2023**, *349*, 128586. [CrossRef]

8. Mohammadi, S.; Shahbazi, K.A. comprehensive review on acid-induced sludge formation during matrix acidizing: Nature, mechanism, and effective parameters. *Geoenergy Sci. Eng.* **2023**, *229*, 212150. [CrossRef]
9. Li, Y.; Deng, Q.; Zhao, J.; Liao, Y.; Jiang, Y. Simulation and analysis of matrix stimulation by diverting acid system considering temperature field. *J. Pet. Sci. Eng.* **2018**, *170*, 932–944. [CrossRef]
10. Zhao, L.; Zhou, W.; Zhong, Y.; Guo, R.; Jin, Z.; Chen, Y. Control factors of reservoir oil-bearing difference of cretaceous mishrif formation in the h oilfield, Iraq. *Pet. Explor. Dev.* **2019**, *46*, 314–323. [CrossRef]
11. Zhao, L.; Chen, X.; Zou, H.; Liu, P.; Liang, C.; Zhang, N.; Li, N.; Luo, Z.; Du, J. A review of diverting agents for reservoir stimulation. *J. Pet. Sci. Eng.* **2020**, *187*, 106734. [CrossRef]
12. Abdollahi, R.; Esfandyari, H.; Pari, M.N.; Davarpanah, A. Conventional diverting techniques and novel fibr-assisted self-diverting system in carbonate reservoir acidizing with successful case studies. *Pet. Res.* **2021**, *6*, 247–256. [CrossRef]
13. Jafarpour, H.; Moghadasi, J.; Khormali, A.; Petrakov, D.; Ashena, R. Increasing the stimulation efficiency of heterogeneous carbonate reservoirs by developing a multi-bached acid system. *J. Pet. Sci. Eng.* **2019**, *172*, 50–59. [CrossRef]
14. Zhu, D.; Hu, Y.; Cui, M.; Chen, Y.; Liang, C.; He, Y.; Wang, X.; Wang, D. Feasibility analysis on the pilot test of acid fracturing for carbonate reservoirs in halfaya oilfield, Iraq. *Energy Sci. Eng.* **2019**, *7*, 721–729. [CrossRef]
15. Dong, R.; Wheeler, M.F.; Su, H.; Ma, K. Modeling multistage acid fracturing treatments in carbonate reservoirs. In Proceedings of the SPE Hydraulic Fracturing Technology Conference and Exhibition, Virtual, 4–6 May 2021.
16. Zhu, D.; Wang, Y.; Cui, M.; Zhou, F.; Wang, Y.; Liang, C.; Zou, H.; Yao, F. Acid system and stimulation efficiency of multistage acid fracturing in porous carbonate reservoirs. *Processes* **2022**, *10*, 1883. [CrossRef]
17. Liu, F.; Fan, Y.; Li, L.; Chen, Y.; Lv, Z.; He, T. Case study of successfully staged acid fracturing on the ultra-deep horizontal well for the Qixia Fm HTHP tight carbonate gas reservoir in China. *Front. Energy Res.* **2022**, *10*, 917740. [CrossRef]
18. Wang, Y.; Zhou, F.; Su, H.; Li, Y.; Yu, F.; Dong, R.; Wang, Q.; Li, J. Performance evaluation of microemulsion acid for integrated acid fracturing in middle eastern carbonate reservoirs. *Pet. Explor. Dev.* **2023**, *50*, 1196–1205. [CrossRef]
19. Cui, M.; Ding, Y.; Zhu, D.; Chen, Y.; Cai, W.; Liang, C.; Hu, Z.; Yan, J.; Wang, D.; Zhang, T. A case study: The first hydraulically fractured well in carbonate reservoirs in Iraq. In Proceedings of the SPE Asia Pacific Oil and Gas Conference and Exhibition, Brisbane, Australia, 23–25 October 2018.
20. Zhu, D.; Cui, M.; Chen, Y.; Wang, Y.; Ding, Y.; Xiong, C.; Liang, C.; Yao, F.; Wang, X.; Cai, W.; et al. First successful application of multi-stage proppant fracturing on horizontal well in carbonate reservoirs in Iraq. In Proceedings of the SPE International Hydraulic Fracturing Technology Conference and Exhibition, Muscat, Oman, 11–13 January 2022.
21. Qiao, J.; Tang, X.; Hu, M.; Rutqvist, J.; Liu, Z. The hydraulic fracturing with multiple influencing factors in carbonate fracture-cavity reservoirs. *Comput. Geotech.* **2022**, *147*, 104773. [CrossRef]
22. Martyushev, D.A.; Ponomareva, I.N.; Filippov, E.V. Studying the direction of hydraulic fracture in carbonate reservoirs: Using machine learning to determine reservoir pressure. *Pet. Res.* **2023**, *8*, 226–233. [CrossRef]
23. Liu, S.; Liu, Z.; Zhang, Z. Numerical study on hydraulic fracture-cavity interaction in fractured-vuggy carbonate reservoir. *J. Pet. Sci. Eng.* **2022**, *213*, 110426. [CrossRef]
24. Saberhosseini, S.E.; Ahangari, K.; Mohammadrezaei, H. Optimization of the horizontal-well multiple hydraulic fracturing operation in a low-permeability carbonate reservoir using fully coupled XFEM model. *Int. J. Rock Mech. Min. Sci.* **2019**, *114*, 33–45. [CrossRef]
25. Shi, F.; Wang, D.; Chen, X. A Numerical Study on the Propagation Mechanisms of Hydraulic Fractures in Fracture-Cavity Carbonate Reservoirs. *Comput. Model. Eng. Sci.* **2021**, *127*, 575–598. [CrossRef]
26. Zhong, Y.; Zhou, L.; Tan, X.; Guo, R.; Zhao, L.; Li, F.; Jin, Z.; Chen, Y. Characteristics of depositional environment and evolution of upper cretaceous mishrif formation, halfaya oil field, Iraq based on sedimentary microfacies analysis. *J. Afr. Earth Sci.* **2018**, *140*, 151–168. [CrossRef]
27. Ismail, M.J.; Ettensohn, F.R.; Handhal, A.M.; Al-Abadi, A. Facies analysis of the middle cretaceous mishrif formation in southern Iraq borehole image logs and core thin-sections as a tool. *Mar. Pet. Geol.* **2021**, *133*, 105324. [CrossRef]
28. Chen, X.; Lu, Y. A perspective review on degradable polylactic acid diverters for well stimulations. *Fuel* **2023**, *348*, 128557. [CrossRef]
29. Mou, J.; Zhang, S. Modeling acid leakoff during multistage alternate injection of pad and acid in acid fracturing. *J. Nat. Gas Sci. Eng.* **2015**, *26*, 1161–1173. [CrossRef]
30. Petriz Munguia, J.M.; Gonzalez Valtierra, B.E.; Trujillo Hernandez, J.; Santos, S.; Monroy, K.C. Acid-fracturing techniques as a good alternative to help improve field development assets. In Proceedings of the Abu Dhabi International Petroleum Exhibition and Conference, Abu Dhabi, United Arab Emirates, 11–14 November 2019.
31. AlOtaibi, F.; Dahlan, M.; Khaldi, M.; AlGhamdi, A.; Al-Hulail, I. Evaluation of inorganic-crosslinked-based gelled acid system for high-temperature applications. In Proceedings of the SPE International Conference and Exhibition on Formation Damage Control, Lafayette, LA, USA, 19–21 February 2020.
32. Kong, X.; Liu, B.; Wan, X.; Li, S.; Liu, Z.; Chen, M.; Shen, J. Experimental evaluation and optimization of improved conductivity of gelling acid etched fractures in deep, low-permeability reservoirs. *Fuel* **2023**, *348*, 128468. [CrossRef]

processes

MDPI

Article

# Simulation of Key Influencing Factors of Hydraulic Fracturing Fracture Propagation in a Shale Reservoir Based on the Displacement Discontinuity Method (DDM)

Pengcheng Ma [1,2,3,*] and Shanfa Tang [1,2,*]

1 Cooperative Innovation Center of Unconventional Oil and Gas, Yangtze University, Wuhan 430100, China
2 Hubei Key Laboratory of Drilling and Production Engineering for Oil and Gas, Yangtze University, Wuhan 430100, China
3 China Urban-Rural Energy Co., Ltd., Wuhan 430071, China
* Correspondence: pengcheng_ma2024@126.com (P.M.); tangsf2005@126.com (S.T.)

**Abstract:** In the process of the large-scale hydraulic fracturing of a shale gas field in the Weiyuan area of Sichuan province, the quantitative description and evaluation of hydraulic fracture expansion morphology and the three-dimensional distribution law are the key points of evaluation of block fracturing transformation effect. Many scholars have used the finite element method, discrete element method, grid-free method and other numerical simulation methods to quantitatively characterize hydraulic fractures, but there are often the problems that the indoor physical simulation results are much different from the actual results and the accuracy of most quantitative studies is poor. Considering rock mechanics parameters and based on the displacement discontinuity method (DDM), a single-stage multi-cluster fracture propagation model of horizontal well was established. The effects of Young's modulus, Poisson's ratio, the in situ stress difference, the approximation angle, the perforation cluster number and the perforation spacing on the formation of complex fracture networks and on the geometrical parameters of hydraulic fractures were simulated. The research results can provide theoretical reference and practical guidance for the optimization of large-scale fracturing parameters and the quantitative post-fracturing evaluation of horizontal wells in unconventional reservoirs such as shale gas reservoirs.

**Keywords:** shale gas; hydraulic fracturing; fracture parameters; numerical simulation; discontinuity of displacement

## 1. Introduction

The exploration and development of shale gas in China is still in the early stage of rapid development. The shale gas reservoir structure is complex. However, as a common unconventional reservoir development technology, hydraulic fracturing has been widely used in Weiyuan, Changning and other areas [1]. The morphology and geometry parameters of hydraulic fracturing are mainly affected by natural fractures, rock mechanic parameters of the reservoir, perforation spacing, the number of perforation clusters and other operation parameters [1–3]. When natural fractures develop, the probability of communication between hydraulic fractures and natural fractures increases, and reservoirs have the potential to form complex fracture networks. The larger the Young's modulus, the less likely the rock is to undergo strain under the same stress, and the simpler the crack morphology formed [4,5]. The heterogeneity of the reservoir and the mutual interference between fractures are the root causes of uneven fracture propagation, and the number of perforation clusters and the perforation spacing have important effects on the distribution of the induced stress field [1,6]. In the process of multi-cluster fracturing in horizontal wells, the negative effects of the uneven expansion of fractures include the inability to achieve optimal reservoir transformation and the expansion of advantageous fractures, which leads to serious inter-well interference caused by communication with adjacent wellbore or hydraulic fractures [7–9]. In addition, due to the different properties and filling methods of different natural fracture cements, the cementation strength of natural fractures is different, so the cementation strength of natural fractures also has a great influence on the opening of fractures. As an important factor, the approaching angle also affects the opening of natural cracks.

When extracting unconventional oil and gas resources by hydraulic fracturing, the numerical simulation method has become the best method to study the fracture growth pattern in complex media. With the improvement in research theory and computing power, the numerical simulation of hydraulic fracturing has changed from the early simple two-dimensional model to the complex quasi-three-dimensional model. In recent decades, a series of numerical simulation calculation methods such as finite element method, boundary element method, discrete element method, meshless method and finite difference method have appeared [10]. Cheng et al. used the finite element method to deal with the elastic and flow equations, studied the problem of multi-layer hydraulic crack propagation in a plane homogeneous longitudinal direction, and proposed a finite element solution for plane arbitrarily shaped cracks perpendicular to the interface of multi-layer media [11–13]. However, this method has limitations in simulating the crack intersection and non-planar propagation. By establishing a finite element model of the dynamic competitive propagation of multiple fractures, Kresse et al. proposed an optimization method of perforation cluster spacing to ensure the balanced propagation of fractures by strengthening stress interference between fractures [14]. Zou et al. applied the boundary element method to discretized wellbore wall and fracture trace, solving the problem of the initiation and propagation of hydraulic fractures [15]. The model can simulate the non-plane propagation of hydraulic fractures on a horizontal plane, which is of great significance for the study of the hydraulic fracture propagation mechanism. Above, there are many studies on multi-layer hydraulic fracture propagation, non-plane hydraulic fracture propagation on the horizontal plane, induced stress field and geostress field and other factors influencing fracture propagation, fluid–structure coupling, and dynamic fracture network simulation of horizontal wells. However, there are two defects in the sensitivity analysis of these studies on related influencing factors: ① Indoor experiments are limited by indoor conditions and core size, resulting in significant deviations between physical simulation results and actual conditions; ② Most studies only analyze the corresponding qualitative trends or have poor accuracy in quantitative characterization.

In this work, a fracturing model is established using the key core technology research project software Frsmart 2.0 platform of CNPC. The fracture model simulates fracture propagation, fluid flow and proppant migration in a fully coupled fluid mechanics manner. By considering the parameters of rock mechanics and based on the displacement discontinuity method, a single-segment multi-cluster fracture propagation model for horizontal wells is established. The effects of Young's modulus, Poisson's ratio, in situ stress difference, approximation angle, perforation cluster number and perforation spacing on the formation of complex fracture networks and on the geometrical parameters of hydraulic fractures are simulated. Contrary to traditional numerical simulation methods such as finite element, the DDM method is used in this work, and the calculation accuracy was higher in the quantitative analysis of the sensitivity of related influencing factors. Its main principles include the following: ① the dimension of the calculation model is reduced by one dimension, so the calculation amount is reduced. The number of grids is reduced, and the input data preparation is simple. Under the same conditions, boundary elements tend to be more accurate than finite element and finite difference methods. ② DDM unit adopts a 1/2 displacement mode to simulate the singular stress field of the crack tip well, and the integral of its coefficient is precisely obtained, so it has high precision. This study provides theoretical reference and practical guidance for the optimization of large-scale fracturing parameters and quantitative post-fracturing evaluation of horizontal wells in unconventional reservoirs such as shale gas.

## 2. The Establishment of Hydraulic Fracturing Model

In this work, a simulation of hydraulic fracture propagation based on DDM was established to solve the problem of a strange stress field at the fracture tip, and sensitivity quantification characterization was, respectively, conducted on the reservoir and construction factors that affect the expansion of multiple fractures. The high-precision simulation results provide theoretical reference and practical guidance for the optimization of large-scale fracturing parameters and quantitative evaluation of hydraulic fracturing in shale reservoirs. The fracture simulation adopted in this work is a non-planar 3D fracture model, which follows the following assumptions:

(1) All hydraulic fractures are vertical (vertical stress equal to the maximum principal stress or at least greater than the minimum horizontal principal stress).
(2) Hydraulic cracks can turn in the horizontal plane.
(3) There are parallel formations within each fracturing stage.
(4) The fracture opening follows a linear elastic fracture mechanics theory.

In the process of the initiation and expansion of hydraulic fractures, the fracture stress interferes, and the fracture growth path will be distorted. If hydraulic fractures meet natural fractures in the process of expansion, the hydraulic fractures may directly penetrate the natural fractures or be captured by natural fractures. Fracture stress interference and fracture propagation intersection behavior should meet the corresponding criteria. Under the premise of meeting the above assumptions, displacement discontinuity method (DDM) is used to describe the fracture initiation and propagation intersection behavior of a hydraulic fracture, and the fracture stress interference and fracture propagation intersection behavior criteria are restricted. It is used to simulate the formation process of fracture network and the sensitivity analysis of each related variable.

The basic equations of DDM and the basic criteria of crack propagation are as follows (the flow chart as Figure 1).

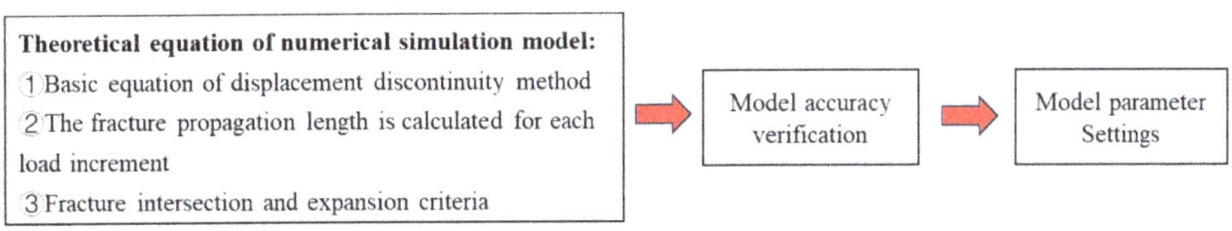


**Figure 1.** The flow chart of simulation steps.

### 2.1. Displacement Discontinuity Method

The relative displacement between two surfaces on a fracture surface is called displacement discontinuity, and its magnitude is the displacement discontinuity quantity. The displacement discontinuity method takes displacement discontinuity as the basic variable and deduces the stress value and displacement value of any point calculated based on displacement discontinuity, which is called the basic solution [5,14].

The deformation of a crack is described by the three-dimensional displacement discontinuity method:

$$\begin{aligned} -\sigma_{sl}^{i} &= \sum_{K=1}^{N} K_{sl,sl}^{ik} D_{sl}^{k} + \sum_{K=1}^{N} K_{sl,sl}^{ik} D_{sh}^{k} + \sum_{K=1}^{N} K_{sl,nn}^{ik} D_{nn}^{k} \\ -\sigma_{sh}^{i} &= \sum_{K=1}^{N} K_{sh,sl}^{ik} D_{sl}^{k} + \sum_{K=1}^{N} K_{sh,sh}^{ik} D_{sh}^{k} + \sum_{K=1}^{N} K_{sh,nn}^{ik} D_{nn}^{k} \\ p-\sigma_{nn}^{i} &= \sum_{K=1}^{N} K_{nn,sl}^{ik} D_{sl}^{k} + \sum_{K=1}^{N} K_{nn,sh}^{ik} D_{sh}^{k} + \sum_{K=1}^{N} K_{nn,nn}^{ik} D_{nn}^{k} \end{aligned} \tag{1}$$

where $D_{nn}$—normal displacement discontinuity surface; $D_{sl}$, $D_{sh}$—a surface with discontinuous shear displacement in the direction of crack length and crack height; $P$—fluid pressure; $\sigma_{nn}$, $\sigma_{sl}$, $\sigma_{sh}$—effective formation normal and shear stresses, which can be obtained from the distant stress field by the superposition principle. The nine matrices *Ks* are boundary influence matrices related to the element size, position, orientation, and elastic constant. Superscripts *i* and *k*—element index; *N*—the total number of elements.

Fluid flow and proppant migration in fractures are controlled by the volume conservation equation:

$$\begin{aligned} \frac{\partial w}{\partial t} &= \nabla \cdot \left[ w(1-c)v^f + wcv^p \right] + \delta(x,y)q_0 - q_l \\ \frac{\partial (cw)}{\partial t} &= \nabla \cdot (cwv^p) + \delta(x,y)c_0 q_0 \end{aligned} \tag{2}$$

where *T*—time; w—crack width, $D_{nn}$—normal displacement discontinuity surface; *c*—volume concentration of thc proppant; $q_o$—the injection fluid rate; $C_0$—the specified volume concentration of proppant in the fluid injection; $\delta(x, y)$—the Dirac function; and $q_l$ corresponds to the frac fluid filtration velocity of the one-dimensional Carter filtration model:

$$q_l = \frac{2C_L}{\sqrt{t - \tau\,(x,y)}} \tag{3}$$

where $\tau(x, y)$—the time the surface position is first exposed to fracturing fluid; $C_L$—Carter's filtration coefficient.

The average frac fluid volume velocity ($V^f$), average proppant volume velocity ($V^p$), and proppant settling velocity ($V^s$) are, respectively, expressed as

$$V^f = \frac{w^2}{12\mu^f} Q^f\left(c, \frac{w}{a}\right)\left(\nabla p - \rho^f g\right) \tag{4}$$

$$V^p = Q^p\left(c, \frac{w}{a}\right)\left(v^f + v^s\right) \tag{5}$$

$$V^s = \frac{a^2}{12\mu^f}\left(\rho^p + \rho^f\right)g \tag{6}$$

where *p*—fluid pressure, *g*—gravity vector, *a*—proppant radius, $\mu^f$—apparent viscosity of the fracturing fluid, $\rho^f$—fracturing fluid density, $v^s$—proppant settling velocity, and $\rho^p$—proppant density.

The following function is used in a planar three-dimensional fracture to reflect the effect of proppant concentration on the viscosity of the fracturing fluid. This function can be used to describe the transition from the Poisson flow state to the Darcy flow state when the proppant concentration reaches its maximum value.

$$Q^f(c, \frac{w}{a}) = \left(1 - \frac{c}{c_{\max}}\right)^{\beta} + \frac{a^2}{w^2}\frac{c}{c_{\max}}\overline{D} \tag{7}$$

where $\beta = 1.5$, $\overline{D} = 8(1 - c\alpha c_{\max_{\max}})$, $\alpha = 4.1$, and $c_{\max} = 0.585$.

When the sand plug or proppant concentration reaches the maximum, the function value is 0; otherwise it is 1. The proppant function expression is different; all this can be obtained from the experimental data.

The equation of perforating pressure drop:

$$P_r = \frac{8\rho}{\pi^2 C_d^2 d_p}\left(\frac{q_{ri}}{n}\right) \tag{8}$$

where $C_d$ is the dimensionless flow coefficient; $d_p$ is the perforation diameter; $q_{ri}$ is the volumetric injection rate at a specific injection location; *n* is the number of holes at a specific perforation; $\rho$ is the mud density, and $\rho = (1-c)\ \rho^f + c\rho^p$. It was assumed that the pressure at the wellbore perforation is equal everywhere.

### 2.2. The Fracture Propagation Length under Load Increment

Through the analysis of energy angle, we found that the structure will release energy continuously during the process of crack propagation, and only when the energy is greater than the energy required to maintain the crack propagation can the crack continue to expand. In most cases, after the same load increment is applied, the length of the crack expansion will be different due to the different energy released by the crack expansion. Since the equivalent stress intensity factor $K_e$ is proportional to the energy release rate $G_e$, the following steps can be used to calculate the crack growth length after each load increment is applied [16,17]:

(1) Apply an additional incremental load on the basis of the existing load, and finally obtain the equivalent stress intensity factor $K_I$ and $K_{II}$ by calculating the stress intensity factor $K_e$ at the crack tip under the current total load.

(2) Conduct judgment on crack propagation. If the calculated equivalent stress intensity factor $K_e$ is less than the critical stress intensity factor $K_c$, then the crack will not expand under the current load increment and need to go back to step (1). If $K_e$ is greater than or equal to $K_c$, then the crack begins to expand. Based on the calculated crack expansion angle $\theta_1$, the original crack extends a short length counterclockwise along the $\theta_1$ direction.

(3) Under the condition that the current total load remains unchanged, the stress field after the fracture extension for a short length is calculated, and the equivalent stress intensity factor $K_e$ at the new crack tip is obtained.

(4) Conduct judgment on crack propagation. If the equivalent stress intensity factor $K_e$ is greater than or equal to the critical stress intensity factor $K_c$, the crack will continue to expand. According to the calculated new crack propagation angle $\theta_2$, rotate $\theta_2$ counterclockwise along the new crack tip of the previous section and extend it a further distance, as shown in Figure 2, then return to step (3). If $K_e$ is less than $K_c$, then under the action of this incremental load, the crack has already expanded, so return to step (1).

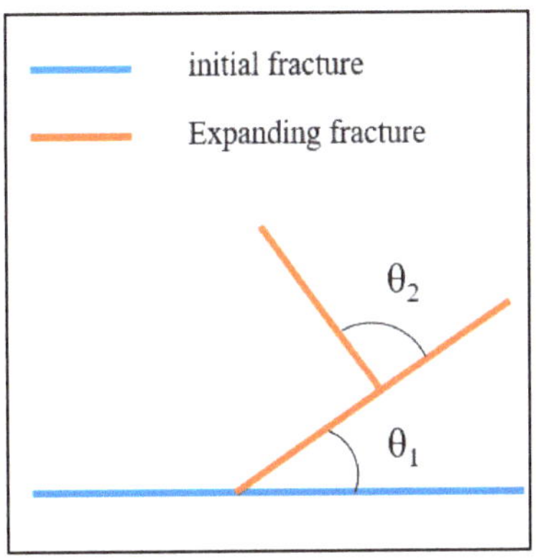


**Figure 2.** Combined with the relationship between the equivalent stress intensity factor and the critical stress intensity factor in step (2), the new fracture propagation angle and propagation distance are determined.

### 2.3. Fracture Intersection and Expansion Criteria

The stress field distribution when the hydraulic fracture extends to the intersection with the natural fracture is shown in Figure 3. $\beta$—the approaching angle between the hydraulic fracture and the natural fracture, $\tau_\beta$ and $\sigma_{\beta,n}$—the shear stress and normal stress acting on the surface of the natural fracture, and $\sigma_H$ and $\sigma_h$—the maximum and minimum horizontal principal stress.

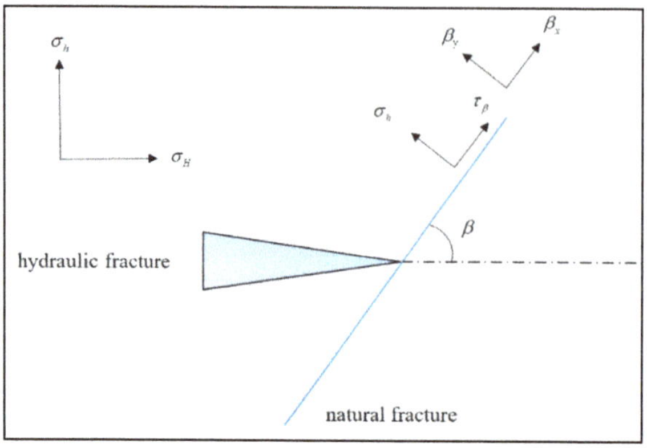


**Figure 3.** The intersection stress field diagram of the hydraulic fracture with the natural fracture.

When the maximum principal stress is greater than the tensile strength of the rock, the hydraulic fracture will expand forward through the natural fracture under the induction of the maximum horizontal principal stress, and the critical conditions are

$$\sigma_1 = T_0 \tag{9}$$

$$\left|\tau_\beta\right| < C - \mu_f \sigma_{\beta,n} = 0 \tag{10}$$

where $C$ is the natural crack interface bonding force, and $\mu_f$ is the natural crack friction factor.

The maximum circumferential stress criterion is used to judge the expansion behavior of cracks. The mutual integral method is used to calculate the stress intensity factor $K_{\mathrm{I}}$ and $K_{\mathrm{II}}$ at the crack tip and obtain the equivalent stress intensity factor $K_e$. By comparing the equivalent stress intensity factor $K_e$ in the fracture direction with the fracture toughness of the rock mass $K_{IC}$, the crack expansion can be judged. The critical conditions are

$$\begin{cases} K_e = \cos\frac{\theta}{2}\left(K_{\mathrm{I}}\cos^2\frac{\theta}{2} - \frac{3K_{\mathrm{II}}}{2}\sin\theta\right) \\ K_e \geq K_{IC} \end{cases} \tag{11}$$

The formula for calculating the crack initiation angle is

$$\theta = 2\mathrm{arctan}\theta \frac{-2K_{\mathrm{II}}/K_{\mathrm{I}}}{1+\sqrt{1+8(K_{\mathrm{II}}/K_{\mathrm{I}})^2}} \tag{12}$$

### 2.4. Fracture Propagation-Induced Stress Field Superposition Criterion

When calculating fracture propagation by the displacement discontinuity method, it is necessary to divide the fracture into multiple line elements. The total displacement and stress of the operation object are calculated by superposition of all discontinuous elements. The influence of discontinuous components on the line element $i$ ($D_s^i$ and $D_n^i$) on the displacement and stress at any point in the operand can be calculated by the following equation [15,18].

The expression of the displacement and stress of line element $j$ to the midpoint of line element $i$ is as follows:

$$\begin{aligned} u_s^j = {} & D_s^j\left[(1-2\nu)\sin\theta\overline{F}_2 + 2(1-\nu)\cos\overline{F}_3 - \overline{y}(\sin\theta\overline{F}_4 + \cos\theta\overline{F}_5)\right] \\ & + D_n^j\left[-(1-2\nu)\cos\theta\overline{F}_2 + 2(1-\nu)\sin\theta\overline{F}_3 - \overline{y}(\cos\theta\overline{F}_4 + \sin\theta\overline{F}_5)\right] \end{aligned} \tag{13}$$

$$\begin{aligned} u_n^j = {} & D_s^i\left[(1-2\nu)\cos\theta\overline{F}_2 + 2(1-\nu)\cos\theta\overline{F}_3 - \overline{y}(\sin\theta\overline{F}_4 + \cos\theta\overline{F}_5)\right] \\ & + D_n^i\left[(1-2\nu)\sin\theta\overline{F}_2 + 2(1-\nu)\cos\theta\overline{F}_3 + \overline{y}(\sin\theta\overline{F}_4 + \cos\theta\overline{F}_5)\right] \end{aligned} \tag{14}$$

$$\sigma_s^j = 2GD_s^i\left[\sin 2\theta\overline{F}_4 - \cos 2\theta\overline{F}_5 + \overline{y}(\sin 2\theta\overline{F}_6 + \cos 2\theta\overline{F}_7)\right] \\ +2GD_s^i\left[-\overline{y}(\cos\theta\overline{F}_6 + \cos\theta\overline{F}_7)\right] \tag{15}$$

$$\sigma_s^j = 2GD_s^i\left[\sin 2\theta\overline{F}_4 - \sin 2\theta\overline{F}_5 + \overline{y}(\cos 2\theta\overline{F}_6 + \sin 2\theta\overline{F}_7)\right] \\ +2GD_s^i\left[-\overline{F}_5 - \overline{y}(\sin 2\theta\overline{F}_6 + \cos 2\theta\overline{F}_7)\right] \tag{16}$$

### 2.5. Model Accuracy Verification

In order to verify the accuracy of the hydraulic fracturing model, as shown in Figure 4, fracturing data from several horizontal wells in Weiyuan X well area were selected, and data from 10 samples were input into the model for calculation. Based on the comprehensive filtration coefficient of the formation and the total volume of fracturing fluid injected into the formation, the volume of fracturing fluid used for fracturing, namely SRV, can be obtained. By simulating single-well SRVs with a different pumping flow rate at the same operation time, when the displacement is low, the net pressure inside the joint is insufficient to activate natural fractures. As the pumping flow rate increases, the net pressure inside the joint continues to rise. When the net pressure is increased to the required net pressure to activate the natural fracture, the natural fracture around the main fracture will be activated and opened, forming a complex fracture system. At this time, the SRV increases sharply. It can be seen from the sample data that when the displacement increases to 8 $m^3/min$, hydraulic fractures begin to induce a large number of secondary fractures to open around them. Moreover, the numerical simulation results of the selected 10 sample wells have a high fit with the SRV of the sample wells in the field, which verifies the accuracy of the hydraulic fracturing model.

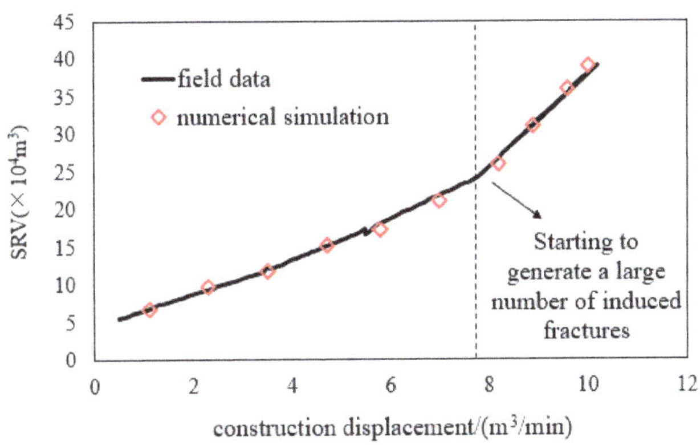


**Figure 4.** Change curve of SRV with displacement.

### 2.6. Model Parameter Settings

It is assumed that the reservoir has three layers: upper, middle and lower; the upper and lower layers are interlayers, the middle reservoir is 40 m thick, the fixed fracture height is the reservoir thickness, and the model is 2000 m long and 500 m wide. The basic parameter settings of fracking model are shown in Table 1.

Here, a coefficient of difference in inflow liquid volume was introduced to quantitatively evaluate the difference in inflow liquid volume into each cluster crack. This coefficient is defined by the standard deviation of the distribution of inflow liquid volume in each cluster:

$$S_d = \sqrt{\frac{1}{N_f - 1}\sum_{i=1}^{N_f}\left(\frac{v_i}{v_t} - \frac{1}{N_f}\right)^2} \tag{17}$$

where $V_t$—total liquid volume, $m^3$; $S_d$—the differential coefficient of the inflow of each cluster of liquid in the section, without a dimension.

**Table 1.** Hydraulic fracturing model parameter settings.

| Parameter | Top Interlayer | Middle Interlayer | Bottom Interlayer |
|---|---|---|---|
| Top depth (m) | 2850 | 2900 | 2940 |
| Bottom depth (m) | 2900 | 2940 | 2990 |
| Rock type | | Shale | |
| Porosity | | 0.01 | |
| Permeability (mD) | | 0.01 | |
| Young's modulus (GPa) | | 20 | |
| Poisson's ratio | | 0.25 | |
| Maximum horizontal principal stress (MPa) | 39 | 43 | 44 |
| Minimum horizontal principal stress (MPa) | 40 | 44 | 45 |
| Vertical stress (MPa) | 41 | 45 | 46 |
| Fracturing fluid | | Slicwater | |
| Propping agent | | 40/70 mesh quartz sand | |
| Hole diameter (mm) | | 10 | |
| Single cluster perforation density (/m) | | 16 | |
| Flow rate ($m^3$/min) | | 12 | |
| Filtration coefficient (m/$s^{0.5}$) | | $2 \times 10^{-5}$ | |
| Fracture toughness (MPa·$m^{0.5}$) | | 1 | |

The size of this coefficient can characterize the degree of difference in the amount of liquid in each cluster fracture during multi-cluster fracturing, and the amount of liquid in each cluster fracture determines the expansion form of each cluster fracture. The greater the difference in the amount of liquid in the cluster, the more uneven the crack expansion. Generally, 4.4 is taken as the limit, and if $S_d$ is lower than 4.4, it is considered that the amount of liquid flowing into each cluster fracture is uniform, and the fracture expansion morphology is more uniform.

## 3. Simulation of Influencing Factors of Hydraulic Fracturing Fracture Parameters

There are two main types of factors affecting the formation of a complex fracture network by hydraulic fracturing, internal reservoir factors and external operation factors. The internal reservoir factors are mainly Young's modulus, Poisson's ratio, the in situ stress difference, the natural fracture approaching angle, etc. The external operation factors are mainly divided into the perforation cluster number, perforation spacing, pumping flow rate, fracturing fluid viscosity and so on. Only when the above two kinds of factors are considered together can a complex fracture network system be formed to carry out sufficient volumetric fracturing transformation of the reservoir. The following will carry out the quantitative research on the sensitivity of the relevant parameters from these two aspects.

### *3.1. Reservoir Influencing Factors*

#### 3.1.1. Young's Modulus

Young's modulus ($E$) is one of the important parameters affecting the propagation of hydraulic fractures, which reflects the difficulty of rock deformation under the action of external forces. The higher the brittleness of the rock, the easier it is to be pressed open. We simulated crack morphology using a single segment with three clusters of perforations and a cluster spacing of 30 m, with Young's modulus set at 10 GPa, 20 GPa, 30 GPa, and 40 GPa. The simulation results are shown in Figure 5. With the same basic parameter settings, as the Young's modulus of the reservoir increased, the length and width of each cluster of fractures in a single-stage three-cluster fracturing increased. The fracture lengths were 169.8 m, 189.7 m, 209.7 m, and 229.7 m, respectively. The deflection angle of the two sides of the fractures increased, and the middle fracture was squeezed by the induced stress of the two sides of the fractures, with a width smaller than that of the two sides and almost no deflection. According to the simulation results, under the condition of a certain fracture height, the integral area of each curve and coordinate was the fracture bottom area under the control of the same fracture height. That is, in Figure 6, although the elastic modulus was different, the scale of fracture 1 and 2 was the same. From the geometric meaning of the simulation results, it can be shown that fractures formed by reservoirs with a low elastic modulus are usually short and wide, while the fractures formed by reservoirs with a high elastic modulus are usually longer and narrower.

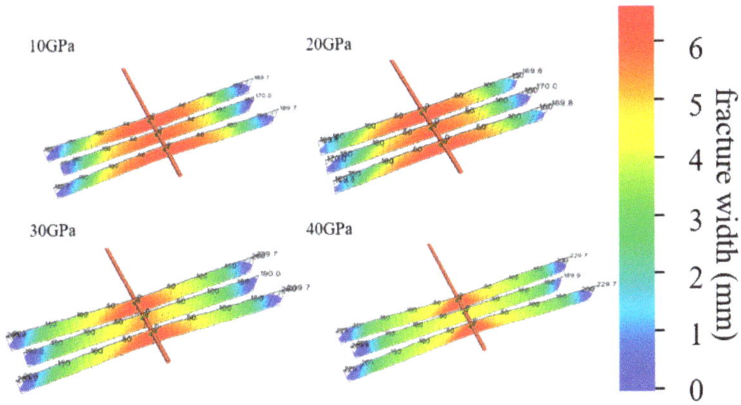


**Figure 5.** Fracture morphology and width under different Young's modulus.

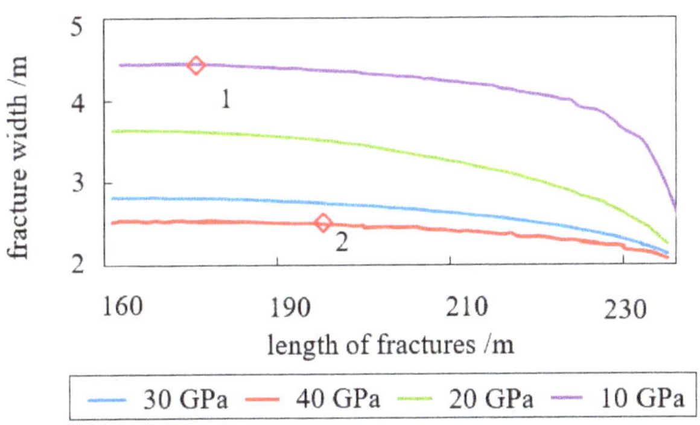


**Figure 6.** Hydraulic fracture length and average width under different Young's modulus. (No 1 and red square represent left side fracture 1, No 2 and red square represent right side fracture 2).

### 3.1.2. Poisson's Ratio

Poisson's ratio, which is the ratio of the transverse strain to axial strain of rock, has no effect on the final shape of hydraulic fracture, but Poisson's ratio has a certain effect [19,20]. Generally speaking, the smaller the Poisson's ratio is, the more likely the hydraulic fracture formed in the half length direction is to deform in unit time when the rock is subjected to shear force, that is, the faster the expansion speed. The fracture morphology was simulated by using a single section with three clusters of perforations, a cluster spacing of 30 m, and Poisson's ratio of 0.1, 0.2, 0.3, 0.4, as shown in Figure 7. Under the same setting of basic parameters, with the increase in reservoir Poisson's ratio, the change in the final fracture morphology was not significant, but the simulation end time was 30 s, 35 s, 38 s, 40 s, respectively. By extracting the fracture pressure values of four fracturing models in the simulation process and drawing them into curves (Figure 8), it can be seen from the curve change trend that with the increase in Poisson's ratio, the fracture pressure of the shale reservoir will also increase, and the difficulty of rock cracking will also increase. In other words, under the same operation conditions, reservoirs with low Poisson's ratio will show a certain lag in fracture expansion; that is, the smaller the Poisson's ratio, the faster the fracture expansion speed.

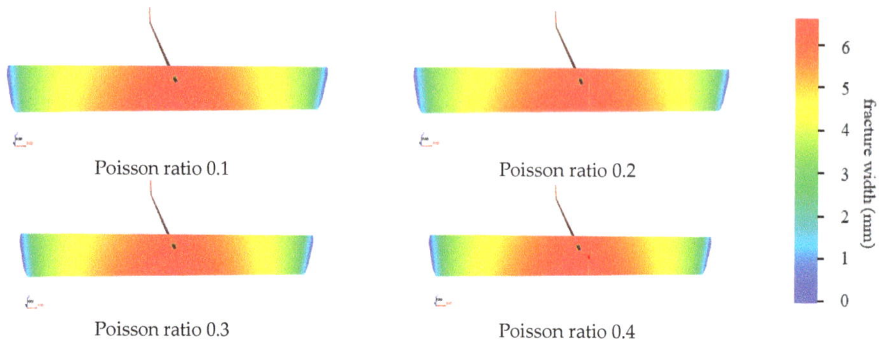


**Figure 7.** Influence of Poisson's ratio on crack growth pattern.

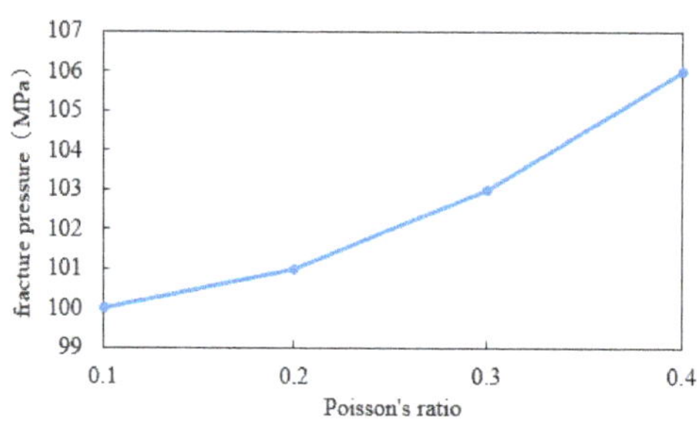


**Figure 8.** Variation in fracturing pressure with Poisson's ratio in shale reservoirs.

#### 3.1.3. In Situ Stress Difference and Approximation Angle

The crack approach angle refers to the angle between a newly generated crack and a natural crack when they intersect. Generally speaking, the smaller the geostress difference and approximation angle, the easier hydraulic fractures are to be captured by natural fractures. However, the larger the geostress difference and approximation angle, the easier hydraulic fractures are to extend forward through natural fractures [19,20]. According to the maximum likelihood diagram of well block X, the natural fractures in well block X in the Weiyuan area of Sichuan Basin were mainly low-angle (30°), and the distribution of in situ stress field had strong heterogeneity (Figure 9). Therefore, this work considered the comprehensive effect of in situ stress difference and fracture approach angle, and conducted orthogonal numerical simulation.

A single segment and four cluster horizontal well fracturing model with a cluster spacing of 30 m was established. The natural fracture strength was randomly distributed at 3 MPa and the standard deviation was 2 MPa to simulate the heterogeneity of specific reservoirs in the Weiyuan area. The natural fracture density was 0.002 pieces/m$^2$. By changing the combination of the approximation angle of natural fractures and the horizontal principal stress difference, orthogonal numerical simulations were conducted when the natural fracture azimuth angle was 30°, 45°, 60°, and the ground stress difference was 0, 5, and 10 MPa.

**Figure 9.** Maximum likelihood map (**left**) and geostress difference distribution map ((**right**), the color code from bottom to top indicates the increase in geostress difference) in Weiyuan X well area, Sichuan Province (The red and blue lines on the left both represent natural cracks, while the red lines are only highlighted in certain areas).

The simulation results show that when the approach angle is 30 ° and the geostress difference is 0 MPa, almost all hydraulic fractures are captured by natural fractures (Figure 10). When the hydraulic fractures extend to the natural fractures, they will extend along the weak bonding surface of the natural fractures [19,20], thereby activating the natural fractures. As the horizontal stress difference and approach angle increase, the hydraulic fractures tend to extend forward through the natural fractures, and the number of transverse fractures formed gradually increases. When the main stress difference is 10 MPa and the approach angle is 60°, almost all natural cracks are penetrated, and the formed crack network is similar to a grid, forming a complex fracture network.

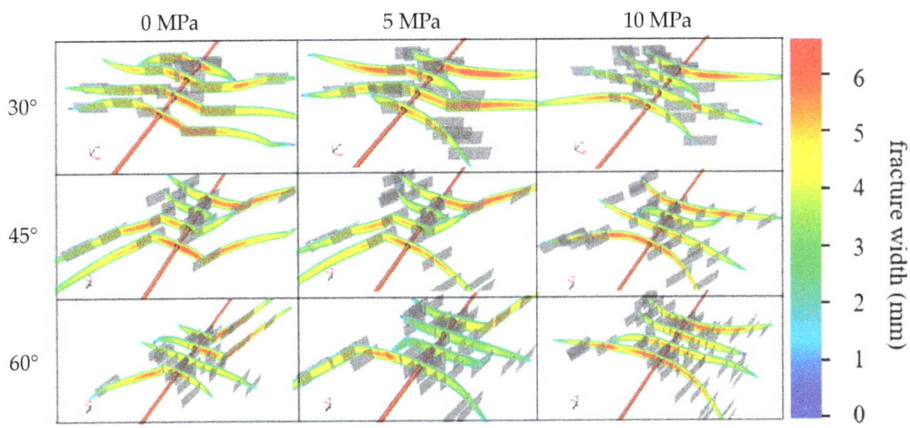


**Figure 10.** Simulation results of hydraulic fracture propagation under different stress differences and approach angles.

The local stress difference gradually increased from 0 to 20 MPa, and the coefficient of difference in liquid inflow of the three different positions of the perforation cluster within the section increased from 4 to 8.2 (Figure 11). This indicates that as the geostress difference increases, the difference in liquid inflow into each cluster crack gradually increases. Specifically, the amount of liquid inflow into the outer crack is more than that into the inner crack, and the development of the inner crack is suppressed, with the size of the crack being smaller than that of the outer crack. After the local stress difference exceeds 12 MPa, the coefficient of difference in the liquid inflow of the perforation cluster at three locations within the section reaches 8.2, which is not conducive to the uniform expansion of cracks and makes it difficult to form a complex crack network.

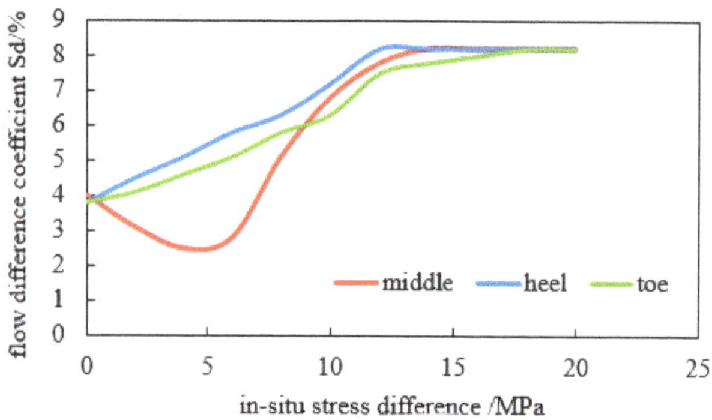


**Figure 11.** Difference coefficient of fluid inflow for each fracture cluster under different in situ stress difference. Notes: Flow difference coefficient—the liquid flow difference between different perforating clusters under different ground stress difference.

### 3.2. Operation Factors

#### 3.2.1. Number of Perforating Clusters

Changing the number of perforation clusters will change the morphology, density and length of the fracture, and affect. With the same amount of injected fluid, as the number of fracturing clusters increases, the size of a single fracture will continue to shrink (Figure 12). When conducting single section and two clusters of perforations, during the extension process of the two cracks, due to mutual interference of stress between the cracks, they will gradually deflect towards the direction away from each other. The shear forces on the two cracks are basically equal, so the deflection angle is also the same [19,20]. When single-stage three-cluster perforating was carried out, the middle crack was inhibited by the induced stress generated by the expansion of cracks on both sides, resulting in its length and width being smaller than the cracks on both sides. At the same time, there was almost no deflection due to the equal and opposite shear forces on both sides. When four clusters of fractures were fractured within a segment, the middle two clusters of fractures are severely disturbed by stress, and the half length of the fractures decreased from 190 m in a single segment with three clusters of perforations to 150 m. Compared to three clusters of perforations, the width of the middle fractures was also smaller; the pattern of five cluster perforations within the segment was consistent with the previous plan.

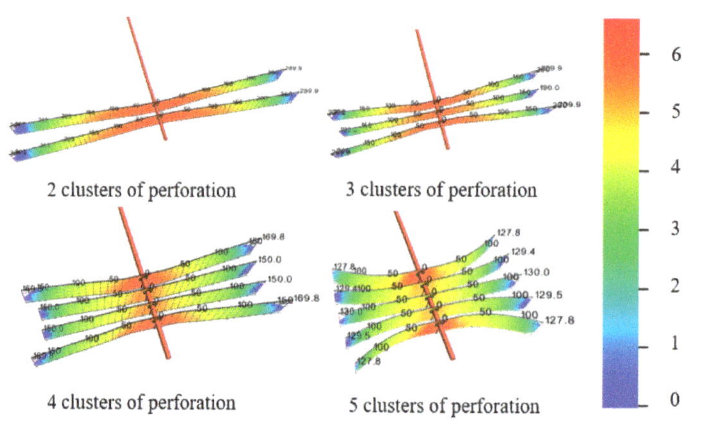


**Figure 12.** Fracture morphology and fracture width under different fracture cluster.

Figure 13 depicts the inflow of liquid into each crack cluster under different cluster numbers in a single segment. When the number of clusters in a single segment was less than 4, the interference of stress shadow between clusters was small, and the cracks expanded uniformly. When the number of clusters exceeded 4, the difference in liquid content between each cluster reached 4%, and the cracks began to expand unevenly. The outer cracks were deflected and became larger due to induced stress interference. Meanwhile, the simulation results in Figure 14 show that both the average control area of the mesh and the average deflection angle of the crack increased with the increase in the number of clusters, and the increased amplitude continued to increase. The number of clusters increased from 4 to 5, and the maximum increase in the control area of the mesh was 62.3%; the maximum increase in average deflection angle was 59%.

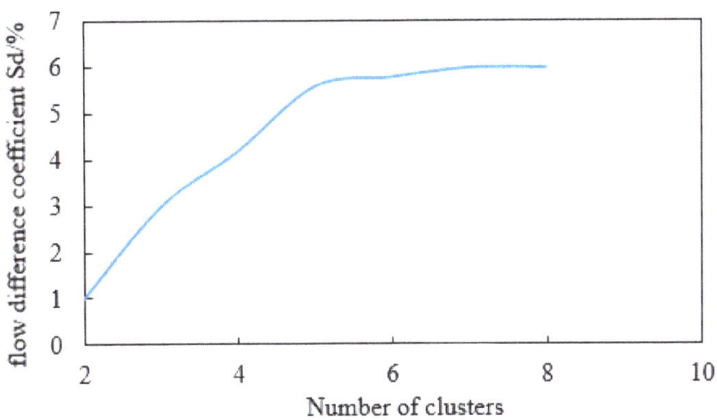


**Figure 13.** Difference coefficient of fluid inflow of each fracture cluster with different cluster.

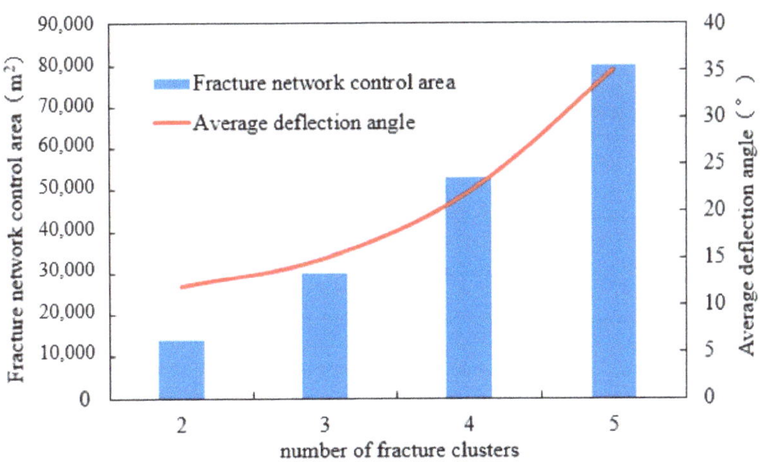


**Figure 14.** Influence of fracture cluster number on the control area and average deflection angle of the seam network.

### 3.2.2. Perforation Spacing

The determination of cluster spacing is influenced by two factors: inter-cluster stress interference and seepage interference. When the distance between clusters was large, the interference of stress shadows between clusters was relatively small, and the crack propagation in each cluster of a single segment was more uniform. Due to the competition between clusters, there were differences in the amount of liquid entering each cluster crack.

Specifically, the amount of liquid entering the outer crack was greater than that of the inner crack. However, the inner crack, due to the shear stress on both sides, did not deflect, but the crack size was smaller than that of the outer crack. Due to the more fracturing fluid entering the outer crack, its size was larger, resulting in an imbalance in the shear force on both sides of the outer crack, which can cause the crack to deflect [15]. Other factors being equal, we set the cluster spacing as 20 m, 30 m, 40 m and 50 m, and conducted single-section three-cluster perforation (Figure 15). When the cluster spacing was 20 m, the deflection angle of hydraulic fractures was relatively large, and the expansion of hydraulic fractures was mainly affected by induced stress. As the cluster spacing increased, the deflection angle of outer fractures decreased, and the fractures tended to be straight. The expansion of hydraulic fractures was mainly controlled by the original geostress field, and the fractures expanded along the direction of the maximum horizontal principal stress. At the same time, with the increase in the cluster spacing, the length of the hydraulic fracture (209.7 m) had little overall change, while the width gradually decreased and the height became larger (Figure 16). This is because in the basic setting of this simulation, the vertical fracture toughness was small: 0.5 MPa·m$^{0.5}$. In addition, the formation filtration coefficient was $2 \times 10^{-5}$ m/s$^{0.5}$, which made the fracturing fluid filtration loss small and was basically used for fracture-making. In addition, the small vertical fracture toughness made the fracture easier to expand vertically.

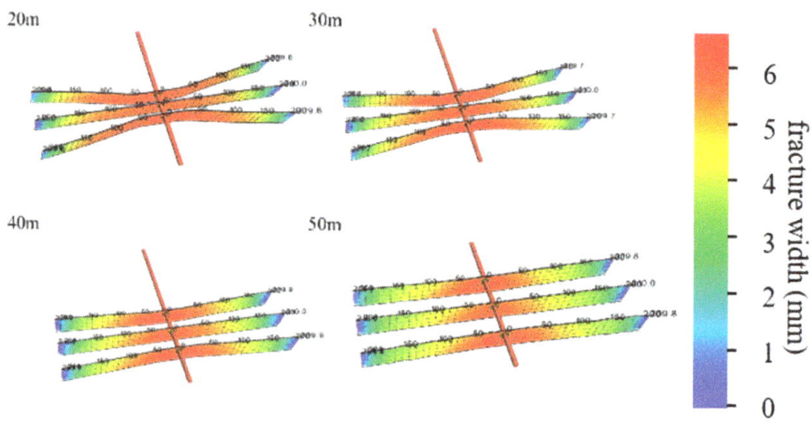


**Figure 15.** Fracture morphology and fracture width under the condition of cluster spacing in different sections.

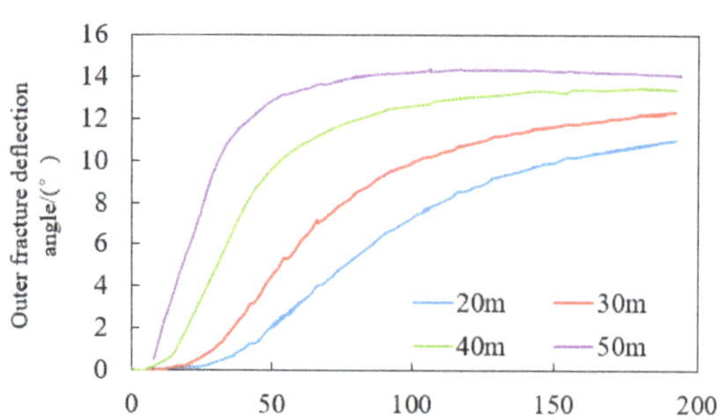


**Figure 16.** Variation curves of fracture length and width under different degrees of cluster spacing in different segments.

### 3.2.3. Fracturing Fluid Viscosity

In large-scale hydraulic fracturing operations, the viscosity of fracturing fluid has an important influence on the complexity of fracture propagation. The lower the viscosity of the fracturing fluid, the higher the complexity of the fracture network. When high-viscosity fracturing fluid was used, there was obvious main fracture expansion, and the hydraulic fracture hardly intersected with the natural fracture, which was more likely to form a single fracture, while the low viscosity fracturing fluid was more likely to form a complex fracture network. Other parameters remained unchanged. The viscosity of the fracturing fluid was set as 5 MPa·s, 10 MPa·s, 20 MPa·s and 40 MPa·s, respectively (Figure 17). The density of natural fractures was 0.004 pieces/m$^2$, the length was 40 m and the azimuth angle was 30°. According to the simulation results, when the viscosity of fracturing fluid was low, secondary fractures were more developed. Hydraulic fracturing and natural fractures not only formed longitudinal fractures along the direction of maximum principal stress, but also transversed fractures perpendicular to the wellbore. When the viscosity of the fracturing fluid was high, the development trend of the main fracture gradually appeared, the hydraulic fracture was relatively straight, and the interaction with the natural fracture was less. The fracturing fluid viscosity had little influence on the hydraulic fracture width, and the change curve of influence on length is shown in Figure 18.

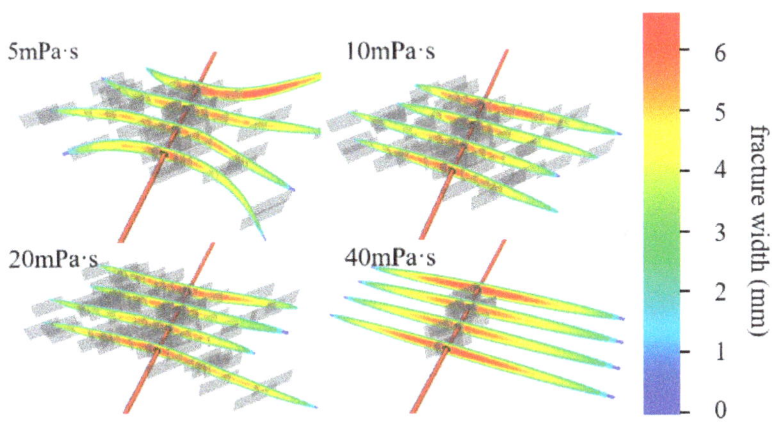


**Figure 17.** Fracture morphology and fracture width under different fracturing fluid viscosity.

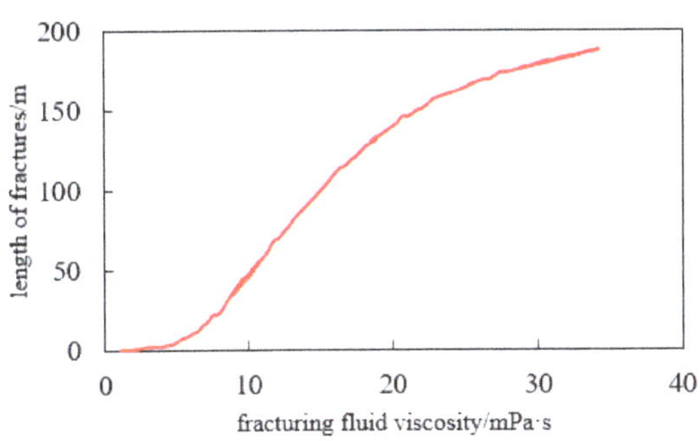


**Figure 18.** Influence of different fracturing fluid viscosity on hydraulic fracture length.

### 3.2.4. Pumping Flow Rate

With the increase in the pumping flow rate, the net pressure in the hydraulic fracture also increased; the smaller the pumping flow rate, the more fracturing fluid loss per unit time. When the displacement was low, the net pressure in the fracture was not enough to activate the natural fracture. However, with the increase in displacement, the net pressure in the seam gradually increases. When the net pressure increases enough to activate the natural fractures, the natural fractures are activated and interact with the hydraulic fractures to form a complex fracture network. Other parameters remain unchanged. Set the total injection volume of fracturing to 500 $m^3$, the pumping flow rate to 3 $m^3$/min, 8 $m^3$/min, 12 $m^3$/min, 16 $m^3$/min, fracturing fluid viscosity to 10 MPa·s, natural fracture density to 0.004 pieces/$m^2$, length to 40 m, azimuth angle to 45° (Figure 19). According to the simulation results, it can be seen that when the pumping flow rate is 3 $m^3$/min, the fracturing fluid displacement is less than the formation filtration rate during the pumping flow rate, and the fracturing fluid filtration loss is serious, and a large-scale fracture network cannot be formed. When the pressure fluid displacement increases to 12 $m^3$/min, the optimal sewing mesh size has been formed. When the pressure fluid displacement continues to increase to 16 $m^3$/min, the increase in seam mesh control area is minimal, but the operation cost increases more. The changes in crack length and width are shown in Figure 20. The larger the displacement, the longer and wider the crack. Therefore, the optimal pumping flow rate in Weiyuan area is 12 $m^3$/min.

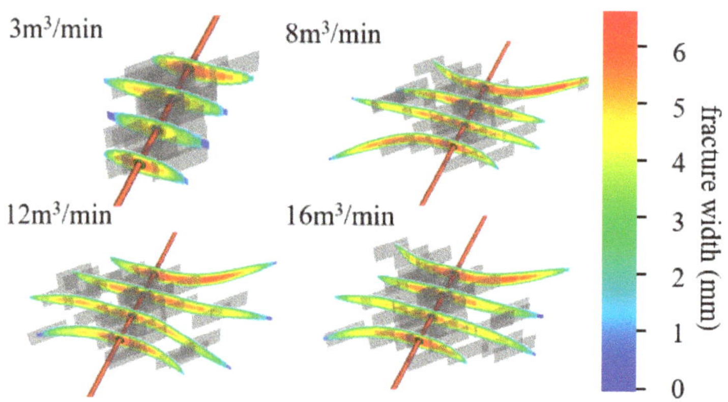


**Figure 19.** Fracture morphology and width under different displacement.

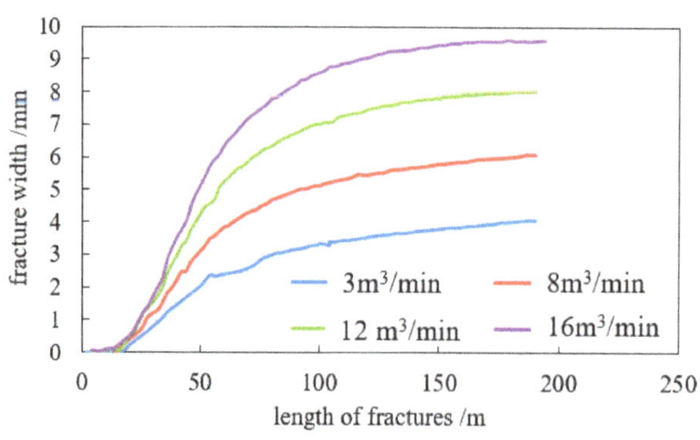


**Figure 20.** Change curves of crack length and crack width under different displacement.

### 3.2.5. Fracturing Fluid Consumption

The amount of fracturing fluid determines the scale of fractures, and optimizing and researching the amount of fracturing fluid can provide a basis for the construction design of large-section and multi-cluster horizontal Wells in Weiyuan area of Sichuan Province. A horizontal well model with four clusters in a single section and cluster spacing of 30 m was set, other parameters remain unchanged and the fracturing fluid dosage was set as

2000 $m^3$, 3000 $m^3$ and 4000 $m^3$, to study the influence of fracturing fluid dosage on fracture scale. According to Figure 21 of the simulation results, the length and width of fractures increased linearly with the increase in fracturing from 2000 $m^3$ to 4000 $m^3$. Considering that the well spacing of horizontal Wells in Weiyuan area of Sichuan is generally 400 m, excessive hydraulic fractures usually cause inter-well interference, it is recommended that the amount of fracturing fluid in a single stage should be controlled within 3000 $m^3$ (Figure 22).

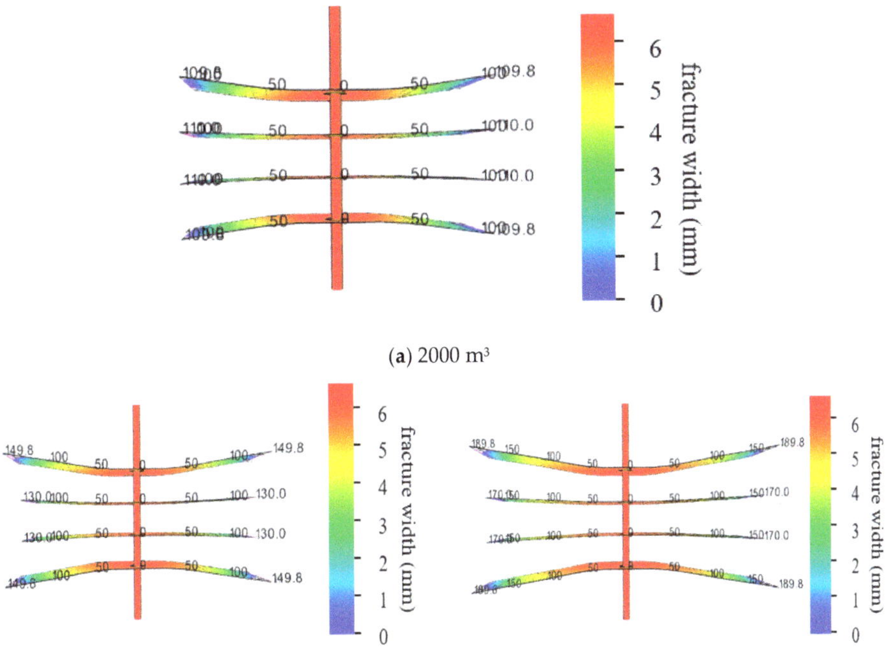


(**a**) 2000 m³

(**b**) 3000 m³

(**c**) 4000 m³

**Figure 21.** Influence of fracturing fluid dosage on fracture morphology.

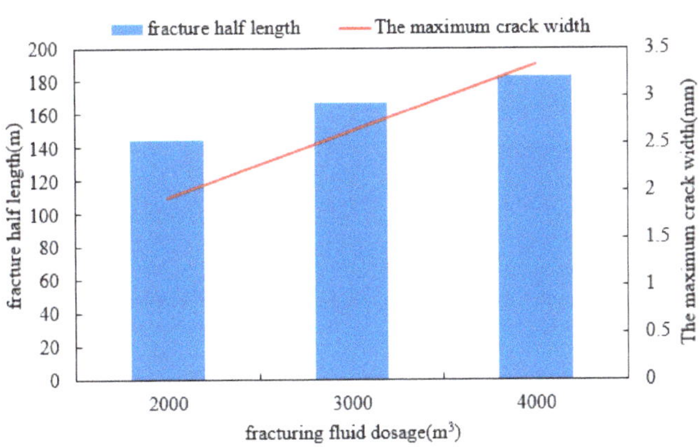


**Figure 22.** Influence of fracturing fluid dosage on fracture size.

## 4. Conclusions

(1) Both rock modulus and Poisson's ratio are important indicators of rock brittleness, and are also the main parameters for formulating reservoir reconstruction schemes. The larger the Young's modulus and Poisson's ratio are, the more brittle the rock is, and the easier it is to crack during hydraulic fracturing. In the reservoir with higher Poisson's ratio, the propagation rate of hydraulic fracture is faster.

(2) When the in situ stress difference and approximation angle are small, the hydraulic fractures will not only induce the natural fractures to extend in the direction of the maximum principal stress, but also form transverse fractures, and finally form a grid-like fracture network. The low-angle (30°) natural fractures in Weiyuan X well area, Sichuan Province have the potential to form complex fracture networks. At the same time, the strong heterogeneity of geostress difference distribution brings many uncertain factors to the formation of fracture network. Therefore, when selecting the well location, it is necessary to combine the maximum likelihood chart and preferentially select the area with small in situ stress difference.

(3) When the number of perforating clusters in a single stage is more than four clusters, the interfracture interference is more serious. When the cluster spacing is less than 30 m, the fracture shape is distorted, which is the premise of forming a complex fracture network.

(4) The viscosity of fracturing fluid has no obvious influence on the width of fracture expansion, but has great influence on the length of fracture. The higher the viscosity of the fracturing fluid, the more obvious the development trend of the main fracture. When the fracturing fluid with low viscosity (10 MPa·s) is used, the more obvious the development trend of the main fracture is or does not show the development trend of the main fracture, and more secondary fractures are activated to form a complex fracture network with the main fracture.

(5) The construction displacement is related to the net pressure inside the fracture, and only enough net pressure inside the fracture can open the natural fracture. When the construction displacement is 3 $m^3$/min, the filtration loss is more serious, and when it is above 12 $m^3$/min, the fracture network can reach the best scale. It is suggested that the average construction displacement of large-scale hydraulic fracturing operations in Weiyuan shale gas Wells is 12 m/min.

(6) The amount of fracturing fluid determines the scale of the fracture. In order to avoid inter-well interference, it is recommended that the maximum amount of fracturing fluid used in single section of horizontal well construction in Weiyuan area should be 3000 $m^3$.

**Author Contributions:** Writing—original draft, P.M.; writing—review & editing, S.T. All authors have read and agreed to the published version of the manuscript.

**Funding:** This research received no external funding.

**Data Availability Statement:** Data are contained within the article.

**Conflicts of Interest:** Author Pengcheng Ma was employed by the company China Urban-Rural Energy Co., Ltd. The remaining authors declare that the research was conducted in the absence of any commercial or financial relationships that could be construed as a potential conflict of interest. The China Urban-Rural Energy Co., Ltd. had no role in the design of the study; in the collection, analyses, or interpretation of data; in the writing of the manuscript, or in the decision to publish the results.

## References

1. Wang, J.; Xu, H.; Yang, K.; Jiang, H.; Zhou, F. Process simulation and performance evaluation of plugging cakes during temporary plugging and diverting fracturing. *Pet. Sci. Technol.* **2022**, *40*, 2504–2524. [CrossRef]
2. Tang, H.; Winterfeld, P.H.; Wu, Y.-S.; Huang, Z.-Q.; Di, Y.; Pan, Z.; Zhang, J. Integrated simulation of multi-stage hydraulic fracturing in unconventional reservoirs. *J. Nat. Gas Sci. Eng.* **2016**, *36*, 875–892. [CrossRef]

3. Heister, T.; Wheeler, M.F.; Wick, T. A primal-dual active set method and predictor-corrector mesh adaptivity for computing fracture propagation using a phase-field approach. *Comput. Methods Appl. Mech. Eng.* **2015**, *290*, 466–495. [CrossRef]
4. Li, Y.; Yu, T.; Xing, C.; Natarajan, S. Crack growth in homogeneous media using an adaptive isogeometric fourth-order phase field model. *Comput. Methods Appl. Mech. Eng.* **2023**, *413*, 116122. [CrossRef]
5. Lecampion, B.; Desroches, J. Simultaneous initiation and growth of multiple radial hydraulic fractures from a horizontal wellbore. *J. Mech. Phys. Solids* **2015**, *82*, 235–258. [CrossRef]
6. Yang, C.-X.; Yi, L.-P.; Yang, Z.-Z.; Li, X.-G. Numerical investigation of the fracture network morphology in multi-cluster hydraulic fracturing of horizontal wells: A DDM-FVM study. *J. Pet. Sci. Eng.* **2022**, *215*, 110723. [CrossRef]
7. Wang, D.-B.; Zhou, F.-J.; Li, Y.-P.; Yu, B.; Martyushev, D.; Liu, X.-F.; Wang, M.; He, C.-M.; Han, D.-X.; Sun, D.-L. Numerical simulation of fracture propagation in Russia carbonate reservoirs during refracturing. *Pet. Sci.* **2022**, *19*, 2781–2795. [CrossRef]
8. Umar, I.A.; Negash, B.M.; Quainoo, A.K.; Ayoub, M.A. An outlook into recent advances on estimation of effective stimulated reservoir volume. *J. Nat. Gas Sci. Eng.* **2021**, *88*, 103822. [CrossRef]
9. Chen, M.; Zhang, S.; Xu, Y.; Ma, X.; Zou, Y. A numerical method for simulating planar 3D multi-fracture propagation in multi-stage fracturing of horizontal wells. *Pet. Explor. Dev.* **2020**, *47*, 171–183. [CrossRef]
10. Chen, X.; Li, Y.; Zhao, J.; Xu, W.; Fu, D. Numerical investigation for simultaneous growth of hydraulic fractures in multiple horizontal wells. *J. Nat. Gas Sci. Eng.* **2018**, *51*, 44–52. [CrossRef]
11. Cheng, W.; Jin, Y.; Chen, M.; Xu, T.; Zhang, Y.; Diao, C. A criterion for identifying hydraulic fractures crossing natural fractures in 3D space. *Pet. Explor. Dev.* **2014**, *41*, 371–376. [CrossRef]
12. Cong, Z.; Li, Y.; Liu, Y.; Xiao, Y. A new method for calculating the direction of fracture propagation by stress numerical search based on the displacement discontinuity method. *Comput. Geotech.* **2021**, *140*, 104482. [CrossRef]
13. Taleghani, A.D.; Gonzalez-Chavez, M.; Yu, H.; Asala, H. Numerical simulation of hydraulic fracture propagation in naturally fractured formations using the cohesive zone model. *J. Pet. Sci. Eng.* **2018**, *165*, 42–57. [CrossRef]
14. Kresse, O.; Weng, X. Numerical Modeling of 3D Hydraulic Fractures Interaction in Complex Naturally Fractured Formations. *Rock Mech. Rock Eng.* **2018**, *51*, 3863–3881. [CrossRef]
15. Zou, Y.S.; Ma, X.F.; Zhang, S.C.; Zhou, T.; Li, H. Numerical investigation into the influence of bedding plane on hydraulic fracture network propagation in shale formations. *Rock Mech. Rock Eng.* **2016**, *49*, 3597–3614.
16. Wang, R.; Yang, Y.; Pan, J.; Li, G. Network fracturing technology of hydraulic fracturing in coalbed methane reservoir based on induced stress. *Arab. J. Geosci.* **2021**, *14*, 351. [CrossRef]
17. Wang, X.; Shi, F.; Liu, C.; Lu, D.; Liu, H.; Wu, H. Extended finite element simulation of fracture network propagation in formation containing frictional and cemented natural fractures. *J. Nat. Gas Sci. Eng.* **2018**, *50*, 309–324. [CrossRef]
18. Zou, Y.; Zhang, S.; Ma, X.; Zhou, T.; Zeng, B. Numerical investigation of hydraulic fracture network propagation in naturally fractured shale formations. *J. Struct. Geol.* **2016**, *84*, 1–13. [CrossRef]
19. Wang, B.; Zhou, F.; Zou, Y.; Liang, T.; Wang, D.; Xue, Y.; Gao, L. Quantitative investigation of fracture interaction by evaluating fracture curvature during temporarily plugging staged fracturing. *J. Pet. Sci. Eng.* **2019**, *172*, 559–571. [CrossRef]
20. Wang, B.; Zhou, F.; Yang, C.; Xu, C.; Liu, J.; Han, S.; Wang, D.; Ren, Z.; Liang, T. A novel experimental method to investigate the plugging characteristics of diversion agents within hydro-fracture. *J. Pet. Sci. Eng.* **2019**, *183*, 106354. [CrossRef]

*processes* MDPI

*Article*

# Further Investigation of $CO_2$ Quasi-Dry Fracturing in Shale Reservoirs—An Experimental Study

**Bo Zheng [1], Weiyu Tang [2,*], Yong Wang [1], Yipeng Li [1], Binbin Shen [1], Yongkang Wang [1], Longqiao Hu [2], Yougen Deng [1], Mingjiang Wu [1], Shangyong Xi [1] and Xiongfei Liu [2,*]**

1 Tuha Underground Operation Company, CNPC Western Drilling Engineering Co., Ltd., Turpan 838000, China; zbjx@cnpc.com.cn (B.Z.); wytlmjx@cnpc.cpm.cn (Y.W.); liypjx@cnpc.com.cn (Y.L.); sbbjx@cnpc.com.cn (B.S.); wangykjx@cnpc.com.cn (Y.W.); dengygjx@cnpc.com.cn (Y.D.); wmjjxgs@cnpc.com.cn (M.W.); xisyjxgs@cnpc.com.cn (S.X.)

2 State Key Laboratory of Petroleum Resources and Prospecting, China University of Petroleum-Beijing, Beijing 102249, China; 2022316206@student.cup.edu.cn

* Correspondence: 2021316212@student.cup.edu.cn (W.T.); lxfei@cup.edu.cn (X.L.)

**Abstract:** The physical properties of shale reservoirs are typically poor, necessitating the use of fracturing technology for effective development. However, the high clay content prevalent in shale formations poses significant challenges for conventional hydraulic fracturing methods. To address this issue, $CO_2$-based fracturing fluid has been proposed as an alternative to mitigate the damage caused by water-based fracturing fluids. In this paper, the applicability of quasi-dry $CO_2$ fracturing in shale reservoirs is examined from three key perspectives: the viscosity of $CO_2$ fracturing fluid, the fracture characteristics induced by the $CO_2$ fracture fluid, and the potential reservoir damage caused by the fracturing fluid. Firstly, the viscosity of $CO_2$ fracturing fluid was determined by a rheological experiment. Rheological tests revealed that the viscosity of $CO_2$ fracturing fluid was significantly influenced by the water–carbon ratio. Specifically, when the water–carbon ratio was 30:70, the maximum viscosity observed could reach 104 mPa·s. Moreover, increasing reservoir temperature resulted in decreased fracturing fluid viscosity, with a 40 °C temperature rise causing a 20% viscosity reduction. Secondly, matrix permeability tests were conducted to investigate permeability alteration during $CO_2$ fracturing fluid invasion. Due to the weak acidity of $CO_2$-based fracturing fluid, the permeability reduction induced by clay hydration was inhibited, and an increase in permeability was observed after a 3-day duration. However, the matrix permeability tends to decrease as the interaction time is prolonged, which means prolonged soaking time can still cause formation damage. Finally, triaxial fracturing experiments facilitated by a three-axis servo pressure device were conducted. The fracture properties were characterized using computed tomography (CT), and 3D reconstruction of fractured samples was conducted based on the CT data. The results demonstrate that $CO_2$ fracturing fluid effectively activates weak cementation surfaces in the rock, promoting the formation of larger and more complex fractures. Hence, $CO_2$ quasi-dry fracturing technology emerges as a method with significant potential, capable of efficiently stimulating shale reservoirs, although a reasonable soaking time is necessary to maximize hydrocarbon production.

**Keywords:** shale; $CO_2$ quasi-dry fracturing; carbon dioxide–water–shale interaction; permeability

## 1. Introduction

$CO_2$ fracturing technology has emerged as a promising method for enhancing hydrocarbon recovery while mitigating environmental impacts associated with conventional hydraulic fracturing. Unlike traditional hydraulic fracturing that relies on water-based fluids, $CO_2$ fracturing utilizes supercritical or liquid $CO_2$ as the fracturing fluid [1]. This approach offers several advantages, including reduced water usage, a reduced environmental footprint, and the potential for enhanced reservoir stimulation [2–5]. Hence, $CO_2$-based fracturing methods have steadily garnered attention within the oil industry.

Supercritical $CO_2$ fluid boasts a plethora of unique physical and chemical properties, setting it apart in various applications [6–9]. Its low viscosity, high diffusion coefficient, and near-zero surface tension make it a standout choice [10–12]. Since its introduction into the oil industry, $CO_2$ fracturing has showcased numerous advantages. Notably, it readily forms intricate fracture networks within reservoirs [13]. Moreover, its low water content mitigates the risk of reservoir clay expansion. One of its standout benefits lies in the rapid flowback after fracturing, significantly reducing non-production periods for oil and gas, thereby enhancing economic returns [14]. Furthermore, the robust absorption capabilities enable $CO_2$ to displace methane molecules adsorbed in shale, thereby amplifying production and recovery efficiency while facilitating permanent $CO_2$ storage [12,15–17]. Consequently, $CO_2$-based fracturing technology is lauded as a promising new approach with expansive potential. The effectiveness of $CO_2$-induced fracture propagation hinges on several key mechanisms [18]. Firstly, thermal stress coupled with pore pressure alleviates effective stress, instigating shear failure. Secondly, the absence of surface tension diminishes the net pressure necessary for crack propagation. Thirdly, adsorption phenomena decrease the critical stress required for crack growth. Lastly, localized phase transformations drive dynamic fracture growth. These mechanisms collectively underscore the multifaceted nature of $CO_2$-induced fracture propagation, shedding light on its efficacy and versatility in various engineering contexts. However, liquid $CO_2$ encounters challenges due to its significant friction resistance, which hinders its ability to meet the demands of high-displacement injection during field application [19]. Moreover, its low viscosity exacerbates issues such as rapid settling of proppants, resulting in limited sand-carrying ability and increased filtration loss [1,20].

To tackle these challenges, a thickening agent is introduced into the liquid $CO_2$ [21,22]. This additive serves to enhance the suspended sand capacity, effectively slowing down the settling process and reducing filtration loss. By incorporating a thickening agent, the performance of liquid $CO_2$ is improved, making it better suited for high-displacement injection tasks while mitigating issues associated with proppant settling and filtration loss. In theory, the molecular structure of carbon dioxide thickeners typically comprises two components: it is a polymer of a $CO_2$-philic group, which can fully dissolve in carbon dioxide, and a $CO_2$-phobic group, which enhances the viscosity of carbon dioxide. $CO_2$ exhibits physical properties compatible with organic agents. In this mixed system, water molecules attract $CO_2$ to form hydrophilic groups of rod-like or worm-like micelles, while $CO_2$-philic molecules attract $CO_2$ groups. Through a series of interactions, the intermolecular forces strengthen, consequently elevating the viscosity of $CO_2$. Many scholars have dedicated efforts to studying the factors influencing the viscosity of gelled $CO_2$ and have established criteria for evaluating its rheological properties. Through systematic experimentation and theoretical analysis, researchers have identified key parameters that impact the viscosity of gelled $CO_2$. These parameters include the concentration and type of thickening agent, the temperature, the pressure, the shear rate, and the presence of additives such as surfactants or polymers. Luo et al. [23] investigate the rheological properties and friction performance of thickened $CO_2$ fracturing fluid. They establish correlations for rheological parameters, noting a significant decrease in effective viscosity with increasing shear rate and temperature, albeit with a modest increase linked to thickener concentration. Frictional resistance decreases with flow velocity, temperature, and pressure, but exhibits a smaller change range concerning pressure and thickener concentration. Zhao et al. [21] investigate a novel supercritical $CO_2$ fracturing fluid system composed of 5% polydimethylsiloxane, 5% kerosene, and 90% supercritical $CO_2$. Through systematic evaluation experiments, it demonstrates significantly enhanced rheological properties, an improvement in the sand-carrying capacity by 38.17%, and a reduction in the filtration loss coefficient by up to 71.71%. Zhao et al. [24] propose that a siloxane-based polymer significantly enhances $CO_2$ viscosity while minimizing damage and chemical residue in the stratum after fracturing.

In addition to viscosity, the formation damage caused by fracturing-fluid intrusion is another crucial parameter for evaluating the suitability of fracturing fluid. The potential for reservoir damage is a significant concern in hydraulic fracturing operations, as it can affect well productivity and reservoir performance in the long term. The intrusion of fracturing fluid into the reservoir rock can lead to various forms of damage, including formation matrix swelling, clay dispersion, fines migration, and pore plugging. Researchers have developed various methods to assess reservoir damage induced by fracturing-fluid intrusion. These methods often involve laboratory experiments, core analysis, and numerical simulations to quantify the extent of damage and its impact on reservoir properties such as permeability, porosity, and fluid flow behavior. Lu et al. [25] investigate the impact of supercritical $CO_2$–slickwater hybrid fracturing on shale pore structures using various analytical techniques. The treatment leads to an increase in micropores (<2 nm) and mesopores (2–50 nm), expanding total specific surface area (TSSA) and total pore volume (TPV) while reducing average pore size. Tian et al. [26] investigate the impact of exposure to supercritical $CO_2$ on the porosity and permeability evolution of both dry and wet shale samples over varying periods. Results indicate an initial increase followed by a decrease in porosity and permeability, with a distinct inflection point observed. Therefore, evaluating the reservoir damage resulting from fracturing-fluid intrusion is crucial for optimizing fracturing fluid formulations and designing effective fracture treatments.

Pure $CO_2$ fluid suffers from low viscosity, resulting in poor sand-carrying capacity and inadequate reduction of filtration loss, which hampers its effectiveness in fracturing operations. In this paper, a novel $CO_2$ thickener system (VIC-2, VIC-6) was proposed to enhance $CO_2$ viscosity, and the following research was carried out. Firstly, high-temperature and high-pressure rheological experiments are conducted to evaluate the viscosity of $CO_2$ fracturing fluid across varying water–carbon ratios. Secondly, the impact of $CO_2$–water–rock reactions on shale matrix permeability during fracturing remains ambiguous despite the presence of a small amount of water and polymer in the $CO_2$ fracturing fluid system. Hence, permeability tests are conducted on shale samples subjected to different soaking times to elucidate this influence. Finally, $CO_2$ triaxial fracturing experiments are conducted using downhole cores, and fracture reconstruction is facilitated through CT imaging. These experiments provide a theoretical foundation for the further application of $CO_2$ quasi-dry fracturing in shale reservoirs. By systematically investigating these aspects, the research aims to enhance the understanding of $CO_2$ fracturing fluid behavior and optimize its performance in shale reservoir stimulation, paving the way for more efficient and effective fracturing operations.

## 2. Methodology

In this paper, the experimental section primarily encompasses 3 key components:

1. Shale damage test: This aspect aimed to assess the permeability reduction induced by water-based fracturing fluid and $CO_2$ quasi-dry fracturing fluid. In this part, the PDP instrument was used to determine the shale permeability. By subjecting cores to both types of fluids, we evaluated the extent of damage inflicted on the cores. This evaluation provides insights into the comparative performance of the two types of fracturing fluids in terms of core preservation and permeability retention.
2. Rheological experiment: This involved studying the rheological behavior of $CO_2$ quasi-dry fracturing fluid under varying water–carbon ratios and temperatures.
3. True triaxial fracturing experiment: Quasi-dry $CO_2$ fracturing fluid and slickwater were applied to conduct the fracturing experiment. By subjecting shale rock samples to different hydraulic fracturing, the fracture characteristics were further evaluated.

### 2.1. Rock Sample Preparation

The experimental setup involved rock samples sourced from the Jimsar Block in Xinjiang, extracted from depths ranging from 3045 to 3172 m underground (Figure 1). To conduct the shale damage test, cylindrical samples with a diameter of 25.4 mm and a height of 30 mm were prepared for permeability testing. Meanwhile, the cylinder samples were encapsulated in epoxy resin and molded into cube-shaped samples measuring 100 mm × 100 mm × 100 mm (Figure 2) to facilitate fracturing tests.

**Figure 1.** Full-diameter shale cores.

**Figure 2.** Cemented core with stainless steel wellbore.

The $CO_2$ fracturing fluid used in the experiments comprised formation water, the $CO_2$ thickeners VIC-6 and VIC-2 (Xian Victory Co., Ltd., Xi'an, China), and 99.9% pure $CO_2$ (Beijing Huatong Jingke Co., Ltd., Beijing, China). For shale damage tests and triaxial fracturing tests, water-based fracturing fluids containing a 0.1% concentration of polyacrylamide were employed.

### *2.2. Experimental Setup and Procedures*

#### 2.2.1. Rheological Test of $CO_2$ Fracturing Fluid

Liquid viscosity is the main factor affecting fracturing effects, which has a direct influence on fracture morphology and proppant-carrying effects. In this part, the rheology of $CO_2$ fracturing fluid was studied. The $CO_2$ fracturing fluid preparation device consists of a heating system, a $CO_2$ pumping system, a pressurization system, a stirring system, a viscosity testing system, a control system, and a reaction kettle with a sapphire window (Figure 3). The equipment was provided by Tochuang Co., Ltd., Ningbo, China. The characteristics and main parameters were as follows:

- Maximum torque: 200 mN·m;
- Volume: 400 mL;
- Minimum rotation speed CR: 10 rpm;
- Maximum rotation speed: 1500 rpm;
- Temperature range: high-temperature electric heating system: 20 degrees to 400 degrees; able to withstand pressure of 40 MPa;
- Maximum lateral loading capacity: 6 Kn.

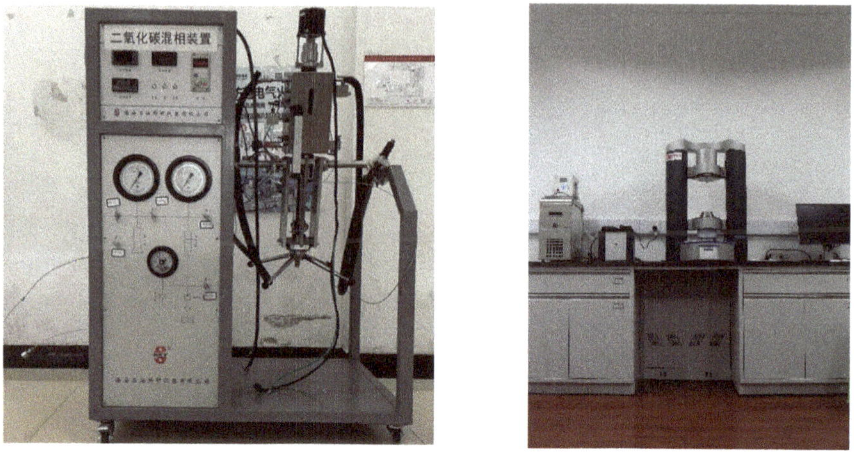

**Figure 3.** $CO_2$ fracturing fluid preparation and rheological testing device.

The experimental setup for configuring and testing the $CO_2$ fracturing fluid involved the following steps:

1. Base fluid preparation: The experimental water was prepared with NaCl at a salinity of 600 ppm. The base fluid was then configured with various water–carbon ratios. For instance, if the ratio was 30/70 (water/$CO_2$), 132 mL of water and 268 mL of $CO_2$ were combined. The thickeners, VIC-6 and VIC-2, were added in appropriate proportions. Specifically, the VIC-6 volume was 1.2% of the water volume (1.584 mL), while the VIC-2 volume was 1.5% of the $CO_2$ volume (4.02 mL). After adding the thickeners, the mixture was stirred thoroughly to ensure uniform distribution before transferring it to the miscible device.
2. $CO_2$ fracturing fluid configuration: $CO_2$ was introduced into the miscible device using a gas booster pump to increase the pressure inside the device. Once the pressure stabilized, mechanical stirring was employed to disperse and mix the miscible system. The stirring was performed in both the forward and reverse directions for 15 min at a speed of 2500 rpm to ensure thorough dispersion and homogeneity of the fracturing fluid.
3. Rheological testing: The prepared $CO_2$ fracturing fluid is then subjected to rheological testing using a Harkar rheometer (Thermo Fisher, Waltham, MA, USA) to evaluate its rheological characteristics at different temperatures.

#### 2.2.2. Shale Permeability Damage Test

When the shale matrix is exposed to $CO_2$ fracturing fluid, a series of chemical reactions will occur, resulting in changes in the physical properties of the shale matrix. In this part, the shale permeability alteration induce by $CO_2$-based fracturing fluid is studied, and a PDP instrument was applied to conduct the permeability determination. The experimental process was as follows:

1. The column samples were dried, and the permeability of the untreated shale was measured to establish a baseline.
2. The core sample comprising both the core plug and fragments was placed in the high-temperature and high-pressure reactor. The reactor was then evacuated and subsequently immersed in a water bath device set to a constant temperature of 60 °C.
3. $CO_2$-based fracturing fluid was pumped into the high-temperature and high-pressure reactor until the pressure within the reactor increased by 10 MPa. The core sample was allowed to react with the fracturing fluid for designated durations of 3, 5, and 7 days, respectively.

4. After the specified reaction periods, the core plug and fragments were removed from the reactor. The core plug was dried and subjected to permeability testing to assess any changes in permeability resulting from the interaction with the fracturing fluid. Meanwhile, the fragments underwent further treatment to meet the requirement of the XRD test, and the alterations in mineral composition were evaluated.

In an experimental setup similar to the one described above, a control group utilizing slickwater was employed to conduct shale hydration experiments. This control was included to provide a comparative baseline and further elucidate the changes in shale permeability under the influence of $CO_2$-based fracturing fluid. The experimental procedure closely resembled that of the $CO_2$ fracturing fluid experiments, with the main difference lying in the choice of fluid used. In this case, slickwater was utilized instead of $CO_2$ fracturing fluid.

### 2.2.3. True Triaxial Fracturing Experiment

The experimental instruments was provided by Tochuang Co., Ltd., Ningbo, China. The experimental equapment are shown in Figure 4, including a stress servo control system, a rock high-temperature heating system, and an injection pump. After fracturing, a Computed Tomography (CT) device was applied to reconstruct the fracture distribution in rock samples. In this study, a GE Brivo CT385 (General Electric Company, Shanghai, China) was used (Figure 5). The experimental process was as follows:

1. A stainless steel shaft was drilled into the center of the rock sample, and the well was cemented using high-strength cement to ensure stability.
2. The prepared sample was placed in the triaxial loading chamber, ensuring the device's airtightness. $CO_2$ fracturing fluid was prepared in an intermediate container, and the pipeline was connected. Triaxial stress is gradually applied until it reaches stability. Fluid is injected into the sample at a set flow rate, and fracturing tests are conducted. Pressure changes are recorded during the test, and injection ceases when a significant pressure drop occurs, indicating completion of fracturing.
3. After fracturing, the sample is placed in a high-precision CT device. Cracks in the rock are reconstructed using the CT device and post-processing software. This step allows for detailed visualization and analysis of fracture distribution within the rock sample, providing valuable insights into the fracturing process and fracture characteristics.

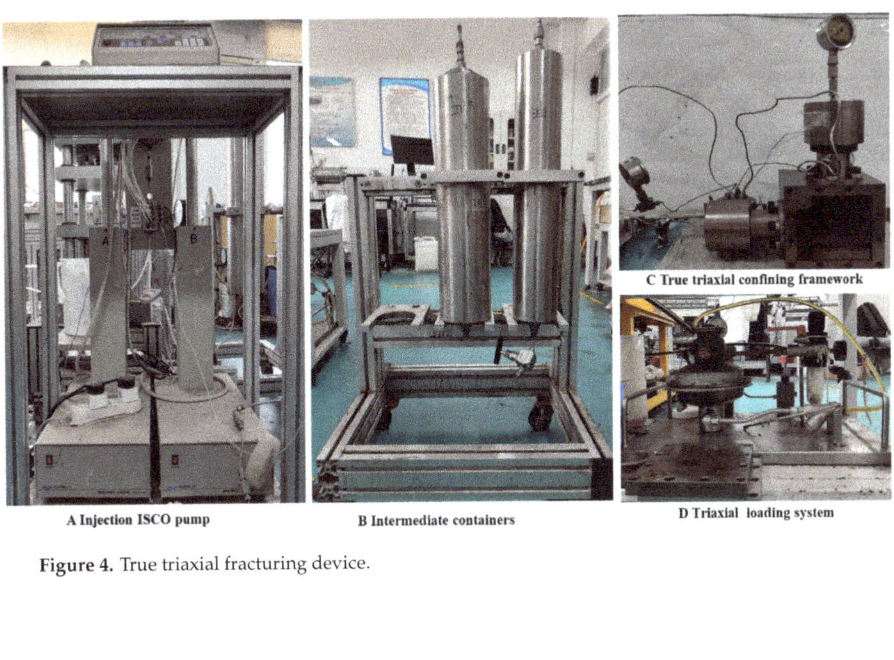


**Figure 4.** True triaxial fracturing device.

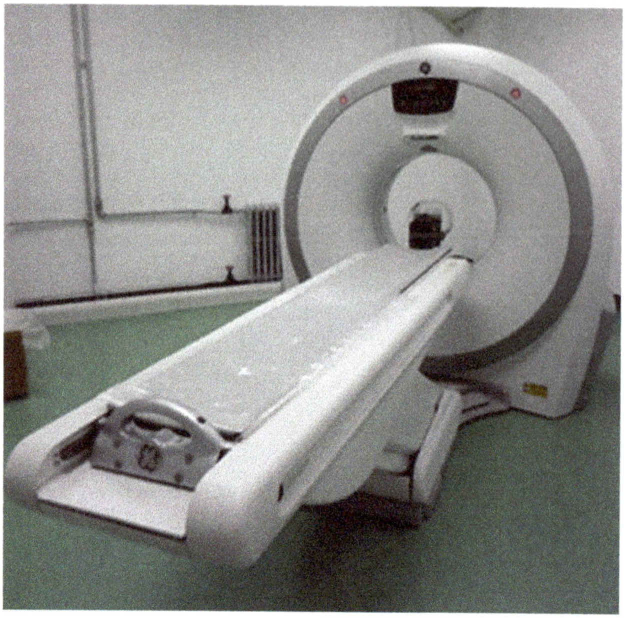

**Figure 5.** CT equipment.

## 3. Results and Discussion

### *3.1. High-Pressure and High-Temperature Rheological Experiment*

Conventional $CO_2$ fluid exhibits low viscosity, posing challenges in meeting sand-carrying requirements during fracturing operations. To address this issue, viscosity enhancement of $CO_2$ was achieved herein through the use of thickeners, specifically the water-based thickener VIC-6 and the $CO_2$ thickener VIC-2 (Figure 6). These two thickeners are miscible and, when mixed, undergo copolymerization of monomers such as ethyl propenyl ether, methyl methacrylate, octadecyl methacrylate, and allyl carbonate in a $CO_2$-philic solvent. This process leads to increased structural viscosity through molecular chain stretching and an enhanced probability of entanglement with neighboring molecules.

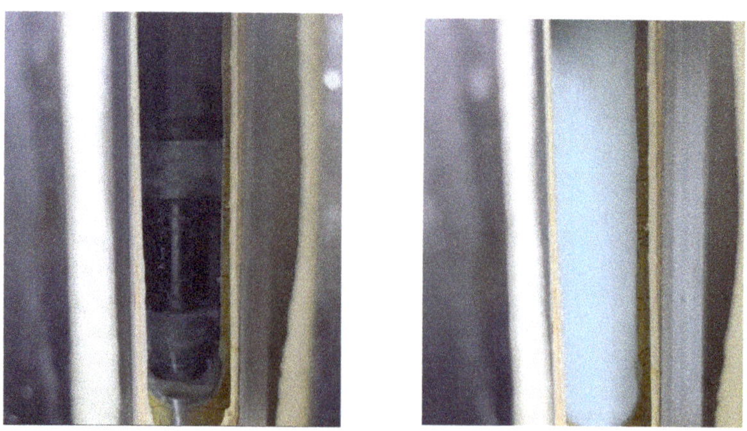

(**a**) (**b**)

**Figure 6.** Mixture of $CO_2$ and water-based thickener (**a**) before stirring and (**b**) after stirring.

However, as liquid $CO_2$ is a weak solvent and almost immiscible with water, a small amount of water based fracturing fluid is necessary for sand carrying in $CO_2$ quasi-dry fracturing. Optimizing the water–carbon ratio of the liquid $CO_2$ and water base is essential to minimize damage while ensuring stable fracturing fluid performance. $CO_2$ fracturing fluid comprises a thickener base fluid and $CO_2$, with varying rheological properties at different water–carbon ratios. Rheological experiments were conducted for different water–carbon ratios (30:70, 40:60, and 35:65). The viscosity of $CO_2$ fracturing fluid is summarized in Table 1. The results indicate that the fluid viscosity is sensitivity to the temperature. Specifically, the liquid viscosity decreases with increasing temperature, with a maximum drop of approximately 50 mPa·s observed. Moreover, the viscosity of the fracturing fluid initially increases and then decreases with increasing water content. The viscosity is 104 mPa·s at a water–carbon ratio of 30:70, and further increases in water content result in a rapid viscosity decrease (Figures 7–9). This phenomenon suggests that 30:70 is the optimal ratio.

**Table 1.** Viscosity of $CO_2$ fracturing fluid.

| Water–Carbon Ratio | Temperature, °C | | |
|---|---|---|---|
| | 50 | 70 | 90 |
| 30:70 | 104.25 | 93.37 | 78.70 |
| 35:65 | 63.50 | 45.63 | 22.56 |
| 40:60 | 19.43 | 16.77 | 15.62 |

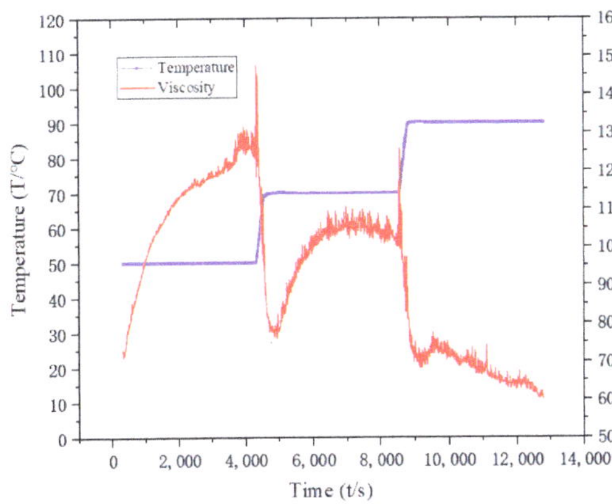


**Figure 7.** Viscosity–temperature relationship at a water–carbon ratio of 30:70.

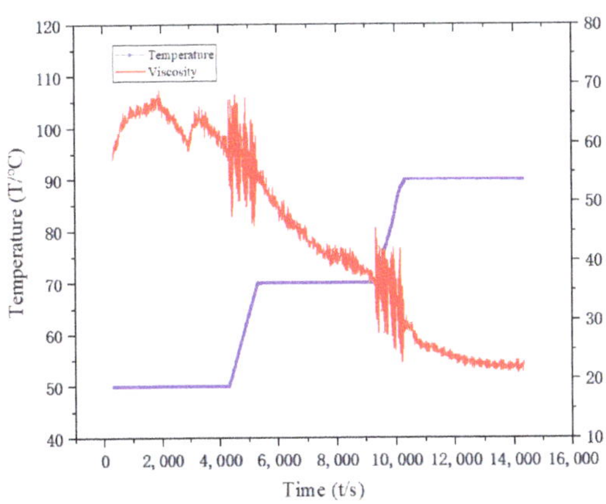


**Figure 8.** Viscosity–temperature relationship at a water–carbon ratio of 35:65.

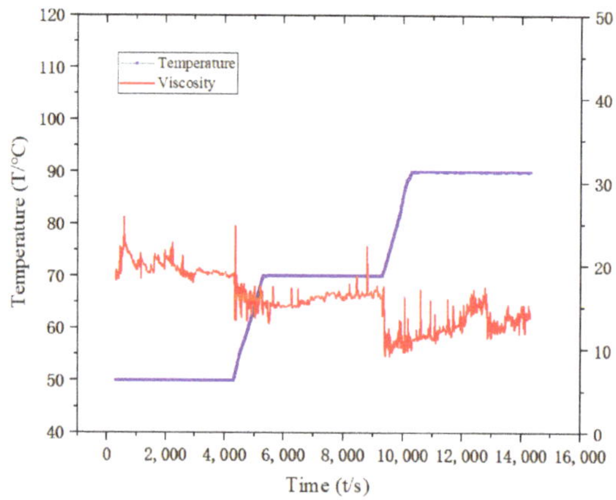


**Figure 9.** Viscosity–temperature relationship at a water–carbon ratio of 40:60.

On the basis of the optimal water–carbon ratio, the viscosity of $CO_2$ fracturing fluid at different concentrations of water-based thickener was studied. Testing was conducted using different concentrations (0.5%, 0.6%, 0.8%, and 1.2%) of an aqueous-phase viscosifier at temperatures of 50 °C, 70 °C, and 90 °C through the quasi-dry method. As the concentration of liquid thickener decreases, the viscosity of the miscible liquid experiences a rapid decline. This reduction in viscosity poses challenges in achieving the requisite viscosity for sand-carrying purposes (Figures 10–12).

### 3.2. Mineral Composition and Permeability Variation

Shale is a type of sedimentary rock primarily composed of fine-grained mineral particles, with a composition that typically includes clay minerals, quartz, and a variety of accessory minerals. XRD results revealed that the shale samples primarily consisted of quartz and clay minerals, constituting approximately 30% and 25% of the composition, respectively. Additionally, carbonate minerals such as dolomite and calcite made up around 20% of the composition (Figure 13).

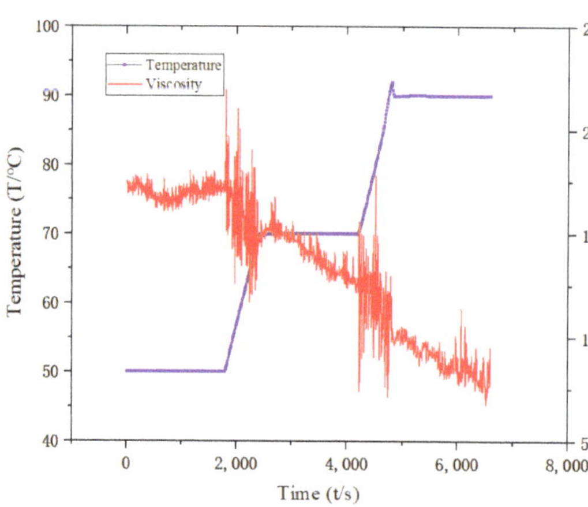


**Figure 10.** Viscosity–temperature relationship at a VIC-6 concentration of 0.5%.

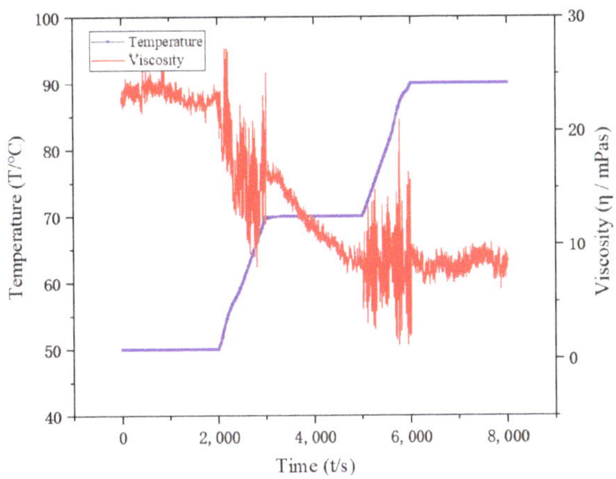


**Figure 11.** Viscosity–temperature relationship at a VIC-6 concentration of 0.6%.

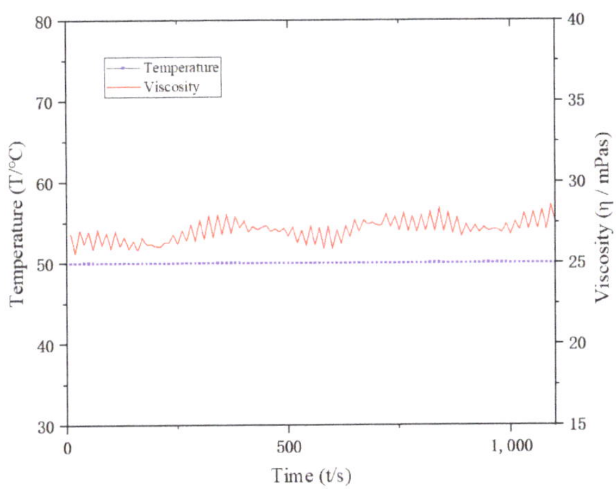


**Figure 12.** Viscosity–temperature relationship at a VIC-6 concentration of 0.7%.

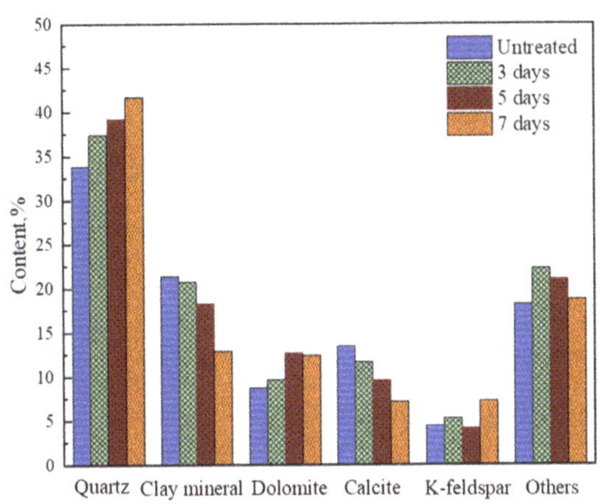


**Figure 13.** Mineral composition of shale under $CO_2$ fracturing fluid treatment.

In this section, a $CO_2$ fracturing fluid was prepared with a water–carbon ratio of 30:70. When the fracturing-fluid base is not fully mixed with $CO_2$, its pH remains neutral at 7. However, upon complete mixing with $CO_2$, the fracturing fluid becomes acidic, initiating a series of chemical reactions with the mineral composition of the rock [27]. These reactions involve mineral dissolution and the formation of new minerals, which are the primary factors influencing the change in shale permeability. In an acidic environment, feldspar, illite, and chlorite undergo partial dissolution, leading to the formation of more stable minerals such as quartz and kaolinite. This transformation process results in an increase in quartz content as the reaction progresses. Additionally, calcite, being a typical carbonate mineral, is consumed in this acidic environment, leading to a decrease in its content over time.

$$KAl_2(Si_3, Al_{10}(OH)_2\ (Illite) + 1.1H^+ \leftrightarrow 0.77kaolinite + 0.6K^+ + 0.25Mg^{(2+)} + 1.2quartz + 1.35H_2O$$

$$2KAlSi_3O_8\ (K - feldspar) + 2H^+ + H_2O \leftrightarrow 2K^+ + Al_2Si_2O_5(OH)_4\ (Kaolinite) + 4SiO_2\ (Quartz)$$

$$chlorite + 16H^+ \leftrightarrow 5Fe^{2+} + 2.3Al^{3+} + 3SiO_2 + 12H_2O$$

The permeability testing results are depicted in Figure 14. The permeability of the untreated shale was 474 nD. The permeability of shale in a slickwater environment exhibited a clear decreasing trend over time. After 3 days, 5 days, and 7 days of hydration reaction, the shale permeability values were recorded as 313 nD, 249 nD, and 217 nD, respectively. The absorption of water may induce swelling and expansion of clay minerals [28]. Water absorption causes shrinkage of pore spaces and reduces pore connectivity, ultimately resulting in decreased permeability. In contrast, the variation of shale permeability is more complicated when it is exposed to a $CO_2$-based fracturing fluid environment. After 3 days of reaction, the permeability of the shale matrix increased to 541 nD, indicating an enlargement effect due to chemical reactions, leading to improved pore connectivity and increased permeability. Subsequently, after 5 days of reaction, the matrix permeability exhibited a downward trend, dropping to 314 nD. This suggested that the chemical reaction was nearing completion, and while a small amount of water in the fracturing fluid might cause hydration expansion leading to core damage, the permeability changes were relatively minor with further extension of the reaction time. Finally, after 7 days of reaction, the permeability stabilized, indicating that $CO_2$–water–rock interaction had minimal influence on permeability at this stage. This stability suggested that the chemical reactions had reached equilibrium, resulting in a relatively stable permeability state in the shale matrix. Indeed, while $CO_2$ fracturing fluid can still cause some damage to the reservoir, the extent of damage is generally less severe than that induced by water-based fracturing fluids. This is primarily due to the chemical properties of $CO_2$ and its interaction with the reservoir rock. One advantage of $CO_2$ fracturing fluid is its ability to undergo a phase change, transitioning from a supercritical state to a gaseous state upon depressurization. This phase change facilitates the rapid flowback of the fracturing fluid from the reservoir, minimizing the duration of fluid–rock interaction and reducing potential damage to the formation. Efficient flowback of the $CO_2$ fracturing fluid is key to further reducing reservoir damage and enabling efficient reservoir development.

### 3.3. True Triaxial Fracturing Experiment Results

In this section, fracture morphology and propagation characteristics with different fracturing media were studied. To ensure experimental consistency, all rock samples were subjected to the same stress state, and both fracturing fluids were injected into the core at the same rate through the wellbore.

The shale samples exhibited prominent bedding development and typical layered stacking characteristics, indicative of their sedimentary origin. The presence of distinct bedding planes is essential for understanding fracture propagation behavior, as these planes can influence the path and orientation of fractures within the rock matrix (Figures 15 and 16).

1. Fractures induced by slickwater

The injection of water into the core resulted in the formation of a single symmetrical two-winged slit fracture. This fracture initiated along the borehole and propagated radially outward, creating a distinct fracture pattern. Analysis of CT results revealed that the average crack width generated by water injection measured 0.27 mm, with the resulting crack volume calculated at 1296 mm$^3$.

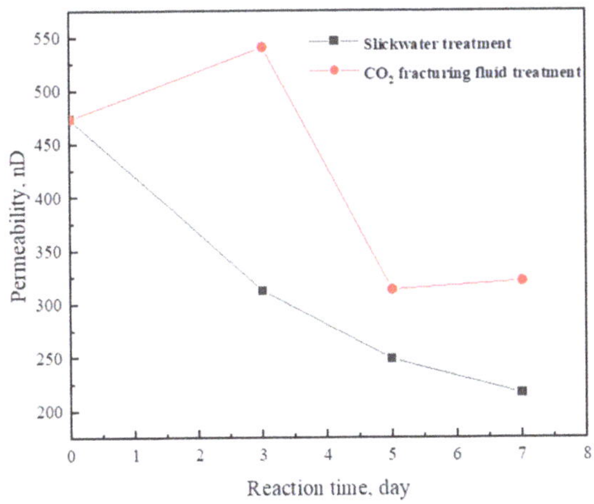


**Figure 14.** Permeability variation with different fracturing fluid treatments.

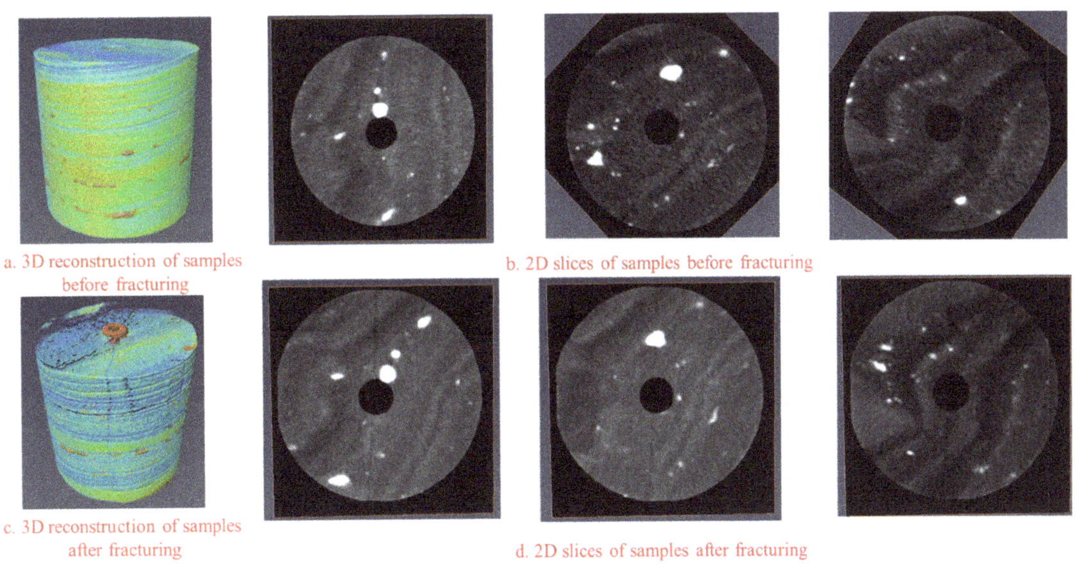


**Figure 15.** Reconstruction image of hydraulically fractured shale sample.

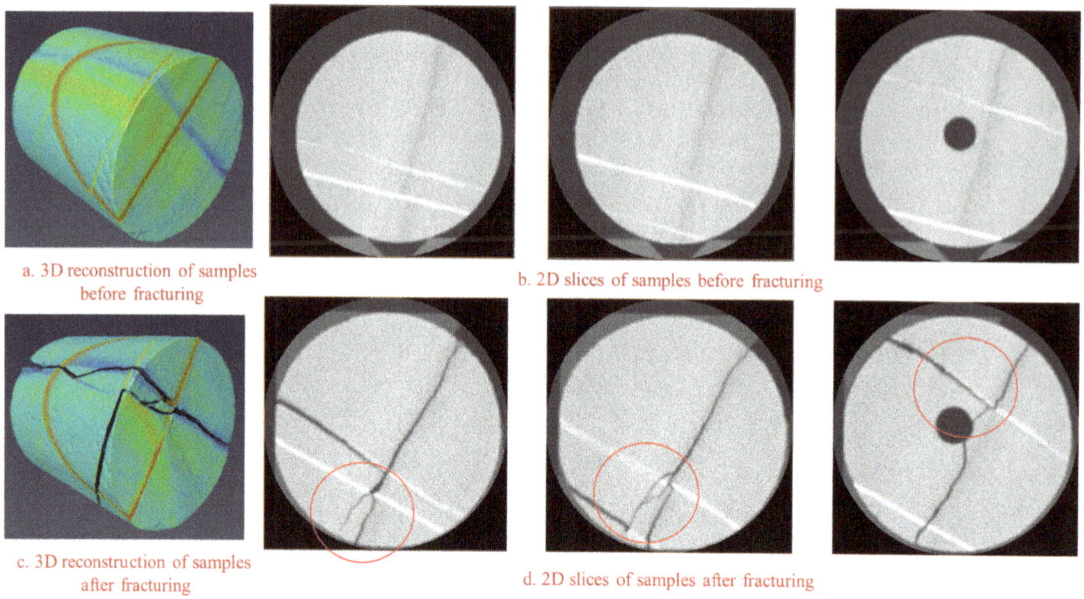


**Figure 16.** Reconstruction image of $CO_2$ quasi-dry fractured shale sample.

2. Fractures induced by $CO_2$ fracturing fluid

As depicted in Figure 16, fractures formed by $CO_2$ injection exhibited greater complexity than those induced by water. The $CO_2$-induced fractures displayed a variety of characteristics, including fractures along bedding planes and fractures cutting through the bedding. Some fractures expanded rapidly along weak surfaces such as filling joints within the core, while others extended vertically through original veins and weak surfaces. Additionally, some fractures changed direction upon encountering weak surfaces, leading to the complete opening of these weak surfaces and resulting in the complete fracturing of the core. This multifaceted behavior culminated in the formation of a more intricate fracture network.

CT results revealed that although the fractures induced by $CO_2$ fracturing fluid were narrower than water-induced fractures, with an average width of 0.14 mm, the resulting fracture volume was significantly higher. Specifically, the fracture volume generated by $CO_2$ hydraulic fracturing was approximately 7.6 times that induced by water, totaling about 9873 $mm^3$.

## 4. Conclusions

In this paper, a comprehensive investigation of $CO_2$ quasi-dry fracturing in shale reservoirs was conducted. The rheological characteristics of $CO_2$ fracturing fluid under high temperature and pressure were determined through laboratory experiments, and the fracture distribution characteristics and matrix permeability changes under $CO_2$ fracturing fluid treatment were evaluated. Several key findings were obtained:

1. Compared to conventional water-based fracturing fluids, quasi-dry $CO_2$ fracturing fluid causes less damage to the reservoir. Under the influence of $CO_2$ fracturing fluid, core permeability initially increases before decreasing. Short-term exposure to the fracturing fluid enhances reservoir permeability, but prolonged exposure is detrimental.
2. The acidic environment of $CO_2$ quasi-dry fracturing fluid promotes the dissolution of feldspar, illite, and chlorite minerals, which initially compensates for the decrease in permeability caused by shale expansion. However, as soaking time increases, the

weakening effect of chemical reactions allows shale hydration expansion to regain dominance, leading to a decrease in shale permeability.

3. The viscosity of conventional $CO_2$ is insufficient for proppant transport into fractures, and water-based thickeners are necessary to improve $CO_2$ viscosity. Rheological tests demonstrate that a water–carbon ratio of 3:7 yields a maximum viscosity of 104 mPa·s for $CO_2$ fracturing fluid. Adjusting the water proportion affects viscosity, with higher proportions leading to decreased viscosity. Additionally, temperature significantly influences fracturing fluid viscosity.
4. Fractures formed by conventional water-based fracturing fluids tend to be relatively singular, with small volumes but larger widths (approximately 0.27 mm). In contrast, fractures induced by $CO_2$ fracturing fluid exhibit larger volumes, with an average width of only 0.14 mm.
5. $CO_2$ fracturing fluid exhibits a high rock-breaking ability. Particularly, in cores with numerous weak surfaces, the fluid tends to rapidly open along these weak surfaces, leading to the formation of a complex fracture network.

**Author Contributions:** Conceptualization, Y.W. (Yong Wang); software, Y.L.; validation, Y.W. (Yongkang Wang); formal analysis, B.S.; investigation, Y.D.; resources, S.X.; data curation, L.H.; writing—original draft, B.Z.; writing—review and editing, X.L.; visualization, M.W.; supervision, W.T. All authors have read and agreed to the published version of the manuscript.

**Funding:** This research was funded by National Natural Science Foundation of China (No. 52174045).

**Data Availability Statement:** The data presented in this study are available on request from the corresponding author.

**Conflicts of Interest:** Authors Bo Zheng, Yong Wang, Yipeng Li, Binbin Shen, Yongkang Wang, Yougen Deng, Mingjiang Wu, and Shangyong Xi were employed by the Tuha Underground Operation Company, CNPC Western Drilling Engineering Co., Ltd. The remaining authors declare that the research was conducted in the absence of any commercial or financial relationships that could be construed as a potential conflict of interest. The Underground Operation Company, CNPC Western Drilling Engineering Co., Ltd. had no role in the design of the study; in the collection, analyses, or interpretation of data; in the writing of the manuscript, or in the decision to publish the results.

## References

1. Tudor, R.; Vozniak, C.; Peters, W.; Banks, M.-L. Technical advances in liquid $CO_2$ fracturing. In Proceedings of the Annual Technical Meeting, Calgary, AB, Canada, 11–14 June 1994.
2. Jia, B.; Tsau, J.-S.; Barati, R. A review of the current progress of $CO_2$ injection EOR and carbon storage in shale oil reservoirs. *Fuel* **2019**, *236*, 404–427. [CrossRef]
3. Johnson, J.M.-F.; Franzluebbers, A.J.; Weyers, S.L.; Reicosky, D.C. Agricultural opportunities to mitigate greenhouse gas emissions. *Environ. Pollut.* **2007**, *150*, 107–124. [CrossRef] [PubMed]
4. Metcalf, G.E. On the economics of a carbon tax for the United States. *Brook. Pap. Econ. Act.* **2019**, *2019*, 405–484. [CrossRef]
5. Tang, W.; Sheng, J.J. Huff-n-puff gas injection or gas flooding in tight oil reservoirs? *J. Pet. Sci. Eng. Fract. Mech.* **2022**, *208*, 109725. [CrossRef]
6. Alemu, B.L.; Aagaard, P.; Munz, I.A.; Skurtveit, E. Caprock interaction with $CO_2$: A laboratory study of reactivity of shale with supercritical $CO_2$ and brine. *Appl. Geochem.* **2011**, *26*, 1975–1989. [CrossRef]
7. Asadi, M.; Scharmach, W.; Jones, T.; Sampayo, A.; Chesney, E.; Unruh, D. Water-Free Fracturing: A Case History. In Proceedings of the SPE/CSUR Unconventional Resources Conference, Calgary, AB, Canada, 20–22 October 2015.
8. Rempel, K.U.; Liebscher, A.; Heinrich, W.; Schettler, G. An experimental investigation of trace element dissolution in carbon dioxide: Applications to the geological storage of $CO_2$. *Chem. Geol.* **2011**, *289*, 224–234. [CrossRef]
9. Zhi, Y.; Caineng, Z.; Songtao, W.; Senhu, L.; Songqi, P.; Xiaobing, N.; Guangtian, M.; Zhenxing, T.; Guohui, L.; Jiahong, Z.; et al. Formation, distribution and resource potential of the "sweet areas (sections)" of continental shale oil in China. *Mar. Pet. Geol.* **2019**, *102*, 48–60. [CrossRef]
10. Fu, C.; Liu, N. Waterless fluids in hydraulic fracturing–A review. *J. Nat. Gas Sci. Eng.* **2019**, *67*, 214–224. [CrossRef]
11. Speight, J.G. *Shale Oil Production Processes*; Gulf Professional Publishing: Houston, TX, USA, 2012.
12. Tang, W.; Peng, Z.; Sheng, J.J. Evaluation of $CO_2$ utilization and storage potential in the Jimsar shale play from an optimization study. *Geoenergy Sci. Eng.* **2023**, *224*, 211607. [CrossRef]

13. Middleton, R.S.; Carey, J.W.; Currier, R.P.; Hyman, J.D.; Kang, Q.; Karra, S.; Jiménez-Martínez, J.; Porter, M.L.; Viswanathan, H.S. Shale gas and non-aqueous fracturing fluids: Opportunities and challenges for supercritical $CO_2$. *Appl. Energy* **2015**, *147*, 500–509. [CrossRef]
14. Xia, Y.; Li, L.; Wang, Z. Experimental and numerical study on influencing factors of replacement capacity and slickwater flowback efficiency using pre-$CO_2$ fracturing in tight oil reservoirs. *J. Pet. Sci. Eng.* **2022**, *215*, 110697. [CrossRef]
15. Abedini, A.; Torabi, F. On the $CO_2$ storage potential of cyclic $CO_2$ injection process for enhanced oil recovery. *Fuel* **2014**, *124*, 14–27. [CrossRef]
16. Gozalpour, F.; Ren, S.R.; Tohidi, B. $CO_2$ EOR and storage in oil reservoir. *Oil Gas Sci. Technol.* **2005**, *60*, 537–546. [CrossRef]
17. Kim, T.H.; Cho, J.; Lee, K.S. Evaluation of $CO_2$ injection in shale gas reservoirs with multi-component transport and geomechanical effects. *Appl. Energy* **2017**, *190*, 1195–1206. [CrossRef]
18. Hou, L.; Zhang, S.; Elsworth, D.; Liu, H.; Sun, B.; Geng, X. Review of fundamental studies of $CO_2$ fracturing: Fracture propagation, propping and permeating. *J. Pet. Sci. Eng.* **2021**, *205*, 108823. [CrossRef]
19. Michels, A.; Botzen, A.; Schuurman, W. The viscosity of carbon dioxide between 0 °C and 75 °C and at pressures up to 2000 atmospheres. *Physica* **1957**, *23*, 95–102. [CrossRef]
20. Middleton, R.; Viswanathan, H.; Currier, R.; Gupta, R. $CO_2$ as a fracturing fluid: Potential for commercial-scale shale gas production and $CO_2$ sequestration. *Energy Procedia* **2014**, *63*, 7780–7784. [CrossRef]
21. Zhao, M.; Li, Y.; Gao, M.; Wang, T.; Dai, C.; Wang, X.; Guan, B.; Liu, P. Formulation and performance evaluation of polymer-thickened supercritical $CO_2$ fracturing fluid. *J. Pet. Sci. Eng.* **2021**, *201*, 108474. [CrossRef]
22. Li, Q.; Wang, Y.; Wang, F.; Li, Q.; Kobina, F.; Bai, H.; Yuan, L. Effect of a modified silicone as a thickener on rheology of liquid $CO_2$ and its fracturing capacity. *Polymers* **2019**, *11*, 540. [CrossRef]
23. Luo, X.; Wang, S.; Wang, Z.; Jing, Z.; Lv, M.; Zhai, Z.; Han, T. Experimental investigation on rheological properties and friction performance of thickened $CO_2$ fracturing fluid. *J. Pet. Sci. Eng.* **2015**, *133*, 410–420. [CrossRef]
24. Li, Q.; Wang, Y.; Wang, X.; Yu, H.; Li, Q.; Wang, F.; Bai, H.; Kobina, F. An application of thickener to increase viscosity of liquid $CO_2$ and the assessment of the reservoir geological damage and $CO_2$ utilization. *Energy Sour. Part A Recovery Util. Environ. Eff.* **2019**, *41*, 368–377.
25. Lu, Y.; Liu, J.; Tang, J.; Ao, X.; Li, H.; Zhou, J.; Sun, X. Pore changes of slickwater-containing shale under supercritical $CO_2$ treatment. *Fuel* **2022**, *312*, 122775. [CrossRef]
26. Tian, S.; Zhou, J.; Xian, X.; Gan, Q.; Yang, K.; Zheng, Y.; Deng, G.; Zhang, F. Impact of supercritical $CO_2$ exposure time on the porosity and permeability of dry and wet shale: The influence of chemo-mechanical coupling effects. *Energy* **2023**, *270*, 126905. [CrossRef]
27. Li, S.; Zhang, S.; Xing, H.; Zou, Y. $CO_2$–brine–rock interactions altering the mineralogical, physical, and mechanical properties of carbonate-rich shale oil reservoirs. *Energy* **2022**, *256*, 124608. [CrossRef]
28. Muhammed, N.S.; Olayiwola, T.; Elkatatny, S. A review on clay chemistry, characterization and shale inhibitors for water-based drilling fluids. *J. Pet. Sci. Eng.* **2021**, *206*, 109043. [CrossRef]

MDPI AG
Grosspeteranlage 5
4052 Basel
Switzerland
Tel.: +41 61 683 77 34

*Processes* Editorial Office
E-mail: processes@mdpi.com
www.mdpi.com/journal/processes

www.ingramcontent.com/pod-product-compliance
Lightning Source LLC
Chambersburg PA
CBHW061828210926
51850CB00006B/836